Applications of Fractional-Order Calculus in Robotics

Applications of Fractional-Order Calculus in Robotics

Editors

Abhaya Pal Singh
Kishore Bingi

Basel • Beijing • Wuhan • Barcelona • Belgrade • Novi Sad • Cluj • Manchester

Editors
Abhaya Pal Singh
Norwegian University of Life
Sciences (NMBU)
Ås
Norway

Kishore Bingi
Universiti Teknologi
PETRONAS
Seri Iskandar
Malaysia

Editorial Office
MDPI AG
Grosspeteranlage 5
4052 Basel, Switzerland

This is a reprint of articles from the Special Issue published online in the open access journal *Fractal and Fractional* (ISSN 2504-3110) (available at: https://www.mdpi.com/journal/fractalfract/special_issues/fractional_order_calculus_in_robotics).

For citation purposes, cite each article independently as indicated on the article page online and as indicated below:

Lastname, A.A.; Lastname, B.B. Article Title. *Journal Name* **Year**, *Volume Number*, Page Range.

ISBN 978-3-7258-1847-1 (Hbk)
ISBN 978-3-7258-1848-8 (PDF)
doi.org/10.3390/books978-3-7258-1848-8

Contents

About the Editors . . . . **vii**

Abhaya Pal Singh and Kishore Bingi
Applications of Fractional-Order Calculus in Robotics
Reprinted from: *Fractal Fract.* **2024**, *8*, 403, doi:10.3390/fractalfract8070403 . . . . **1**

Kishore Bingi, B Rajanarayan Prusty and Abhaya Pal Singh
A Review on Fractional-Order Modelling and Control of Robotic Manipulators
Reprinted from: *Fractal Fract.* **2023**, *7*, 77, doi:10.3390/fractalfract7010077 . . . . **6**

Yixiao Ding, Xiaolian Liu, Pengchong Chen, Xin Luo and Ying Luo
Fractional-Order Impedance Control for Robot Manipulator
Reprinted from: *Fractal Fract.* **2022**, *6*, 684, doi:10.3390/fractalfract6110684 . . . . **35**

Timi Karner, Rok Belšak and Janez Gotlih
Using a Fully Fractional Generalised Maxwell Model for Describing the Time Dependent
Sinusoidal Creep of a Dielectric Elastomer Actuator
Reprinted from: *Fractal Fract.* **2022**, *6*, 720, doi:10.3390/fractalfract6120720 . . . . **50**

Dora Morar, Vlad Mihaly, Mircea Şuşcă and Petru Dobra
Cascade Control for Two-Axis Position Mechatronic Systems
Reprinted from: *Fractal Fract.* **2023**, *7*, 122, doi:10.3390/fractalfract7020122 . . . . **64**

Xuan Liu, He Gan, Ying Luo, Yangquan Chen and Liang Gao
Digital-Twin-Based Real-Time Optimization for a Fractional Order Controller for Industrial
Robots
Reprinted from: *Fractal Fract.* **2023**, *7*, 167, doi:10.3390/fractalfract7020167 . . . . **81**

Banu Ataşlar-Ayyıldız
Robust Trajectory Tracking Control for Serial Robotic Manipulators Using Fractional
Order-Based PTID Controller
Reprinted from: *Fractal Fract.* **2023**, *7*, 250, doi:10.3390/fractalfract7030250 . . . . **97**

**Mohamed Naji Muftah, Ahmad Athif Mohd Faudzi, Shafishuhaza Sahlan and Shahrol
Mohamaddan**
Fuzzy Fractional Order PID Tuned via PSO for a Pneumatic Actuator with Ball Beam (PABB)
System
Reprinted from: *Fractal Fract.* **2023**, *7*, 416, doi:10.3390/fractalfract7060416 . . . . **119**

Bhukya Ramadevi, Venkata Ramana Kasi and Kishore Bingi
Hybrid LSTM-Based Fractional-Order Neural Network for Jeju Island's Wind Farm Power
Forecasting
Reprinted from: *Fractal Fract.* **2024**, *8*, 149, doi:10.3390/fractalfract8030149 . . . . **150**

Likun Li, Liyu Jiang, Wenzhang Tu, Liquan Jiang and Ruhan He
Smooth and Efficient Path Planning for Car-like Mobile Robot Using Improved Ant Colony
Optimization in Narrow and Large-Size Scenes
Reprinted from: *Fractal Fract.* **2024**, *8*, 157, doi:10.3390/fractalfract8030157 . . . . **175**

Weidong Liu, Liwei Guo, Le Li, Jingming Xu and Guanghao Yang
Fractional Active Disturbance Rejection Positioning and Docking Control of Remotely Operated
Vehicles: Analysis and Experimental Validation
Reprinted from: *Fractal Fract.* **2024**, *8*, 354, doi:10.3390/fractalfract8060354 . . . . **194**

About the Editors

Abhaya Pal Singh

Abhaya Pal Singh obtained a Bachelor of Technology in Electronics and Communication Engineering in 2009 from ITM, Gorakhpur, a Master of Technology in Electrical Engineering with a specialization in Control Systems in 2012 from VJTI, Mumbai, and a Ph.D. in 2019 from Symbiosis International University, Pune, India. Currently, he is a Post-Doctoral Fellow at REALTEK, Robotics Department, Norwegian University of Life Sciences (NMBU), Ås, Norway, where his work involves vision-based robotics and control. His research aims to develop autonomous robots for detecting contamination in salmon fillets and measuring nutrient content, leveraging robotic control and smart sensors for the in-line measurement of heterogeneous food items. His professional journey includes serving as a Researcher at NMBU, contributing to the GentleMAN project aimed at enhancing robotic manipulation skills, and being a Visiting Scientist at SINTEF Ocean, working on visual servoing and deformable object handling. With a decade of experience as an Assistant Professor at Symbiosis Institute of Technology, Pune, he has significantly contributed to the advancements of fractional calculus in the field of robotics. His work is also associated with the University of Sannino, Benevento, Italy, where he worked on the VERITAS project. The main goal was to develop a decision support system (DSS) based on crop growth dynamics for treatment and intervention on the farm. The primary goal was to use unmanned aerial vehicles (UAVs) to sense and track the variability in a vineyard (through drones) for effective precision agriculture (viticulture).

Kishore Bingi

Kishore Bingi received his B.Tech. Degree in Electrical and Electronics Engineering from Acharya Nagarjuna University, India, in 2012. He received his M.Tech. Degree in Instrumentation and Control Systems from the National Institute of Technology Calicut, India, in 2014, and his PhD in Electrical and Electronic Engineering from Universiti Teknologi PETRONAS, Malaysia, in 2019. From 2014 to 2015, he worked as an Assistant Systems Engineer at TATA Consultancy Services Limited, India. From 2019 to 2020, he worked as Research Scientist and Post-Doctoral Researcher at the Universiti Teknologi PETRONAS, Malaysia. From 2020 to 2022, he served as an Assistant Professor at the Process Control Laboratory, School of Electrical Engineering, Vellore Institute of Technology, India. Since 2022, he has been working as a faculty member in the Department of Electrical and Electronic Engineering at Universiti Teknologi PETRONAS, Malaysia. His research area is concerned with developing fractional-order neural networks, including fractional order systems and controllers, chaos prediction and forecasting, and advanced hybrid optimization techniques. He is an IEEE and IET Member and a registered Chartered Engineer (CEng) with the Engineering Council UK. He serves as an Editorial Board Member for the *International Journal of Applied Mathematics and Computer Science* and Academic Editor for *Mathematical Problems in Engineering and the Journal of Control Science and Engineering*.

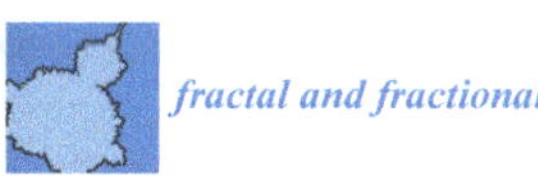

fractal and fractional

Editorial

Applications of Fractional-Order Calculus in Robotics

Abhaya Pal Singh [1] **and Kishore Bingi** [2,*]

[1] Department of Mechanical Engineering and Technology Management, Faculty of Science and Technology, Norwegian University of Life Sciences (NMBU), 1430 Ås, Norway; abhaya.pal.singh@nmbu.no
[2] Department of Electrical and Electronic Engineering, Universiti Teknologi PETRONAS, Seri Iskandar 32610, Malaysia
* Correspondence: bingi.kishore@utp.edu.my

1. Introduction

Fractional calculus, a branch of mathematical analysis, extends traditional calculus that encompasses integrals and derivatives of non-integer orders. This concept provides a robust framework for modelling complex systems, particularly in physics, engineering, and biology, where traditional calculus may fall short of capturing certain behaviours. Unlike integer-order operations in traditional calculus, fractional calculus enables differentiation and integration of arbitrary orders, offering flexibility in describing diverse physical phenomena [1]. The origins of fractional calculus can be traced back to the late 17th century, following the development of classical calculus by Leibniz and Newton. The concept was introduced in a letter from L'H ôpital to Leibniz in 1695, inquiring about the meaning of a half-order derivative. Leibniz's response hinted at the possibility of such an operation, setting the stage for further exploration [2,3]. Despite early interest, fractional calculus remained primarily a theoretical curiosity for over a century, exemplifying the depth and complexity of this mathematical concept. Significant advancements in the 19th century, led by mathematicians such as Liouville and Riemann, solidified the foundations of fractional calculus. Liouville formalized the concept by defining fractional integration and differentiation in terms of definite integrals, while Riemann expanded on these ideas through the Riemann-Liouville integral [4,5]. These pivotal works established a rigorous mathematical basis for fractional calculus, leading to its widespread applications across scientific disciplines today [6,7]. As a result, fractional calculus is now considered an essential mathematical tool, driving research and innovation in various fields [8–12].

The field of robotics has seen a revolutionary shift with the adoption of the fractional calculus framework, which has empowered researchers to develop more resilient and efficient control systems [13]. Conventional control methods often need help to address the intricacies and uncertainties present in robotic systems. Fractional calculus offers a refined approach to control and optimization thanks to its capacity to model systems with memory and hereditary properties [14]. This has led to substantial enhancements in robotics, particularly in motion planning, stability, and adaptive control. A significant advantage of fractional calculus in robotics is its ability to provide more precise descriptions of dynamic systems [15]. Robots frequently operate in environments with unforeseeable disturbances and fluctuating load conditions, where traditional integer-order models may fall short. However, fractional-order controllers deliver enhanced performance and reliability, such as fractional-order proportional integral derivative (FOPID) controllers [16,17]. They excel at managing system non-linearities and parameter variations, leading to smoother operations. Moreover, fractional calculus has ushered in progress in robotic path planning and trajectory optimization, enabling robots to manoeuvre more accurately through intricate and dynamic environments [18–20]. By integrating fractional-order models, researchers have developed algorithms that optimize paths more effectively, ensuring seamless transitions and reduced energy consumption. In conclusion, applying fractional calculus in robotics marks a

Citation: Singh, A.P.; Bingi, K. Applications of Fractional-Order Calculus in Robotics. *Fractal Fract.* **2024**, *8*, 403. https://doi.org/10.3390/fractalfract8070403

Received: 19 June 2024
Accepted: 4 July 2024
Published: 6 July 2024

significant leap forward, providing innovative solutions to persistent challenges and laying the groundwork for more advanced and autonomous robotic systems.

On this note, this Special Issue has captured the diversity of studies focusing on fractional calculus applications in various robotic systems. It contains nine articles and one review, which we will briefly describe in the next section. Please note that the purpose of this editorial is not to elaborate on each of the articles but rather to encourage the reader to explore them.

2. An Overview of Published Articles

Weidong Liu et al.'s article (contribution 1) introduces a fractional active disturbance rejection control scheme for remotely operated vehicles (ROVs). This scheme, which includes a double closed-loop fractional-order PID controller and model-assisted finite-time sliding-mode extended state observer, is more than just a theoretical concept. It has practical significance for ROV applications, as it demonstrates effective resistance to disturbances and independence from accurate model data, enabling high-precision tasks to be achieved despite disturbances and model uncertainties.

Likun Li et al.'s article (contribution 2) presents a novel approach to path planning for car-like mobile robots with suspension. Their fractional-order enhanced path planning method, which uses an improved ant colony optimization (ACO), is a unique solution. It aims to generate smooth and efficient paths in narrow and large-size scenes. It includes an accurate fractional-order-based kinematic modelling method and an improved ACO-based path planning method with dynamic angle constraints, adaptive pheromone adjustment, and fractional-order state-transfer models.

Bhukya Ramadevi et al. (contribution 3) propose a hybrid neural network model that has the potential to revolutionize wind power forecasting. Their model, which uses a long short-term memory model to forecast missing wind speed and direction data and a fractional-order neural network with a fractional arctan activation function to enhance wind power prediction, aims to improve wind power forecasting accuracy by addressing data gaps. The model has shown promising results in the field of wind power prediction.

Mohamed Naji Muftah et al.'s article (contribution 4) focused on enhancing the performance of a pneumatic positioning system by developing a control system based on fuzzy fractional-order proportional integral derivative controllers. The controllers were optimized using a particle swarm optimization algorithm, and real-time experimental results showed improved rapidity, stability, and precision compared to a fuzzy PID controller. The proposed control system effectively controlled a pneumatically actuated ball and beam system.

Banu Ataşlar-Ayyıldız (contribution 5) proposed a fractional-order proportional-tilt integral-derivative controller for a serial robotic manipulator. The controller was designed to achieve high-accuracy trajectory tracking and reduce the impact of disturbances and uncertainties. The controller parameters were optimized using a hybrid Gray Wolf and particle swarm optimization algorithm, demonstrating superior trajectory tracking and increased robustness compared to other controllers. Additionally, it showed reduced energy consumption, confirming its robustness and stability against continuous disturbances.

Xuan Liu et al.'s article (contribution 6) introduces a framework for implementing digital twins in industrial robots to facilitate real-time monitoring and performance optimization. This framework incorporates multi-domain modelling, behavioural matching, control optimization, and parameter updating. A fractional-order controller based on an enhanced particle swarm optimization algorithm improves the system's control performance. Experimental validation demonstrates substantial enhancements in time-domain performance, including reduced overshoot, decreased peak time, and improved settling time.

Dora Morar et al.'s article (contribution 7) introduces two controller design procedures for a mechatronic system. The first method formulates an optimization problem using linear matrix inequalities to determine closed-loop poles and address model uncertainties using linear differential inclusions. The second method involves a cascade controller with an inner P controller and an outer fractional-order FO-ID controller. Both methods offer four degrees of freedom for each axis. The article includes a numerical example and a comparison of performance metrics for the positioning system.

The article by Timi Karner et al. (contribution 8) discusses the use of dielectric elastomer actuators in soft robotics, noting their viscoelastic behaviour. They derived a fully fractional generalized Maxwell model using the Laplace transform to capture this behaviour. Based on the experimental results, they utilized the Pattern Search global optimization procedure to determine the model's optimal parameters and number of branches. This model can be implemented to control dielectric elastomer actuators and applied to various viscoelastic materials in simulations.

Yixiao Ding et al.'s article (contribution 9) introduces a fractional-order impedance controller for robot manipulators. Unlike traditional models, this method employs fractional calculus to describe damping forces more accurately. A systematic tuning procedure is developed based on frequency design, and comparisons with integer-order controllers show the fractional-order controller's superior step response and anti-disturbance performance.

The tenth article by Kishore Bingi et al. (contribution 10) reviews state-of-the-art fractional-order modelling and control strategies for robotic manipulators. Robotic manipulators are crucial in various fields, especially where human access is limited or hazardous. These highly complex systems require effective modelling and robust controllers to handle uncertainties. The review paper presents comprehensive research on modelling and control, aiming to provide the control engineering community with a better understanding and up-to-date knowledge in this area. The paper includes a summary of around 95 related works, focusing on modelling, control strategies, and future research directions.

3. Conclusions

In conclusion, this special issue, "Applications of Fractional-Order Calculus in Robotics", has successfully highlighted fractional calculus's expansive and varied applications in enhancing robotic systems, demonstrating its critical role in modern robotics research and development. By incorporating fractional calculus into their methodologies, researchers have addressed complex problems with greater precision and efficiency, showcasing the versatility and robustness of this mathematical approach. The articles within this issue cover diverse topics, from improving control accuracy and optimizing path planning to enhancing system robustness against disturbances and uncertainties.

The nine articles and one comprehensive review article in this special issue encapsulate various innovative approaches and novel methodologies. These contributions push the boundaries of robotics, emphasizing theoretical advancements and practical implementations. Each article presents unique solutions to longstanding challenges in robotics, highlighting the potential of fractional calculus to revolutionize various aspects of robotic technology. In summary, this special issue collectively underscores the significant impact of fractional calculus in advancing robotic technology and encourages further exploration and development in this promising study area.

Acknowledgments: As the Guest Editors of the Special Issue "Applications of Fractional-Order Calculus in Robotics", we would like to express our deep appreciation to all authors whose valuable work was published in this issue and thus contributed to its success.

Conflicts of Interest: The author declares no conflict of interest.

List of Contributions

1. Liu, W.; Guo, L.; Li, L.; Xu, J.; Yang, G. Fractional Active Disturbance Rejection Positioning and Docking Control of Remotely Operated Vehicles: Analysis and Experimental Validation. *Fractal Fract.* **2024**, *8*, 354. https://doi.org/10.3390/fractalfract8060354.
2. Li, L.; Jiang, L.; Tu, W.; Jiang, L.; He, R. Smooth and Efficient Path Planning for Car-like Mobile Robot Using Improved Ant Colony Optimization in Narrow and Large-Size Scenes. *Fractal Fract.* **2024**, *8*, 157. https://doi.org/10.3390/fractalfract8030157.

3. Ramadevi, B.; Kasi, V.R.; Bingi, K. Hybrid LSTM-Based Fractional-Order Neural Network for Jeju Island's Wind Farm Power Forecasting. *Fractal Fract.* **2024**, *8*, 149. https://doi.org/10.3390/fractalfract8030149.
4. Muftah, M.N.; Faudzi, A.A.M.; Sahlan, S.; Mohamaddan, S. Fuzzy Fractional Order PID Tuned via PSO for a Pneumatic Actuator with Ball Beam (PABB) System. *Fractal Fract.* **2023**, *7*, 416. https://doi.org/10.3390/fractalfract7060416.
5. Ataşlar-Ayyıldız, B. Robust Trajectory Tracking Control for Serial Robotic Manipulators Using Fractional Order-Based PTID Controller. *Fractal Fract.* **2023**, *7*, 250. https://doi.org/10.3390/fractalfract7030250.
6. Liu, X.; Gan, H.; Luo, Y.; Chen, Y.; Gao, L. Digital-Twin-Based Real-Time Optimization for a Fractional Order Controller for Industrial Robots. *Fractal Fract.* **2023**, *7*, 167. https://doi.org/10.3390/fractalfract7020167.
7. Morar, D.; Mihaly, V.; Şuşcă, M.; Dobra, P. Cascade Control for Two-Axis Position Mechatronic Systems. *Fractal Fract.* **2023**, *7*, 122. https://doi.org/10.3390/fractalfract7020122.
8. Karner, T.; Belšak, R.; Gotlih, J. Using a Fully Fractional Generalised Maxwell Model for Describing the Time Dependent Sinusoidal Creep of a Dielectric Elastomer Actuator. *Fractal Fract.* **2022**, *6*, 720. https://doi.org/10.3390/fractalfract6120720.
9. Ding, Y.; Liu, X.; Chen, P.; Luo, X.; Luo, Y. Fractional-Order Impedance Control for Robot Manipulator. *Fractal Fract.* **2022**, *6*, 684. https://doi.org/10.3390/fractalfract6110684.
10. Bingi, K.; Rajanarayan Prusty, B.; Pal Singh, A. A Review on Fractional-Order Modelling and Control of Robotic Manipulators. *Fractal Fract.* **2023**, *7*, 77. https://doi.org/10.3390/fractalfract7010077.

References

1. Tarasov, V.E. On history of mathematical economics: Application of fractional calculus. *Mathematics* **2019**, *7*, 509. [CrossRef]
2. Lischke, A.; Pang, G.; Gulian, M.; Song, F.; Glusa, C.; Zheng, X.; Mao, Z.; Cai, W.; Meerschaert, M.M.; Ainsworth, M.; et al. What is the fractional Laplacian? A comparative review with new results. *J. Comput. Phys.* **2020**, *404*, 109009. [CrossRef]
3. Petráš, I. Fractional-order control: New control techniques. In *Fractional Order Systems*; Elsevier: Amsterdam, The Netherlands, 2022; pp. 71–106.
4. Li, Z.; Liu, L.; Dehghan, S.; Chen, Y.; Xue, D. A review and evaluation of numerical tools for fractional calculus and fractional order controls. *Int. J. Control.* **2017**, *90*, 1165–1181. [CrossRef]
5. Valério, D.; Trujillo, J.J.; Rivero, M.; Machado, J.T.; Baleanu, D. Fractional calculus: A survey of useful formulas. *Eur. Phys. J. Spec. Top.* **2013**, *222*, 1827–1846. [CrossRef]
6. Chen, W.; Liang, Y. New methodologies in fractional and fractal derivatives modeling. *Chaos Solitons Fractals* **2017**, *102*, 72–77. [CrossRef]
7. Podlubny, I.; Petráš, I.; Skovranek, T.; Terpák, J. Toolboxes and programs for fractional-order system identification, modeling, simulation, and control. In Proceedings of the 2016 17th International Carpathian Control Conference (ICCC), High Tatras, Slovakia, 29 May–1 June 2016; IEEE: Piscataway, NJ, USA, 2016; pp. 608–612.
8. Vieira, L.C.; Costa, R.S.; Valério, D. An Overview of Mathematical Modelling in Cancer Research: Fractional Calculus as Modelling Tool. *Fractal Fract.* **2023**, *7*, 595. [CrossRef]
9. Arora, S.; Mathur, T.; Agarwal, S.; Tiwari, K.; Gupta, P. Applications of fractional calculus in computer vision: A survey. *Neurocomputing* **2022**, *489*, 407–428. [CrossRef]
10. Machado, J.T.; Lopes, A.M. Analysis of natural and artificial phenomena using signal processing and fractional calculus. *Fract. Calc. Appl. Anal.* **2015**, *18*, 459–478. [CrossRef]
11. Tapadar, A.; Khanday, F.A.; Sen, S.; Adhikary, A. Fractional calculus in electronic circuits: A review. *Fract. Order Syst.* **2022**, *1*, 441–482.
12. Oliveira, E.C.d.; Machado, J. A review of definitions for fractional derivatives and integral. *Math. Probl. Eng.* **2014**, *2014*, 1–7. [CrossRef]
13. Ahmed, S.; Azar, A.T. Adaptive fractional tracking control of robotic manipulator using fixed-time method. *Complex Intell. Syst.* **2024**, *10*, 369–382. [CrossRef]
14. Saif, A.; Fareh, R.; Sinan, S.; Bettayeb, M. Fractional synergetic tracking control for robot manipulator. *J. Control. Decis.* **2024**, *11*, 139–152. [CrossRef]
15. Angel, L.; Viola, J. Fractional order PID for tracking control of a parallel robotic manipulator type delta. *Isa Trans.* **2018**, *79*, 172–188. [CrossRef] [PubMed]
16. Bąkała, M.; Duch, P.; Ostalczyk, P. New Approach of the Variable Fractional-Order Model of a Robot Arm. *Appl. Sci.* **2023**, *13*, 3304. [CrossRef]
17. Sharma, R.; Rana, K.; Kumar, V. Performance analysis of fractional order fuzzy PID controllers applied to a robotic manipulator. *Expert Syst. Appl.* **2014**, *41*, 4274–4289. [CrossRef]

18. Copot, C.; Burlacu, A.; Ionescu, C.M.; Lazar, C.; De Keyser, R. A fractional order control strategy for visual servoing systems. *Mechatronics* **2013**, *23*, 848–855. [CrossRef]
19. Sharma, R.; Gaur, P.; Mittal, A. Performance analysis of two-degree of freedom fractional order PID controllers for robotic manipulator with payload. *Isa Trans.* **2015**, *58*, 279–291. [CrossRef] [PubMed]
20. Monje, C.A.; Balaguer, C.; Deutschmann, B.; Ott, C. Control of a soft robotic link using a fractional-order controller. *Handb. Fract. Calc. Appl.* **2019**, 321–338.

 *fractal and fractional*

Review

A Review on Fractional-Order Modelling and Control of Robotic Manipulators

Kishore Bingi [1,*], **B Rajanarayan Prusty** [2], **Abhaya Pal Singh** [3,*]

[1] Department of Electrical and Electronics Engineering, Universiti Teknologi PETRONAS, Seri Iskandar 32610, Malaysia
[2] Department of Electrical and Electronics Engineering, Alliance College of Engineering and Design, Alliance University, Bengaluru 562106, India
[3] Department of Mechanical Engineering and Technology Management, Faculty of Science and Technology, Norwegian University of Life Sciences (NMBU), 1430 Ås, Norway
* Correspondence: bingi.kishore@ieee.org (K.B.); abhaya.aps@gmail.com or abhaya.pal.singh@nmbu.no (A.P.S.)

Abstract: Robot manipulators are widely used in many fields and play a vital role in the assembly, maintenance, and servicing of future complex in-orbit infrastructures. They are also helpful in areas where it is undesirable for humans to go, for instance, during undersea exploration, in radioactive surroundings, and other hazardous places. Robotic manipulators are highly coupled and non-linear multivariable mechanical systems designed to perform one of these specific tasks. Further, the time-varying constraints and uncertainties of robotic manipulators will adversely affect the characteristics and response of these systems. Therefore, these systems require effective modelling and robust controllers to handle such complexities, which is challenging for control engineers. To solve this problem, many researchers have used the fractional-order concept in the modelling and control of robotic manipulators; yet it remains a challenge. This review paper presents comprehensive and significant research on state-of-the-art fractional-order modelling and control strategies for robotic manipulators. It also aims to provide a control engineering community for better understanding and up-to-date knowledge of fractional-order modelling, control trends, and future directions. The main table summarises around 95 works closely related to the mentioned issue. Key areas focused on include modelling, fractional-order modelling type, model order, fractional-order control, controller parameters, comparison controllers, tuning techniques, objective function, fractional-order definitions and approximation techniques, simulation tools and validation type. Trends for existing research have been broadly studied and depicted graphically. Further, future perspective and research gaps have also been discussed comprehensively.

Keywords: approximation approaches; fractional calculus; fractional-order control; fractional-order model; industrial manipulators; optimization techniques; robotic manipulators

Citation: Bingi, K.; Prusty, R.; Singh, A. A Review on Fractional-Order Modelling and Control of Robotic Manipulators. *Fractal Fract.* **2023**, *7*, 77. https://doi.org/10.3390/fractalfract7010077

Academic Editor: Inés Tejado

Received: 5 December 2022
Revised: 4 January 2023
Accepted: 7 January 2023
Published: 10 January 2023

1. Introduction

Robotic manipulators are electronically controlled mechanisms consisting of multiple segments that perform tasks by interacting with their environment. They can perform repetitive tasks at speeds and accuracies far exceeding human operators [1]. They can move or handle objects automatically depending upon the given number of DOF. The DOF of industrial robotic manipulators can range from two to ten, or more. As they are capable of automating, many automated applications have recently been seen. The most common include spot welding, assembly, handling, painting, and palletizing [2]. Technological advancements have greatly improved robotic manipulators' accuracy and precision, thus allowing them to automate new applications such as automated 3D printing. Robotic manipulator automation makes manufacturing processes more efficient, reliable, and productive. As a result, considerable attention has been given to modelling the robotic manipulators and designing practical controllers that are easy to implement and provide optimal controlled performance [3–5].

Recently, the fractional-order concept has attracted increasing attention in control research. Fractional-order modelling and control, using fractional-order derivatives/integrals, has been recognized as an alternative strategy to solve many robust control problems effectively [6,7]. This is also true in the case of robotic manipulators. In the last few years, extensive research has been performed on robotic manipulators using fractional-order concepts. Thus, this study thoroughly reviews the application of fractional calculus in modelling and controlling robotic manipulators. Therefore, a comprehensive literature review on fractional-order modelling and control techniques for various robotic manipulators is presented. This study is structured as follows:

- Different conventional and fractional-order modelling strategies for lower and higher DOF robotic manipulators are included in the review.
- A review of developed fractional-order controllers for various robotic manipulators evolved from PID, sliding mode, fuzzy, backstepping, active disturbance rejection control, and impedance control is presented.
- Fractional-order derivative definitions and approximation techniques are also presented.
- Trends for existing research and future developments in this area have been broadly presented and depicted in a graphical layout.

The paper's remaining sections are organized as follows: the preliminaries of fractional calculus, including the derivative definitions, are presented in Section 2. Section 3 summarizes the collected literature review and the graphical trend analysis. Section 4 offers the detailed dynamic modelling of robotic manipulators. The broad overview of fractional-order control strategies developed for various robotic manipulators is presented in Section 5. Finally, the paper concludes in Section 6.

2. Preliminaries of Fractional Calculus

The fractional-order differintegral operator $\mathcal{D}_t^\alpha$ for an order α of a given function $f(t)$ is defined as,

$$\mathcal{D}_t^\alpha f(t) = \begin{cases} \frac{d^\alpha}{dt^\alpha} f(t), & \alpha > 0, \\ f(t), & \alpha = 0, \\ \int_0^t f(\tau)d\tau, & \alpha < 0. \end{cases} \tag{1}$$

The three most frequently used definitions of fractional-order derivative $\mathcal{D}_t^\alpha$ for $\alpha > 0$ are Grünwald–Letnikov, Riemann–Liouville, and Caputo, as given in orange, blue, and grey coloured boxes of Figure 1, respectively. In the definitions, $\Gamma(\cdot)$ is Euler's Gamma function. On the other hand, among the various approximation techniques available in the literature, Oustaloup's technique is the most widely used frequency domain approximation method. The formula for computing the Oustaloup and refined Oustaloup approximations in red and green coloured boxes is in Figure 1. These approximation techniques are valid for estimating the Nth order approximation of order within the lower and higher frequencies of ω_l and ω_h, respectively.

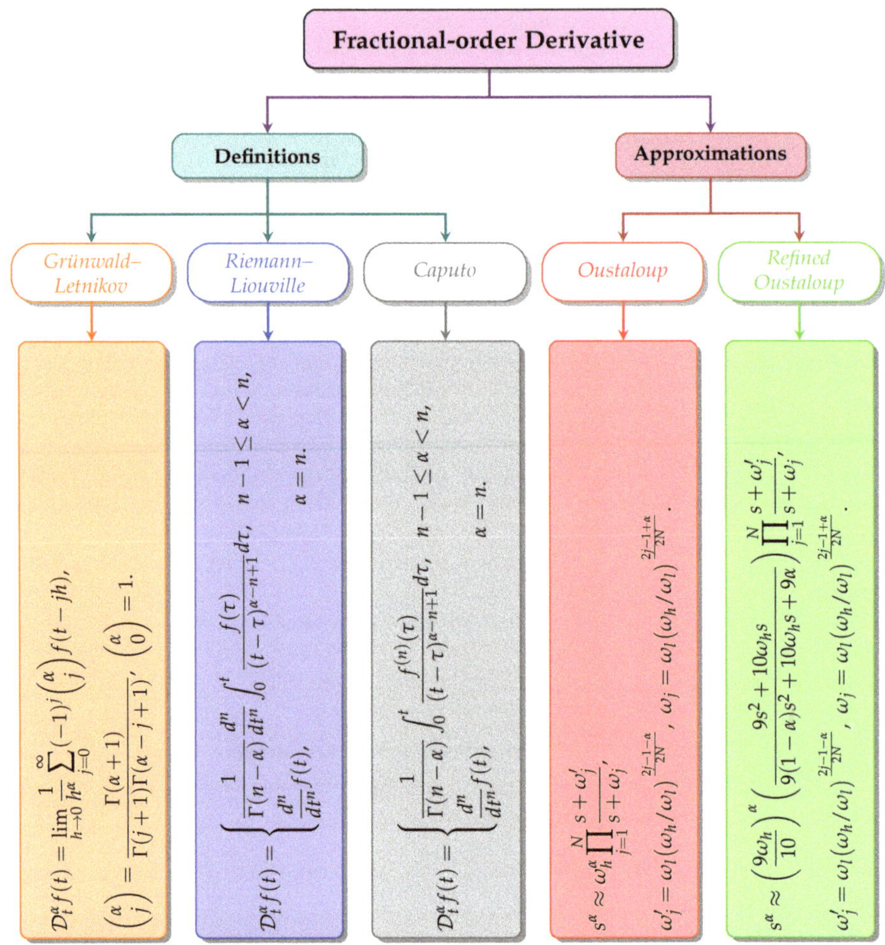

Figure 1. Definitions and approximation techniques of fractional-order derivative.

3. Survey With Trend Analysis

From the collected literature review in Table 1, a graphical trend analysis is made in this section. From the table, the summary of the manipulators' trend is given in Figure 2. As shown in the figure, research has been conducted on various manipulators of DOF ranging from 1 to 7. However, most of the research on developing either fractional-order models or controllers has been conducted on 1, 2, and 3 DOF manipulators, with 2 DOF being the highest, around 60% (see Figure 2a). Moreover, as shown in Figure 2b, about 66% of research has been conducted on robotic manipulators without any payload, and only 34% work with a load. Further, it can be observed from Figure 2c that the research on developing either fractional-order models or controllers has been performed primarily on two-link, rigid planar, and single-link manipulators. It is also worth highlighting that research has been conducted on some industrial manipulators, including PUMA 560, SCARA, Polaris -I, Stewart platform, Staubli RX-60, Robotino-XT, Mitsubishi RV-4FL, KUKA youBot, Fanuc, ETS-MARSE, EFFORT-ERC20C-C10, Delta robot, differential-drive mobile robot [8] and University of Maryland manipulators.

Table 1. Summary of works focussed on fractional-order modelling and controlling of robotic manipulators.

Ref.	Manipulator Details				Modelling Details				Controller Details						Tool S/P	
	Type	DOF	Payload	FOM	Method	Order	FOC	Controller	CP	Tuning Technique	Comparison Controllers	OF	Approximation		
[9]	2R robotic manipulator	2	✗	✗	Mathematical modelling	2	✓	Fractional-order D controller	2	Trial and error	PI and PD controllers	Transient response characteristics	Padé approximation	—	S
[10]	Redundant manipulator	—	✗	✗	Closed-Loop Pseudoinverse	2	✓	Pseudoinverse Algorithm	5	—	—	Tracking error	Grünwald–Letnikov's method	—	S
[11]	Single-link flexible manipulator	1	✓	✗	Mathematical modelling	2	✓	Fractional-order PD controller	3	Trial and error	PD controller	Stability	Digital IIR filter approximation	M	P
[12]	Robotic manipulator	2	✓	✗	Mathematical modelling	2	✓	Fractional fuzzy adaptive sliding mode controller	5	Trial and error	—	Tracking error	CRONE approximations	M	S
[13]	Rotational joints robotic manipulator	2	✓	✗	Mathematical modelling	2	✓	Fractional-order PD-PI controller	5	Trial and error	PD-PI controller	Transient response characteristics	—	—	S
[14]	Two-link robotic manipulator	2	✗	✗	Lagrangian formulation	2	✓	Adaptive fractional-order PID controller	5	Genetic Algorithm	PID controller	ISE	CRONE approximations	—	S
[15]	Polar robotic manipulator	2	✓	✗	State space model	4	✓	Fuzzy Fractional-order PD surface sliding mode controller	8	Genetic Algorithm	Classical PD surface sliding mode controller	RMSE	Caputo derivative	—	S
[16]	Two-link flexible joint manipulator	2	✗	✗	Lagrangian formulation	8	✓	Fractional order fuzzy sliding mode controller	6	Genetic Algorithm	Sliding mode controller, PD surface sliding mode controller, Sliding surfaces through fractional PD controller	IAE, ITAE, ISV	Caputo derivative	—	S
[17]	Two-link planar rigid robotic manipulator	2	✗	✗	Mathematical modelling	2	✓	Fractional-order PID controller	5	Particle Swarm Optimization	Fuzzy and PID controllers	RMSE, MAE, MMFAE	Riemann–Liouville method	—	S
[18]	Mechanical manipulator	2	✗	✗	Mathematical modelling	3	✓	Fractional variable structure control and sliding mode control	6	Trial and error	Integer variable structure control and sliding mode control	Switching activity	Taylor series expansion	—	P
[19]	Two-link planar rigid robotic manipulator	2	✗	✗	Mathematical modelling	2	✓	Fractional-order PID controller	5	Genetic Algorithm, Particle Swarm Optimization	—	RMSE, MAE, MMFAE	—	M	S
[20]	Manipulator robot (Fanuc)	6	✓	✗	Robust disturbance observer	1	✓	Fractional-order PI controller	3	Decentralized tuning	PI controller	Gain Margins	Refined Oustaloup Filter	M	P

Table 1. *Cont.*

Ref.	Manipulator Details				Modelling Details			Controller Details						Tool	S/P
	Type	DOF	Payload	FOM	Method	Order	FOC	Controller	CP	Tuning Technique	Comparison Controllers	OF	Approximation		
[21]	University of Maryland (UMD) manipulator	3	✓	✗	Mathematical modelling	2	✓	Fractional-order PID controller	5	Pattern search optimization	PID controller	MSE	—	—	S
[22]	Flexible link manipulator	2	✓	✗	Euler-Bernoulli method	2	✓	Fractional-order sliding mode controller	6	Particle Swarm Optimization	Sliding mode controller	ISE	Riemann–Liouville method	—	S
[23]	Angular manipulator	3	✗	✗	Lagrange model	2	✓	Fractional-order PID controller	5	Trial and error	—	—	Riemann–Liouville method	M, L	P
[24]	Robotic manipulator	6	✓	✗	Mathematical modelling	6	✓	Fractional-order PD controller	3	Bode tuning	PD controller	Linear and angular velocities	Grünwald–Letnikov method	M	S
[25]	Single-link flexible manipulator	1	✗	✓	Non-commensurate fractional-order model	0.71, 0.92	✓	Fractional order sliding mode controller	4	QR decomposition method	Sliding mode controller	Tracking error	Caputo derivative	M	P
[4]	Two-link planar rigid robotic manipulator	2	✗	✗	Mathematical modelling	2	✓	Fractional-order fuzzy PID controller	6	Cuckoo Search Algorithm	Fuzzy PID, fractional-order PID and PID controllers	IAE, IACCO	Oustaloup's approximation	M	S
[26]	Hydraulic manipulator	2	✓	✗	Mathematical modelling	2	✓	Fractional-order nonsingular terminal sliding mode controller	16	Trial and error	Integer-order nonsingular terminal sliding mode controller	RMSE	Refined Oustaloup filter	M	P
[27]	Single-link flexible manipulator	1	✗	✓	Non-commensurate fractional-order model	0.71, 0.92	✓	Observer-based fractional-order sliding mode controller	8	Stability criterion	Sliding mode controller	Tracking error	Caputo derivative	—	P
[5]	Two-link planar rigid robotic manipulator	2	✓	✗	Mathematical modelling	2	✓	Two-degree of freedom fractional-order PID controller	8	Cuckoo Search Algorithm	Two-degree of freedom PID controller	Weighted sum of ITAE and IACCO	Oustaloup's approximation	M	S
[28]	Two-link robotic manipulator	2	✗	✗	Mathematical modelling	2	✓	Adaptive fractional-order nonsingular fast terminal sliding mode controller	13	Trial and error	Nonsingular terminal, Second-order sliding mode controllers	Error, Reaching time, Chattering effect	Riemann–Liouville method	—	S
[29]	Two-link robotic manipulator	2	✗	✗	Mathematical modelling	2	✓	Fractional-order PID controller	5	Particle swarm optimization, Genetic algorithm and Estimation of distribution algorithm	—	RMSE	Riemann–Liouville method	M	S

Table 1. *Cont.*

Ref.	Manipulator Details				Modelling Details			Controller Details						Tool	S/P
	Type	DOF	Payload	FOM	Method	Order	FOC	Controller	CP	Tuning Technique	Comparison Controllers	OF	Approximation		
[30]	Robotic manipulator (PUMA 560)	2	✗	✗	Mathematical modelling	2	✓	Fractional-order fuzzy PID controller	5	Genetic Algorithm	PID, fractional-order PID and fuzzy PID controllers	ISE	—	M	S
[31]	Two-link planar rigid robotic manipulator (SCARA)	2	✓	✗	Mathematical modelling	2	✓	Two-layered fractional-order fuzzy logic controller	10	Cuckoo Search Algorithm	Two-layered, single-layered fuzzy logic, PID controllers	IAE	Oustaloup's approximation	M	S
[32]	Rotary manipulator	2	✗	✗	Mathematical modelling	2	✓	Fractional-order adaptive backstepping controller	7	Trial and error	Adaptive backstepping controllers	Tracking performance	Caputo derivative	M	P
[33]	Robotic manipulator	4	✗	✓	Pseudoinverse algorithm	0.5, 0.6, 0.8, 0.9, 0.99	✗	—	—	—	—	Tracking accuracy	Grünwald–Letnikov method	M	S
[34]	Inchworm/ Caterpillar robotic manipulator	1	✗	✗	Euler–Lagrange method	2	✓	Neural network-based fraction integral terminal sliding mode controller	5	Trial and error	Sliding mode controller, Integral terminal sliding mode controller, Fraction integral terminal sliding mode controller	Tracking error	—	M	S
[35]	Single-link direct joint driven robotic manipulator	1	✗	✗	Mathematical modelling	2	✓	Sliding mode based fractional-order PD type iterative learning control	5	Trial and error	Sliding mode based fractional-order D type iterative learning control, Higher-order iterative learning control	Tracking error	CRONE approximations	M	S
[36]	Robotic manipulator	2	✓	✗	Mathematical modelling	2	✓	Time delay estimation-based fractional-order nonsingular terminal sliding mode controller	9	Trial and error	Time delay estimation-based, continuous nonsingular terminal, Time delay estimation-based integer-order nonsingular terminal sliding mode controllers	Tracking error	Riemann–Liouville method	M	P
[37]	Inchworm/ Caterpillar robotic manipulator	1	✗	✗	Euler–Lagrange formalism	2	✓	Adaptive fractional-order PID sliding mode controller	5	Bat optimization algorithm	PID, fractional-order PID, sliding mode controller	Weighted sum of IAE and ISV	Oustaloup's recursive approximation	M	S

Table 1. *Cont.*

Ref.	Manipulator Details				Modelling Details			Controller Details						Tool	S/P
	Type	DOF	Payload	FOM	Method	Order	FOC	Controller	CP	Tuning Technique	Comparison Controllers	OF	Approximation		
[38]	Five-bar-linkage robotic manipulator	-	✗	✗	Mathematical modelling	2	✓	Fractional-order PID controller	5	Modified Particle Swarm Optimization	Fractional-order PID controller tuned using standard, constriction factor approach, random inertia weight-based particle swarm optimization algorithms	IAE, ISE, ITSE	Oustaloup's approximation	M	P
[39]	Two-link robotic manipulator	2	✓	✗	Mathematical modelling	2	✓	Interval type-2 fractional-order fuzzy PID controller	6	Artificial Bee Colony-Genetic Algorithm	Interval type-2 fuzzy PID, Type-1 fractional-order fuzzy PID, Type-1 fuzzy PID, PID	ITAE	Oustaloup's approximation	M	S
[40]	Single-link flexible manipulator	1	✗	✗	Mathematical modelling	2	✓	Fractional-order phase-lead compensator	4	Nyquist criterion	PID controller	Gain Margin	Grünwald–Letnikov method	—	P
[41]	Three and five links redundant manipulators	3, 5	✗	✓	Moore-Penrose pseudoinverse	—	✗	—	—	—	—	—	Grünwald–Letnikov method	M	S
[42]	Robotic manipulator	2	✓	✗	State space model	4	✓	Fractional-order global sliding mode controller	10	Trial and error	Sliding mode controller	Tracking error	Riemann–Liouville method	—	S
[43]	Robotic manipulator	2	✗	✗	Mathematical modelling	2	✓	Fractional-order fuzzy pre-compensated fractional-order PID controller	9	Hybrid artificial bee colony-genetic algorithm	Fuzzy pre-compensated PID, fuzzy PID and PID controllers	ITAE	Oustaloup's recursive approximation	M	S
[44]	Two-link planar rigid robotic manipulator	2	✓	✗	Mathematical modelling	2	✓	Non-linear adaptive fractional-order fuzzy PID controller	7	Backtracking search algorithm	Non-linear adaptive fuzzy PID controller	ITAE, ITACO	Grünwald–Letnikov method	L	S
[45]	Two-link robotic manipulator	2	✗	✓	Fractional adaptive neural network	—	✓	Fractional-order PID controller	5	Trial and error	—	Tracking error	Caputo derivative	—	S
[46]	Two-link rigid planar manipulator	2	✗	✗	Mathematical modelling	2	✓	Fractional-order PID controller	5	Genetic Algorithm	PID controller	Weighted sum of IAE and ISCCO	Short memory principle	L	P
[47]	Rotary flexible joint manipulator	1	✗	✗	Mathematical modelling	2	✓	Fractional-order integral controller	2	Gain margins	Integral controller	Tracking accuracy	Oustaloup's approximation	M	P
[48]	Electrically driven three-link rigid robotic manipulator	3	✗	✗	Mathematical modelling	3	✓	Fractional-order fuzzy PD+I controller	4	Cuckoo Search Algorithm	PID, Fractional-order PID, Integer-order fuzzy PD+I	IAE	Grünwald–Letnikov method	M	S

Table 1. *Cont.*

Ref.	Manipulator Details				Modelling Details				Controller Details							Tool	S/P
	Type	DOF	Payload	FOM	Method	Order	FOC	Controller	CP	Tuning Technique	Comparison Controllers	OF	Approximation				
[49]	Robotic manipulator (SCARA)	2	✗	✗	Linear model	2	✓	Fractional-order model reference adaptive controller	3	Trial and error	Model reference adaptive controller	Delay time	Oustaloup's approximation		—	S	
[50]	Robotic manipulator (PUMA 560)	3	✗	✗	Mathematical modelling	2	✓	Fractional-order nonsingular fast terminal sliding mode control based fault tolerant control	7	Trial and error	Adaptive fractional-order nonsingular fast terminal sliding mode controller, Nonsingular fast terminal sliding mode control based active fault tolerant control	Convergence speed	Riemann–Liouville method		—	S	
[51]	Two-link planar electrically-driven rigid robotic manipulator	2	✗	✗	Mathematical modelling	2	✓	Fractional-order self organizing fuzzy controller	6	Cuckoo Search Algorithm	Fractional-order fuzzy PID	IAE	Grünwald–Letnikov method		M	S	
[52]	Serial link manipulator	2	✗	✗	Mathematical modelling	2	✓	Fractional-order PID and auxiliary controllers	5	Trial and error	Torque approach controller	Tracking error	CRONE approximations		M	S	
[53]	Redundant manipulator (SCARA)	5	✗	✗	Mathematical modelling	2	✓	Fuzzy fractional-order PID controller	6	Artificial Bee Colony Algorithm	PID and fuzzy PID controllers	ITAE	—		M	S	
[54]	Three-link robotic manipulator (Staubli RX-60)	6	✗	✗	Mathematical modelling	3	✓	Fractional-order PID controller	5	Cuckoo Search Algorithm	PID controller	IAE, ITAE, ISE and IACCO	—		M	S	
[55]	Robotic manipulator	6	✗	✗	Kinematic modelling	2	✓	Fractional order nonsingular fast terminal sliding mode control	13	Trial and error	—	Tracking error	Riemann–Liouville method		—	S	
[56]	Three-link planar rigid robotic manipulator	3	✗	✗	Euler–Lagrange formalism	3	✓	Fractional-order PID controller	5	Evaporation Rate-Based Water Cycle Algorithm	PID controller	Weighted sum of IAE and IACCO	Grünwald–Letnikov method		M	S	
[57]	Two-link planar rigid robotic manipulator	2	✗	✗	Euler–Lagrange formalism	2	✓	Fractional-order fuzzy sliding mode PD/PID controller	8	Cuckoo Search Algorithm	Integer-order fuzzy sliding mode PD/PID controller	Weighted sum of IAE and chatter	Grünwald–Letnikov method		M	S	
[58]	Two-link planar rigid robotic manipulator	2	✗	✗	Lagrangian-Euler formulation	2	✓	Fractional-order fuzzy sliding mode controller with proportional derivative surface	6	Genetic Algorithm	Integer-order fuzzy SMC with proportional derivative surface	Weighted sum of IAE and chatter	Grünwald–Letnikov method		M	S	
[59]	Parallel robotic manipulators (Delta Robot)	3	✓	✗	Inverse kinematic model	3	✓	Fractional-order PID controller	5	FMINCON (Gradient descent algorithm)	PID controller	RMSE	—		M	P	

Table 1. *Cont.*

Ref.	Manipulator Details				Modelling Details				Controller Details							Tool S/P	
	Type	DOF	Payload	FOM	Method	Order	FOC	Controller	CP	Tuning Technique	Comparison Controllers	OF	Approximation				
[60]	Robotic manipulator (SCARA)	2	✗	✓	Euler–Lagrange and Hamilton formalisms	1.14	✓	Fractional-order PI/PD controller	3	Particle Swarm Optimization	PI/PD controller	ITAE	Grünwald–Letnikov method		M	S
[61]	Serial robotic manipulator	6	✗	✗	Mathematical modelling	2	✓	Fractional-order adaptive nonsingular terminal siding mode controller	8	Trial and error	—	Tracking error	Riemann–Liouville method		M	S
[3]	Cable-driven manipulator (Polaris-I)	2	✓	✗	Mathematical modelling	2	✓	Time delay control scheme-based adaptive fractional-order nonsingular terminal sliding mode controller	15	Trial and error	Time delay estimation-based adaptive, continuous fractional-order nonsingular terminal sliding mode controller	RMSE	Riemann–Liouville method		M	P
[62]	Robotic manipulator	2	✗	✗	Euler–Lagrange formalism	2	✓	Fuzzy fractional-order PID controller	3	Heuristic Tuning	Sliding mode control, Super twisting sliding mode control, Fuzzy PID	ITAE, ISE	Grünwald–Letnikov method		C++	P
[63]	Rigid planar robotic manipulator	2	✓	✗	Mathematical modelling	2	✓	Collaborative fractional order PID and fractional order fuzzy logic controller	9	Cuckoo Search Algorithm	PID, Fractional-order PID, Fractional-order fuzzy PID	ITAE	Oustaloup's recursive approximation		M	S
[64]	Two-link robotic manipulator	2	✗	✗	Mathematical modelling	2	✓	Two-degree-of-freedom fractional-order fuzzy PI-D	16	Multi-objective non-dominated sorting genetic algorithm-II	Two-degree-of-freedom fractional-order PI-D	IAE	Grünwald–Letnikov method		M	S
[65]	Three-link planar rigid robotic manipulator	3	✓	✗	Euler–Lagrange formalism	3	✓	Self-regulated fractional-order fuzzy PID controller	6	Backtracking Search Algorithm	Self-regulated integer-order fuzzy PID controller	IAE, IACCO	Grünwald–Letnikov method		L	S
[66]	Single-link flexible manipulator	1	✓	✗	Lagrangian formulation	2	✓	Sliding fractional order controller	6	Trial and error	PD controller	Tracking error	—		—	S
[67]	Two-link robotic manipulator	2	✗	✗	Mathematical modelling	2	✓	Fractional-order fuzzy PID controller	6	Particle Swarm Optimization	Fractional-order PID controller	IAE, IACCO	Oustaloup's approximation		M	S
[68]	Single-link flexible manipulator	1	✓	✗	State space model	4	✓	Fractional-order sliding mode controller	10	Trial and error	PID, Sliding mode controller	RMSE, MAE	CRONE approximations		M	S
[69]	Cable-driven manipulator (Polaris-I)	2	✓	✗	Mathematical modelling	2	✓	Fractional-order nonsingular terminal sliding mode controller	12	Closed-loop control tuning	Time delay estimation-based and continuous fractional-order nonsingular terminal sliding mode controller	RMSE	Refined Oustaloup filter		M	P

Table 1. *Cont.*

Ref.	Manipulator Details				Modelling Details				Controller Details						Tool S/P	
	Type	DOF	Payload	FOM	Method	Order	FOC	Controller	CP	Tuning Technique	Comparison Controllers	OF	Approximation		
[70]	Serial Flexible Link Robotic Manipulator, Serial Flexible Joint Robotic Manipulator	2	✗	✓	Fractional transfer function model	0.3, 0.9	✓	Fractional-order PID controller	5	Trial and error	PID controller	Transient response characteristics	Oustaloup's approximation	M	P
[71]	Robotic manipulator	2	✗	✗	Kinematic modelling	2	✓	Fractional-order PID controller	5	Particle Swarm Optimization	PID controller	Error	—	—	S
[72]	Two-link flexible robotic manipulator	3	✗	✓	Euler–Lagrange formulation	0.98	✓	Fractional-order adaptive sliding mode controller	13	Trial and error	Adaptive sliding mode controller	Tracking error	—	M	S
[73]	Exoskeleton Robot (ETS-MARSE)	7	✗	✗	Mathematical modelling	2	✓	Adaptive neural network fast fractional integral terminal sliding mode control	6	Trial and error	Fast fractional integral terminal sliding mode controller	Tracking error	Grünwald–Letnikov method	M	P
[74]	Robotic manipulator	2	✓	✗	Mathematical modelling	2	✓	Adaptive fractional high-order terminal sliding mode controller	10	Trial and error	H∞-Adaptive control, intelligent PD, intelligent PID, Adaptive third-order sliding mode controller	Convergence speed and precision	Oustaloup method	M	S
[75]	Robotic manipulator (PUMA 560)	6	✓	✓	Euler–Lagrange formalism	12	✓	Fractional-order PI, PD controllers	9	Cuckoo Search Algorithm	PI, PD controllers	RMSE	Caputo–Fabrizio derivative, Atangana–Baleanu integral	—	P
[76]	3-RRR planar parallel robots	3	✗	✗	Inverse kinematics using Cayley–Menger determinants and bilateration	2	✓	Fractional-order PID controller	5	Bat optimization algorithm	PID controller	Weighted function	—	M	P
[77]	Muscle-actuated manipulator	2	✗	✓	Fractional order describing functions	2	✗	—	—	—	—	—	Grünwald–Letnikov method	—	P
[78]	Rigid robotic manipulator	2	✗	✗	Mathematical modelling	2	✓	Deep convolutional neural network based Fractional-order terminal sliding-mode controller	15	FMINCON (Gradient descent algorithm)	Nonsingular and conventional fractional-order terminal sliding-mode controllers	Fractional-order loss function	Caputo derivative	—	S

Table 1. *Cont.*

Ref.	Manipulator Details				Modelling Details				Controller Details							Tool S/P	
	Type	DOF	Payload	FOM	Method	Order	FOC	Controller	CP	Tuning Technique	Comparison Controllers	OF	Approximation				
[79]	Robotic manipulator	2	✗	✗	Mathematical modelling	2	✓	Fractional-order fuzzy PD and I controller	8	Multi-objective non-dominated sorting genetic algorithm-II, dragonfly algorithm, multi-verse optimization, ant lion optimizer algorithms	PID, fuzzy PID controllers	IAE	Grünwald–Letnikov method			M	P
[80]	Robotic manipulator (SCARA)	2	✗	✗	Mathematical modelling	2	✓	Fractional-order PID and Fractional-order pre-filter	5, 4	Genetic Algorithm, Trial and error	—	Gain Margins	CRONE approximations			M	S
[81]	Two-link robotic manipulator	2	✗	✗	Mathematical modelling	2	✓	Time delay estimation-based adaptive fractional-order nonsingular terminal sliding mode controller	12	Trial and error	Nonsingular fast terminal sliding mode controller, Second order nonsingular fast terminal sliding mode controller	Tracking error	Riemann–Liouville method			M	S
[82]	Parallel robotic manipulator	6	✓	✗	Kinematic modelling	3	✓	Fractional-order active disturbance rejection controller	16	Trial and error	Active disturbance rejection controller	Tracking accuracy	—			M	P
[83]	Single-link robotic manipulator	1	✗	✓	Euler–Lagrange formulation	0.5	✗	Feedback controller	8	Pole placement method	PID, LQR controllers	Tracking accuracy	Oustaloup's approximation			M	P
[83]	Serial-link flexible robotic manipulator, Serial flexible joint robotic manipulator	2	✗	✓	Fractional value selection algorithm	0.3, 0.9	✓	Fractional-order PID controller	5	Trial and error	PID controller	Tracking accuracy	Oustaloup's approximation			M	P
[84]	Rotary flexible joint manipulator	1	✗	✗	Mathematical modelling	2	✓	State-feedback-based fractional-order integral controller	2	Trial and error	Pure state-feedback control scheme and the modified state-feedback-based fractional-order integral controllers	Tracking error	CRONE, Oustaloup's approximations			M	S
[85]	Robotic manipulator (PUMA 560)	3	✓	✗	State space model	2	✓	Fractional-order adaptive backstepping controller	6	Trial and error	PID and Computed torque controllers	Tracking error and convergence speed	Caputo method			M	S

Table 1. *Cont.*

Ref.	Manipulator Details				Modelling Details			Controller Details							Tool S/P	
	Type	DOF	Payload	FOM	Method	Order	FOC	Controller	CP	Tuning Technique	Comparison Controllers	OF	Approximation		
[86]	Two-link robotic manipulator	2	✗	✗	Mathematical modelling	2	✓	Time delay estimation-based adaptive fractional-order nonsingular terminal sliding mode controller	10	Trial and error	—	Tracking performance and speed	Oustaloup's recursive approximation	—	S
[83]	Single Rigid Link Robotic Manipulator, Serial Link Robotic Manipulator	2	✗	✗	Mathematical modelling	2	✓	Adaptive fractional-order controller	5	Trial and error	Integer-order and adaptive controllers	Transient response characteristics	Oustaloup's approximation	M	P
[2]	Cooperative manipulator (Mitsubishi RV-4FL)	6	✓	✗	Kinematic modelling	3	✓	Coupled fractional-order sliding mode control	5	Fuzzy tuning	PI, Sliding mode controllers, fractional-order sliding mode controller	IAE, ISE, STD	Oustaloup's approximation	M	P
[87]	Single flexible link robotic manipulator, Serial flexible joint robotic manipulator	1,2	✗	✓	Euler–Lagrange formulation	0.5	No	Feedback controller	8	Pole placement method	PID, LQR controllers	Tracking accuracy	Oustaloup's approximation	M	P
[88]	Single flexible link robotic manipulator, Serial flexible joint robotic manipulator	1,2	✗	✗	Euler–Lagrange formulation	2	✓	Fractional-order PID controller	5	Trial and error	PID controller	Transient response characteristics	Oustaloup's approximation	M	S
[89]	Stewart Platform	6	✗	✗	Lagrange-Euler approach	3	✓	Fractional order fuzzy PID controller	8	Particle Swarm Optimization	PID, fractional-order PID and fuzzy PID controllers	MAE, RMSE	Oustaloup's approximation	M	P
[90]	Robotic manipulator (PUMA 560)	3	✗	✗	Mathematical modelling	2	✓	Fractional-order backstepping fast terminal sliding mode controller	15	Trial and error	PID, Computed torque controller, Nonsingular fast terminal sliding mode controller	Position tracking error	Oustaloup's approximation	M	S
[91]	Robotic manipulator (EFFORT-ERC20C-C10)	6	✓	✗	Mathematical modelling	2	Yes	Fractional-order impedance control	3	Frequency design method	Impedance control	ITSE	Impulse response method	—	P

Table 1. *Cont.*

| Ref. | Manipulator Details | | | | Modelling Details | | | Controller Details | | | | | | Tool | S/P |
	Type	DOF	Payload	FOM	Method	Order	FOC	Controller	CP	Tuning Technique	Comparison Controllers	OF	Approximation		
[1]	Three-link omnidirectional mobile robot manipulator (KUKA youBot)	5	✗	✗	Lagrangian dynamics equation	3	✓	Adaptive fractional-order nonsingular terminal sliding mode controller	9	Trial and error	Fractional-order terminal sliding mode controller, Nonsingular terminal sliding mode controller	Tracking speed and accuracy	Riemann–Liouville method	M	P
[92]	Two-link Rigid Robotic Manipulator	2	✗	✗	Mathematical modelling	2	✓	Fractional-order fuzzy PID controller	6	Most valuable player algorithm	Integer-order fuzzy PID, One block fractional/Integer order fuzzy PID, Two block Fractional/Integer order fuzzy PID controllers	ITSE	Grünwald–Letnikov method	M	S
[93]	Robotic manipulator	2	✓	✗	Euler–Lagrange method	2	Yes	Fractional-order PID controller	5	Gradient-based optimization	PID controller	ISE	—	M	S
[94]	Single-segment soft continuum manipulator (Robotino-XT)	—	✓	✓	Fractional-order Bouc–Wen hysteresis model	16	—	—	—	—	—	Absolute pose error	Grünwald–Letnikov method	—	P
[95]	Two-link robotic manipulator	2	✓	✗	Mathematical modelling	—	✓	Fractional-order fuzzy PID controller	8	Hybrid grey wolf optimizer and artificial bee colony algorithm	PID	Tracking error	—	M	P
[96]	Robotic manipulator	—	✗	✓	Fractional-order Euler–Lagrange formulation	—	—	—	—	—	—	—	—	—	P
[97]	Stewart Platform	6	✓	✗	Kinematic modelling	2	✓	Fractional-order KDHD impedance control	2	Transient response-based tuning	KD controller	Error	Grünwald–Letnikov method	M	S
[98]	3-PUU parallel robotic manipulator	3	✗	✗	Kinematic modelling	2	✓	PDD1/2 controller	2	Transient response-based tuning	PD controller	Error	Grünwald–Letnikov method	M	S
[99]	Flexible link manipulator	2	✗	✗	Euler–Lagrange formulation	2	✓	Fractional-order phase-lag compensator	3	Optimization process	2DOF PID controller	Tracking error	Grünwald–Letnikov method	M	P

Table 1. *Cont.*

Ref.	Manipulator Details				Modelling Details			Controller Details						Tool S/P	
	Type	DOF	Payload	FOM	Method	Order	FOC	Controller	CP	Tuning Technique	Comparison Controllers	OF	Approximation		
[100]	Single-link flexible manipulator	2	✓	✗	Euler–Bernoull formulation	2	✓	Fractional-order PD	2	Bode Specifications	PD controller	Bode Margins	Grünwald–Letnikov method	M	P
[101]	KUKA LWR IV	7	✓	✓	Inverse Kinematics Model	3.04	✓	Impedance control	4	Genetic Algorithm	—	MSE, MAD	—	—	P
[102]	Single-link flexible manipulator	2	✓	✗	Pseudo-clamped approach	2	✓	Fractional-order PID	2	Bode Specifications	PID controller	Tracking error	Frequency response-based technique	M	P

The notations used in the table header are as follows: DOF—degree of freedom; FOM—fractional-order model; FOC—fractional-order control; CP—controller parameters; OF—objective function; M—MATLAB; L—LabVIEW; S/P—simulation/practical.

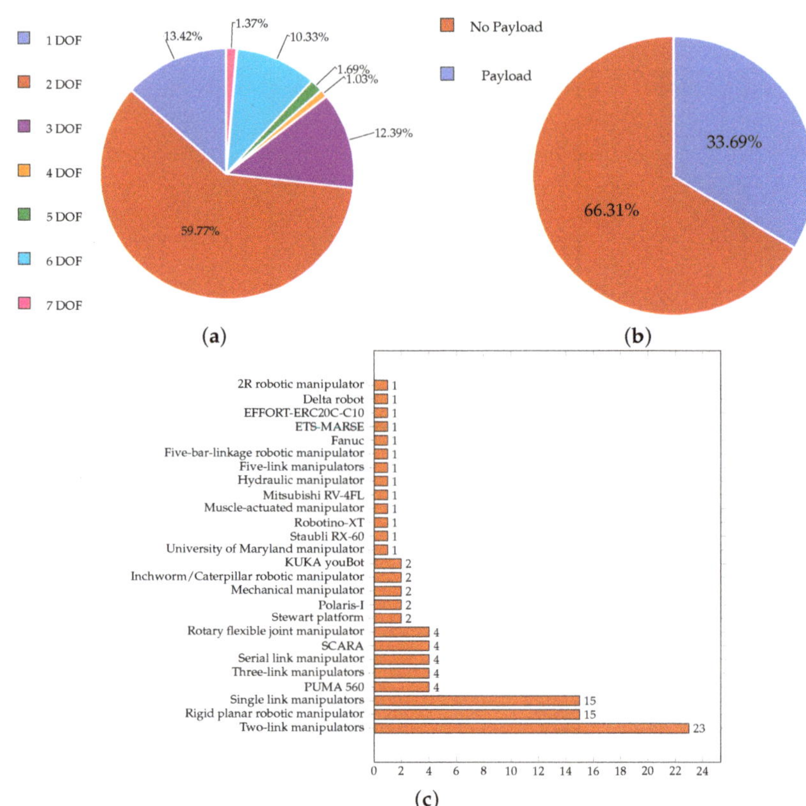

Figure 2. Summary of manipulator details from Table 1. (**a**) Manipulators' DOF trend; (**b**) Payload trend; (**c**) Manipulator's type.

Figure 3 gives a summary of the modelling approach and techniques used for robotic manipulators. As shown in Figure 3a, approximately 85% of modelling approaches used in the literature are conventional/integer-order type only. The remaining 15% of works have developed a fractional-order model of orders 0.3, 0.5, 0.6, 0.71, 0.8, 0.9, 0.92, 0.99, 1.14 and 3.04. Figure 3b shows that Euler–Lagrange relations have often been used to develop the manipulator's dynamic model in the conventional model category. In the fractional-order model category, various approaches, including adaptive neural network, describing functions, value selection algorithm, the Bouc–Wen hysteresis model, and the Euler–Lagrange formulation, have been used to develop commensurate and non-commensurate fractional-order models of manipulators. The following section will give a more detailed review of these modelling stargates.

Similarly, Figure 4 shows the summary of controllers, optimization, and approximation techniques used during the manipulators' control design. As shown in Figure 4a, the most widely developed fractional-order controllers use PID, sliding mode, and fuzzy. This is because PID is often used in the industry due to the advantages of simplicity and easy tuning and implementation. At the same time, the sliding mode offers the benefits of computational simplicity, less sensitivity to parameter uncertainties, being highly robust to disturbances, and fast dynamic response. On the other hand, fuzzy achieves better servo and regulatory response. However, sliding mode and fuzzy requires more controller parameters to be tuned. Researchers have used various optimization algorithms for tuning, as shown in Figure 4b. The figures show that about 70% have used genetic algorithms, cuckoo search, and particle swarm optimization. This is because these are the most popular and widely considered benchmark algorithms. Figure 4c gives the trend of approximation techniques used in manipulator modelling and controller design. The figures show Grünwald–Letnikov, Riemann–Liouville, Caputo, Oustaloup/refined Oustaloup approximations are the most frequently used techniques in the literature. More details regarding these approximation techniques can be found in [7]. A more detailed review of these control and optimization techniques stargates will be given in the following section.

Figure 5 shows the summary of validation type and type of toolbox, collected from Table 1. Figure 5a shows that about 65% of works, either modelling or validating controller, have been performed in the simulation environment. At the same time, the remaining 35% of results have validated the proposed approaches, practically. For these validations, approximately 90% of the researchers have used MATLAB, while others used LabVIEW, C++, and Solidworks. It is also worth highlighting that several researchers have used externally developed MATLAB-based toolboxes such as CRONE, Ninteger, and FOMCON to realize fractional-order systems and controllers [7].

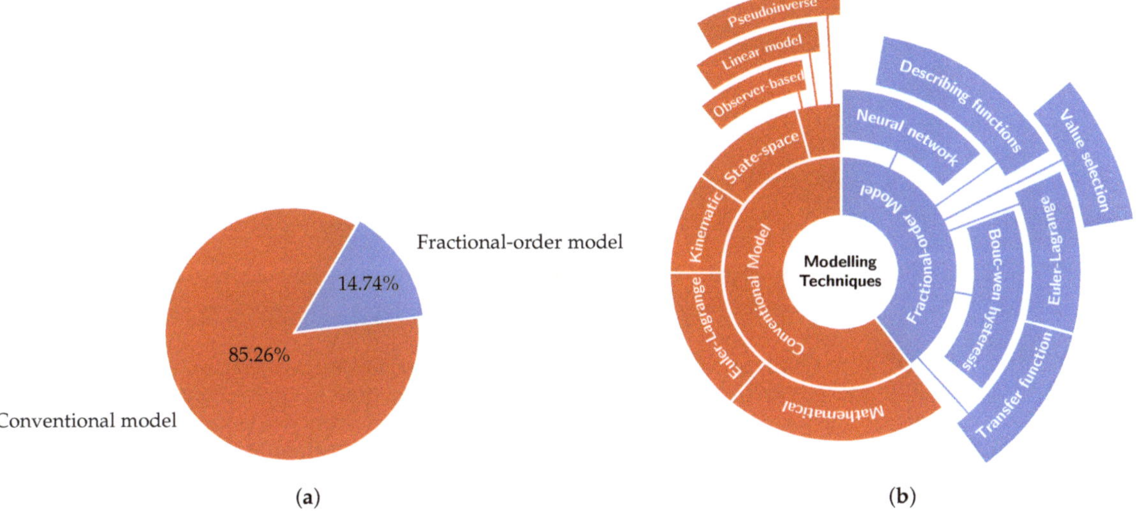

Figure 3. Summary of modelling details from Table 1. (**a**) Type of modelling approach. (**b**) Various types of modelling techniques.

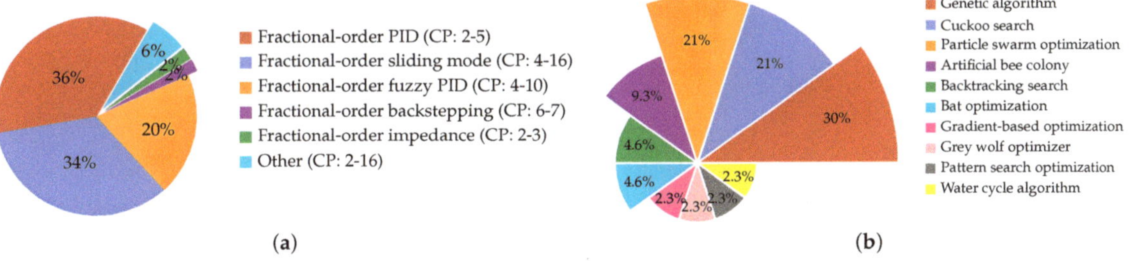

Figure 4. *Cont.*

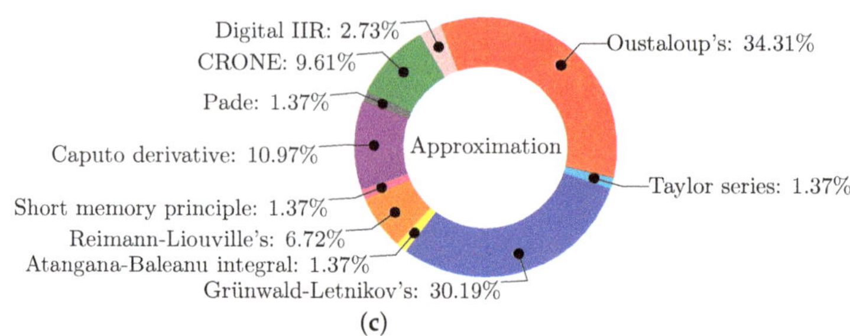

(c)

Figure 4. Summary of controller, optimization and approximation technique details from Table 1. (a) Fractional-order controllers. (b) Optimization techniques. (c) Approximation techniques.

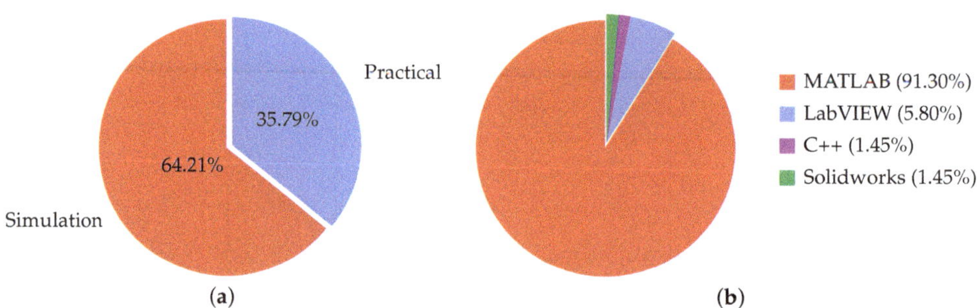

(a)   (b)

Figure 5. Summary of implementation type from Table 1. (a) Validation type. (b) Software toolboxes.

4. Modelling of Robotic Manipulators

As mentioned in Section 3, the Newton–Euler equations and Lagrange-assumed modes methods are most widely used for obtaining the mathematical model of robotic manipulators [103–105]. The Newton–Euler equations are based on Newton's second law of motion, while the Lagrange method derives the motion equations by eliminating interaction forces between adjacent links. In other words, Newton–Euler is a force balance approach, whereas the Lagrange method is an energy-based approach to manipulators' dynamics. Moreover, the Euler–Lagrange relations will produce the same equations as Newton's, which help analyze complicated systems. Additionally, these relations have the advantage of taking the same form in any system of generalized coordinates and are better suited for generalizations. Therefore, for developing the dynamic models of single-, two- and three-link robotic manipulators, Euler–Lagrangian relations are used as explained underneath. Further, the generalized model for the N number of rigid and n number of elastic degrees of freedom using the same technique is also given underneath.

4.1. Single-Link Rigid and Flexible Robotic Manipulators

An ideal single-link planar rigid robotic manipulator is shown in Figure 6. The mathematical relationship between torque τ and position θ using Euler–Lagrangian formulation is given as [66,103,105],

$$ml^2\ddot{\theta} + gml\sin(\theta) + v\dot{\theta} = \tau, \tag{2}$$

where v is the friction coefficient.

Let us assume $x_1 = \theta$ and $x_2 = \dot{\theta}$, then (2) can be rewritten as,

$$\dot{x}_1 = x_2,$$
$$\dot{x}_2 = -\frac{g}{l}\sin(x_1) - \frac{v}{ml^2}x_2 + \frac{1}{ml^2}\tau. \tag{3}$$

The nominal values of robotic manipulator parameters considered in most of the research works are $m = 2$ kg, $v = 6$ kgms, $l = 1$ m and $g = 9.81$ m/s^2. Thus, substituting these nominal values, (3) can be rewritten as,

$$\dot{x}_1 = x_2,$$
$$\dot{x}_2 = -9.81\sin(x_1) - 3x_2 + 0.5\tau. \tag{4}$$

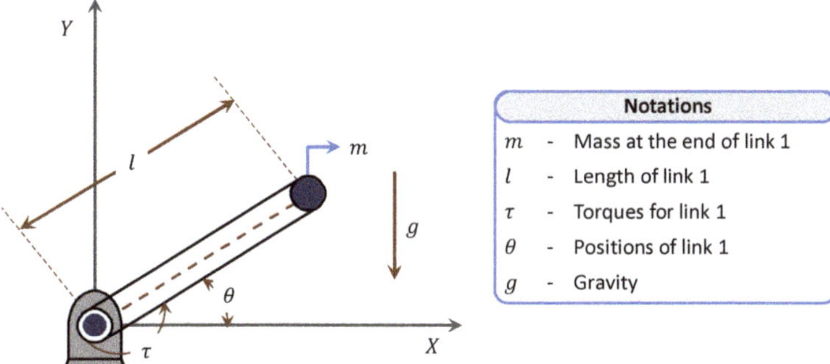

Figure 6. Single-link planar rigid robotic manipulator.

Similarly, the state space representation of an ideal single-link flexible robotic manipulator using Euler–Lagrangian formulation is given as [25,27,70],

$$\ddot{\theta} = -k_1\dot{\theta} + k_2\alpha + k_3 V_m,$$
$$\ddot{\alpha} = k_1\dot{\theta} - k_4\alpha - k_3 V_m, \tag{5}$$

where α is the tip deflection, θ is the motor shaft position, V_m is the motor input voltage and $k_i, i \in (1,4)$ are constants.

Let us assume $x_1 = \theta$, $x_2 = \alpha$, $x_3 = \dot{\theta}$, $x_4 = \dot{\alpha}$ and $V_m = u$, then (5) can be rewritten as,

$$\dot{x}_1 = x_3,$$
$$\dot{x}_2 = x_4,$$
$$\dot{x}_3 = p_2 x_2 - p_1 x_3 + p_3 u,$$
$$\dot{x}_4 = p_4 x_2 + p_1 x_3 - p_3 u. \tag{6}$$

From (6), the fractional-order model of a single-link flexible robotic manipulator in non-commensurate order is given as,

$$\dot{x}_1^\beta = x_3,$$
$$\dot{x}_2^\beta = x_4,$$
$$\dot{x}_3^\alpha = p_2 x_2 - p_1 x_3 + p_3 u,$$
$$\dot{x}_4^\alpha = p_4 x_2 + p_1 x_3 - p_3 u, \tag{7}$$

where α and β are the fractional-orders.

4.2. Two-Link Planar Rigid Robotic Manipulator

An ideal two-link planar rigid robotic manipulator or a SCARA-type manipulator with a payload of mass m_p at the tip is shown in Figure 7. The mathematical relationship between torques (τ_1, τ_2) and positions (θ_1, θ_2) of both the links (1, 2) using Euler–Lagrangian formulation is given as [4,5,28,31,39,44,51,64,103,106,107],

$$
\begin{bmatrix} M_{11} & M_{12} \\ M_{21} & M_{22} \end{bmatrix} \begin{bmatrix} \ddot{\theta}_1 \\ \ddot{\theta}_2 \end{bmatrix} + \begin{bmatrix} -(m_2 l_1 l_{c2} \sin(\theta_2))\dot{\theta}_2 & -(m_2 l_1 l_{c2} \sin(\theta_2))(\dot{\theta}_1 + \dot{\theta}_2) \\ (m_2 l_1 l_{c2} \sin(\theta_2))\dot{\theta}_1 & 0 \end{bmatrix} \begin{bmatrix} \dot{\theta}_1 \\ \dot{\theta}_2 \end{bmatrix}
$$
$$
+ \begin{bmatrix} m_1 l_{c1} g \cos(\theta_1) + m_2 g(l_{c2}\cos(\theta_1 + \theta_2) + l_1 \cos(\theta_1)) \\ m_2 l_{c2} g \cos(\theta_1 + \theta_2) \end{bmatrix} + \begin{bmatrix} v_1 \dot{\theta}_1 \\ v_2 \dot{\theta}_2 \end{bmatrix} + \begin{bmatrix} p_1 sgn(\dot{\theta}_1) \\ p_2 sgn(\dot{\theta}_2) \end{bmatrix} = \begin{bmatrix} \tau_1 \\ \tau_2 \end{bmatrix}, \tag{8}
$$

where

$$
M_{11} = m_1 + l_{c1}^2 + m_2(l_1^2 + l_{c2}^2 + 2l_1 l_{c2}\cos(\theta_2)) + m_p(l_1^2 + l_2^2 + 2l_1 l_2 \cos(\theta_2)) + I_1 + I_2,
$$
$$
M_{12} = m_2(l_{c2}^2 + l_1 l_{c2}\cos(\theta_2)) + m_p(l_2^2 + l_1 l_2 \cos(\theta 2)) + I_2,
$$
$$
M_{21} = m_2(l_{c2}^2 + l_1 l_{c2}\cos(\theta_2)) + m_p(l_2^2 + l_1 l_2 \cos(\theta 2)) + I_2,
$$
$$
M_{22} = m_2 l_{c2}^2 + m_p l_2^2 + I_2.
$$

In (8), v_1, v_2 are the coefficients of viscous friction and p_1, p_2 are the coefficients of dynamic friction of links 1 and 2, respectively. The nominal values of robotic manipulator parameters considered in most of the research works are $m_1 = m_2 = 1.0$ kg, $l_1 = l_2 = 1.0$ m, $l_{c1} = l_{c2} = 0.5$ m, $I_1 = I_2 = 0.2$ kgm^2, $v_1 = v_2 = 0.1$, $p_1 = p_2 = 0.1$, $m_p = 0.5$ kg and $g = 9.81$ m/s^2.

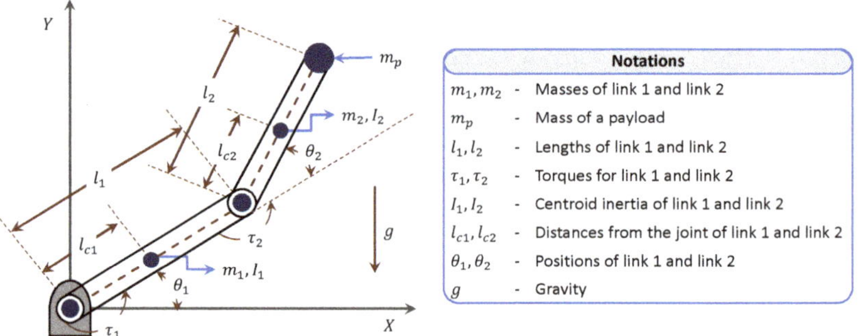

Figure 7. Two-link planar rigid robotic manipulator with a payload.

4.3. Three-Link Planar Rigid Robotic Manipulator

An ideal three-link planar rigid robotic manipulator with no friction, as shown in Figure 8, is where all the masses m_1, m_2 and m_3 exist as a point mass at the end point of each link. The mathematical relationship between torques (τ_1, τ_2, τ_3) and positions (θ_1, θ_2, θ_3) of all the links (1, 2, 3) using Euler–Lagrangian formulation is given as [56,65],

$$
\begin{bmatrix} M_{11} & M_{12} & M_{13} \\ M_{21} & M_{22} & M_{23} \\ M_{31} & M_{32} & M_{33} \end{bmatrix} \begin{bmatrix} \ddot{\theta}_1 \\ \ddot{\theta}_2 \\ \ddot{\theta}_3 \end{bmatrix} +
$$
$$
\begin{bmatrix} -l_1(m_3 l_3 \sin(\theta_2 + \theta_3) + m_2 l_2 \sin(\theta_2) + m_3 l_2 \sin(\theta_2))\dot{\theta}_2^2 - m_3 l_3(l_1 \sin(\theta_2 + \theta_3) + l_2 \sin(\theta_3))\dot{\theta}_3^2 \\ l_1(m_3 l_3 \sin(\theta_2 + \theta_3) + m_2 l_2 \sin(\theta_2) + m_3 l_2 \sin(\theta_2))\dot{\theta}_1^2 - m_3 l_2 l_3 \sin(\theta_3)\dot{\theta}_3^2 \\ m_3 l_3(l_1 \sin(\theta_2 + \theta_3) + l_2 \sin(\theta_3))\dot{\theta}_1^2 + m_3 l_2 l_3 \sin(\theta_3)\dot{\theta}_2^2 \end{bmatrix} + \begin{bmatrix} R_1 \\ R_2 \\ R_3 \end{bmatrix} +, \tag{9}
$$
$$
\begin{bmatrix} (m_1 + m_2 + m_3)gl_1 \cos(\theta_1) + (m_2 + m_3)gl_2 \cos(\theta_1 + \theta_2) + m_3 gl_3 \cos(\theta_1 + \theta_2 + \theta_3) \\ (m_2 + m_3)gl_2 \cos(\theta_1 + \theta_2) + m_3 gl_3 \cos(\theta_1 + \theta_2 + \theta_3) \\ m_3 gl_3 \cos(\theta_1 + \theta_2 + \theta_3) \end{bmatrix} = \begin{bmatrix} \tau_1 \\ \tau_2 \\ \tau_3 \end{bmatrix}
$$

where

$$
M_{11} = (m_1 + m_2 + m_3)l_1^2 + (m_2 + m_3)l_2^2 + m_3 l_3^2 + 2m_3 l_1 l_3 \cos(\theta_2 + \theta_3) + 2(m_2 + m_3)l_1 l_2 \cos(\theta_2) + 2m_3 l_2 l_3 \cos(\theta_3),
$$
$$
M_{12} = (m_2 + m_3)l_2^2 + m_3 l_3^2 + m_3 l_1 l_3 \cos(\theta_2 + \theta_3) + (m_2 + m_3)l_1 l_2 \cos(\theta_2) + 2m_3 l_2 l_3 \cos(\theta_3),
$$

$M_{13} = m_3 l_3^2 + m_3 l_1 l_3 \cos(\theta_2 + \theta_3) + m_3 l_2 l_3 \cos(\theta_3),$

$M_{21} = m_2 l_2^2 + m_3 l_2^2 + m_3 l_3^2 + m_3 l_1 l_3 \cos(\theta_2 + \theta_3) + m_2 l_1 l_2 \cos(\theta_2) + m_3 l_1 l_2 \cos(\theta_2) + 2 m_3 l_2 l_3 \cos(\theta_3),$

$M_{22} = m_2 l_2^2 + m_3 l_2^2 + m_3 l_3^2 + 2 m_3 l_2 l_3 \cos(\theta_3),$

$M_{23} = m_3 l_3^2 + m_3 l_2 l_3 \cos(\theta_3),$

$M_{31} = m_3 l_3^2 + m_3 l_1 l_3 \cos(\theta_2 + \theta_3) + m_3 l_2 l_3 \cos(\theta_3),$

$M_{32} = m_3 l_3^2 + m_3 l_2 l_3 \cos(\theta_3),$

$M_{33} = m_3 l_3^2,$

$R_1 = -2 l_1 (m_3 l_3 \sin(\theta_2 + \theta_3) + (m_2 + m_3) l_2 \sin(\theta_2)) \dot{\theta}_1 \dot{\theta}_2 - 2 m_3 l_3 (l_1 \sin(\theta_2 + \theta_3) + l_2 \sin(\theta_3)) \dot{\theta}_2 \dot{\theta}_3 - 2 m_3 l_3 (l_1 \sin(\theta_2 + \theta_3) + l_2 \sin(\theta_3)) \dot{\theta}_1 \dot{\theta}_3,$

$R_2 = -2 m_3 l_2 l_3 \sin(\theta_3) \dot{\theta}_1 \dot{\theta}_3 - 2 m_3 l_2 l_3 \sin(\theta_3) \dot{\theta}_3 \dot{\theta}_2,$

$R_3 = 2 m_3 l_2 l_3 \sin(\theta_3) \dot{\theta}_1 \dot{\theta}_2.$

In (9), it can be observed that the first, second (i.e., centrifugal), third (i.e., Coriolis) and fourth (i.e., potential energy) terms consist of $\ddot{\theta}_i$, $\dot{\theta}_i^2$, $\dot{\theta}_i \dot{\theta}_j$ and θ_i, respectively, where $i = 1, 2, 3$ and $i \neq j$. The nominal values of robotic manipulator parameters considered in most research works are $m_1 = 0.2$ kg, $m_2 = 0.3$ kg, $m_3 = 0.4$ kg, $l_1 = 0.4$ m, $l_2 = 0.6$ m, $l_3 = 0.8$ m and $g = 9.81$ m/s^2. The payload mass is added to the mass m_3.

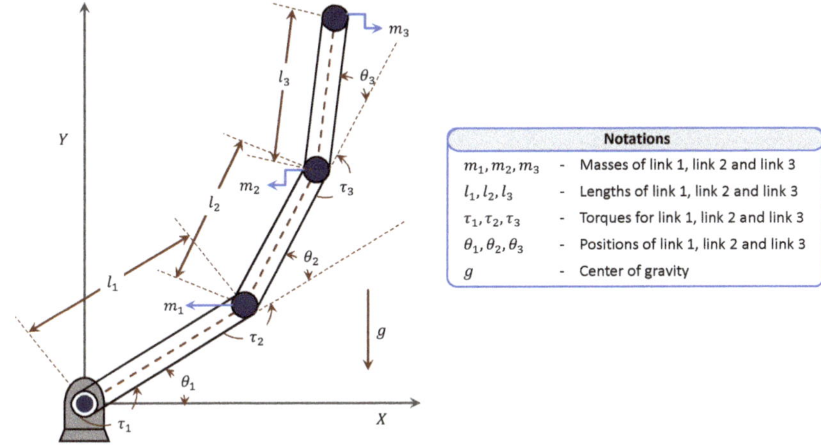

Figure 8. Three-link planar rigid robotic manipulator with a payload.

4.4. Generalized Model of Serial Link Planar Rigid Robotic Manipulator

The mathematical relationship between torques and positions of a robotic manipulator with N number of rigid and n number of elastic degrees of freedom using Euler–Lagrangian formulation is given as [104],

$$
\begin{bmatrix} (M_{rr})_{N \times N} & (M_{rf})_{N \times n} \\ (M_{fr})_{n \times N} & (M_{ff})_{n \times n} \end{bmatrix}_{(N+n) \times (N+n)} \begin{bmatrix} (\ddot{q}_r)_{N \times 1} \\ (\ddot{q}_f)_{n \times 1} \end{bmatrix}_{(N+n) \times 1} +
$$
$$
\begin{bmatrix} (H_r)_{N \times 1} \\ (H_f)_{n \times 1} \end{bmatrix}_{(N+n) \times 1} + \begin{bmatrix} (G_r)_{N \times 1} \\ (G_f)_{n \times 1} \end{bmatrix}_{(N+n) \times 1} = \begin{bmatrix} \tau_{N \times 1} \\ 0_{(n) \times 1} \end{bmatrix}_{(N+n) \times 1},
\tag{10}
$$

where the matrices are defined as,

- M_{rr} and M_{ff} are the mass matrices related to rigid and flexible degrees of freedom, respectively,
- M_{rf} row matrix that defines the coupling between manipulators' rigid and flexible motions,
- M_{fr} row matrix that defines the coupling between manipulators' flexible and rigid motions,
- q_r and q_f are the manipulators' rigid and flexible degrees of freedom representing the motions of joints and elastic motions of flexible links, respectively,

- H_r and H_f are the centrifugal and Coriolis matrix related to rigid and flexible motion, respectively,
- G_r and G_f are the gravity matrix related to rigid and flexible motion, respectively,
- τ is the torque vector.

4.5. Other Robotic Manipulators

The modelling strategies of other robotic manipulators of various degrees of freedom are shown in Figure 9. The figure depicts that the most widely used Euler–Lagrangian formulation has been used to model lower and higher DOF manipulators such as inchworm/caterpillar [34,37], serial/joint manipulators, KUKA youBot [1], and Stewart platforms [89]. Similarly, the kinematic and inverse kinematic modelling approach has also been used for Delta robots [59], parallel manipulators, the Stewart platform [97], KUKA LWR IV [101], and Mitsubishi RV-4FL [2]. The next most widely used is a mathematical model developed for PUMA 560 [50], Quanser manipulators [83,88], Staubli RX-60 [54], Polaris-I [2], and UMD manipulators [21]. On the other hand, the fractional-order models have been developed for only Quanser [83], PUMA 560 [75], and Robotino-XT [94]. Thus, there is broad scope for exploring the concept of fractional-order modelling for various lower DOF manipulators such as inchworm/caterpillar and higher DOF manipulators such as Delta robot, KUKA youBot, Staubli RX-60, Robotino-XT, etc.

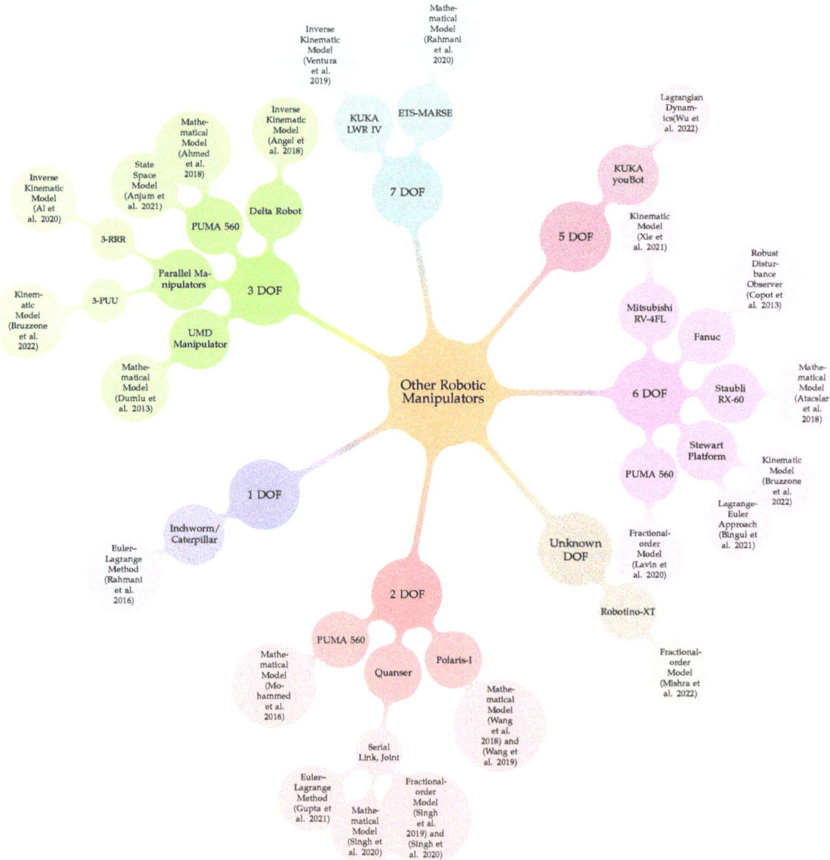

Figure 9. Modelling strategies used for various lower and higher DOF robotic manipulators [1–3,20,21,30,34,37,50,54,59,69,70,73,75,76,76,83,83,85,88,89,94,97,101].

5. Fractional-Order Control of Robotic Manipulators

This section presents a broad overview of fractional-order control strategies developed for various rigid, flexible, and joint robotic manipulators. These control strategies aim to achieve robust and stable performance despite uncertainties, external disturbances, and actual faults. As mentioned in Section 3, the developed fractional-order control strategies for various robotic manipulators are evolved versions of PID, sliding mode, backstepping, fuzzy, active disturbance rejection [82], and impedance control [91,97,98]. A more detailed review of these control strategies will be explained underneath.

5.1. Fractional-Order PID Controllers

The fractional-order PID controller with five parameters is an extension of the PID where the conventional integrator and differentiator are replaced with fractional ones. The serial rigid, flexible, and joint manipulators with DOF varying from 1 to 2 have been effectively controlled in simulation, and practice, using fractional-order PD/PID compared to PI/PD/PID and achieved better tracking accuracy and stability, practically [11,52,70,88,99,100,102,108]. However, the trial and error method has often been used to achieve the controller parameters. However, in the case of a two-link planar rigid robotic manipulator, the optimally tuned fractional-order PID and two-degree of freedom fractional-order PID controllers using the cuckoo search algorithm [4], particle swarm optimization [17,19], genetic algorithm [14,46] have performed better than the conventional and two-degree of freedom PID controllers [29,45]. A similar case has also been seen in a three-link planar rigid robotic manipulator, where fractional-order PID tuned using an evaporation rate-based water cycle algorithm has achieved better performance than the PID [56]. The best fractional-order PI/PD/PID performance is also true for higher DOF robotic manipulators, including Staubli RX-60 [54], UMD manipulator [21], PUMA 560 [75], Fanuc [20,24], Delta robot [59], KUKA LWR IV [101], and 3-RRR planar parallel robots [76]. Moreover, for these higher DOF robotic manipulators, the controller parameters are tuned using rule-based methods including Bode tuning [24] and decentralized tuning [20]. More details regarding the control actions of the fractional-order PID controller family, including two-degree of freedom configuration, can be found in [6,7,109,110].

5.2. Fractional-Order Fuzzy PID Controllers

It is widely known that PID is most often used in industry due to the advantages of simplicity and easy tuning and implementation [111]. As mentioned earlier, the performance of this controller is enhanced using fractional calculus. Moreover, the performance of this fractional-order PID is further enhanced using intelligent fuzzy techniques to achieve better servo and regulatory responses. Therefore, various combinations of fractional-order PID and fuzzy logic are proposed in the literature to form fractional-order fuzzy PID controller for two-link [4,39,43,44,51,62,63,67,79,92,95], three-link manipulators [48,65], SCARA [31,53], PUMA 560 [30], and Stewart platforms [89]. In addition, the authors of [64] have proposed a hybrid two-degree-of-freedom fractional-order fuzzy PID controller by combining two-degree-of-freedom PID, fractional-order concept, and fuzzy logic. These combinations have achieved better performance than the conventional and integer-order ones. Further, to incorporate the self-tuning of controller parameters rather than designing using precise mathematics, researchers have used several optimization techniques where the non-linear controller gains are updated in real-time using error and fractional rate of error. The optimization techniques used in the literature are artificial bee colony [39,43,53,95], genetic algorithm [30,39,43,64,79], cuckoo search [4,31,48,51,63], backtracking search [44,65], dragonfly [79], ant lion optimizer [79], particle swarm optimization [67,89] and grey wolf optimizer [95]. The robustness testing of these self-tuned fractional-order fuzzy PID controllers has shown superior tracking results in comparison to the conventional counterparts. However, in most of the works, the analytical stability analysis of these controllers has yet to be attempted. Thus, the research gap in the analytical proof of stability is noteworthy.

5.3. Fractional-Order Sliding Mode Controllers

Among the non-linear control methods such as an adaptive, fuzzy, neural network, sliding mode, H_∞, and model predictive controllers, the sliding mode control has been widely utilized due to its advantages of being computational simplicity, less sensitive to parameter uncertainties, highly robust to disturbances, and fast dynamic response [2,42]. However, the sliding mode controller has three significant problems: singularity, uncertainties, and chattering effect [78]. The singularity problem in the sliding mode control signal exists because of differentiating the exponential term in the controller equation. Thus, nonsingular sliding mode controllers have been developed to deal with this issue [69]. Moreover, various intelligent and optimization algorithms are hybridized with sliding mode controllers to compensate for the uncertainties issue, which also helps reduce the switching gains [58]. However, the problem of the chattering effect is still a drawback for the sliding mode controller. Therefore, researchers have recently developed fractional-order sliding mode controllers, which help reduce the chattering impact due to their memory and hereditary properties [81]. The two types of sliding mode controllers are given as linear sliding mode and terminal sliding mode controllers. The application of the fractional-order form of these two sliding mode controllers for various robotic manipulators will be explained underneath.

The linear fractional-order sliding mode controller has been developed for a single-link flexible manipulator for DOF varying from 1 to 2, achieving better performance than the conventional sliding mode controller and PID [22,25,42,66,68]. Even though the controller has no chattering effect, the singularity and uncertainties issues still exist. Thus, fuzzy and adaptive sliding mode controllers have been proposed for single-link, two-link, Mitsubishi RV-4FL, polar, and Inchworm/Caterpillar robotic manipulators. In [15,16,37,57,58], the authors have developed fuzzy and adaptive sliding mode controllers using bat optimization, genetic, and cuckoo search algorithms. The adaptive part of the controller will help reduce the uncertainties issue, and the fractional part of the controller will help reduce the chattering effect. On the other hand, the authors of [18] have proposed a fractional variable structure that helps minimize switching actions. However, the singularity problem still exists in these control techniques. Thus, the interest has been shifted towards using nonsingular sliding mode controller configurations.

Various configurations of terminal fractional-order sliding mode controllers have recently been developed for robotic manipulators to deal with singularity, uncertainties, and chattering effects. The authors of [26,55,69] have developed a fractional-order nonsingular terminal sliding mode controller for hydraulic and cable-driven manipulators, where the controller parameters are obtained using the trial and error method. This controller configuration has performed better than the integer-order nonsingular terminal sliding mode controller in both practical and simulation analysis. Even though the chattering and singularity issues have been solved, the controller still has uncertainty issues. Thus, in [1,28,34,61,73,74,78], an adaptive fractional-order nonsingular terminal sliding mode controller has been proposed for serial robotic manipulators, exoskeleton robot, KUKA youBot, and inchworm/caterpillar robotic manipulators. The controller has performed better than all its counterparts, including sliding mode controller, integer-order terminal sliding mode controller, fractional-order terminal sliding mode controller, and fractional-order nonsingular terminal sliding mode controller in solving the singularity issues, uncertainties, and chattering effect. However, this controller configuration is complex and needs more controller parameters to be tuned. Moreover, this controller configuration is further improved using time delay estimation, which forms the time delay estimation-based adaptive fractional-order nonsingular terminal sliding mode controller. In [3,36,81,86], the time delay estimation-based adaptive fractional-order nonsingular terminal sliding mode controller has been proposed for rigid hydraulic manipulators which have performed better than all of its counterparts and solved singularity, uncertainties, and chattering issues. At the same time, the controller configuration is very complex, and around 15 controller parameters need to be tuned. Thus, developing simple evolved versions of fractional-order sliding mode controllers to deal with singularity, uncertainties, and chattering effects are inevitable.

5.4. Fractional-Order Adaptive Backstepping Controller

The adaptive backstepping controller provides an improved tracking performance in the presence of uncertainties and faults, thanks to the controllers' adaptation law. In addition, the controller guarantees closed-loop system stability, which the conventional one failed to achieve. As finite-time convergence is crucial in robotic manipulators, thus, an adaptive backstepping controller is the perfect choice to achieve stable operation even in the presence of uncertainties and external disturbances. Further, to provide better steady-state and transient performances, the authors of [32,85] have proposed a fractional-order adaptive backstepping controller in the presence of actuators' faults and disturbances. The controller achieved adequate performance for PUMA 560 and a rotary manipulator under uncertainties, external load disturbances, and actuator faults. The controller also attained finite-time convergence and asymptotic stability. However, in both works, the controller parameters are chosen using the trial and error method. Thus, there is scope to develop a tuning approach for controller parameters of the fractional-order adaptive backstepping controller.

6. Conclusions

6.1. Findings

A comprehensive review of the application of the fractional-order concept in modelling and control techniques for various robotic manipulators has been discussed, as proposed by previous researchers. This comprehensive review summarizes the research outcomes published from 1998 until 2022 of around 100 works. Firstly, the study includes the conventional and fractional-order modelling strategies for robotic manipulators. Then, a review of developed fractional-order controllers for various robotic manipulators, which evolved from PID, sliding mode, fuzzy, backstepping, active disturbance rejection control, and impedance control, are presented. The graphical trend for existing research has been broadly presented in both cases. Thus, this review is expected to draw the attention of the investigators, experts, and researchers, allowing them to understand the most recent trends and work to advance in this field.

6.2. Future Perspectives

- There is broad scope for exploring the fractional-order modelling concept for various industrial robots, including Delta robot, KUKA youBot, Staubli RX-60, Robotino-XT, etc.
- The performance of fractional-order PID controllers can be further improved using the fractional-order form of predictive PI controllers for achieving robust servo and regulatory responses. Additionally, the performance of fractional-order PID controllers needs to be improved in the presence of uncertainties and faults.
- Even though fractional-order fuzzy PID controllers have achieved better servo and regulatory responses for proper industrial applications, the proof for analytical stability is a considerable research gap.
- The fractional-order nonsingular terminal sliding mode controller has achieved better response and surpassed the issues of singularity, uncertainties, and chattering effects. However, the controller configuration is very complex, and more parameters must be tuned. Thus, research on developing simple, evolved versions of controllers is inevitable.
- The adaptive backstepping controller provided an improved tracking performance in the presence of uncertainties and faults, thanks to the controllers' adaptation law. However, the controller parameters are chosen using the trial and error method. Thus, there is scope to develop a tuning approach for controller parameters.

Author Contributions: Conceptualization, K.B. and A.P.S.; investigation, K.B. and A.P.S.; resources, K.B. and A.P.S.; writing—original draft preparation, K.B. and B.R.P.; writing—review and editing, A.P.S. and B.R.P.; supervision, A.P.S.; project administration, K.B.; funding acquisition, K.B. All authors have read and agreed to the published version of the manuscript.

Funding: This research received no external funding.

Institutional Review Board Statement: Not applicable.

Informed Consent Statement: Not applicable.

Data Availability Statement: Not applicable.

Acknowledgments: The authors also acknowledge the Universiti Teknologi PETRONAS, Malaysia for providing all research facilities.

Conflicts of Interest: The authors declare no conflict of interest.

Abbreviations

The following abbreviations are used in this manuscript:

DOF	Degrees of freedom
FOMCON	Fractional-order modeling and control
IACCO	Integral of absolute change in controller output
IAE	Integral absolute error
ISCCO	Integral square of change in controller output
ISE	Integral square error
ISV	Integral of the square value
ITACO	Integral of time absolute change in controller output
ITAE	Integral time absolute error
ITSE	Integral time square error
LQR	Linear-quadratic regulator
MAD	Mean absolute deviation
MAE	Mean absolute error
MSE	Mean square error
MMFAE	Mean minimum fuel and absolute error
RMSE	Root mean squared error
STD	Standard deviation

References

1. Wu, X.; Huang, Y. Adaptive fractional-order non-singular terminal sliding mode control based on fuzzy wavelet neural networks for omnidirectional mobile robot manipulator. *ISA Trans.* **2022**, *121*, 258–267. [CrossRef]
2. Xie, Y.; Zhang, X.; Meng, W.; Zheng, S.; Jiang, L.; Meng, J.; Wang, S. Coupled fractional-order sliding mode control and obstacle avoidance of a four-wheeled steerable mobile robot. *ISA Trans.* **2021**, *108*, 282–294. [CrossRef]
3. Wang, Y.; Yan, F.; Jiang, S.; Chen, B. Time delay control of cable-driven manipulators with adaptive fractional-order nonsingular terminal sliding mode. *Adv. Eng. Softw.* **2018**, *121*, 13–25. [CrossRef]
4. Sharma, R.; Rana, K.; Kumar, V. Performance analysis of fractional order fuzzy PID controllers applied to a robotic manipulator. *Expert Syst. Appl.* **2014**, *41*, 4274–4289. [CrossRef]
5. Sharma, R.; Gaur, P.; Mittal, A. Performance analysis of two-degree of freedom fractional order PID controllers for robotic manipulator with payload. *ISA Trans.* **2015**, *58*, 279–291. [CrossRef] [PubMed]
6. Shah, P.; Agashe, S. Review of fractional PID controller. *Mechatronics* **2016**, *38*, 29–41. [CrossRef]
7. Bingi, K.; Ibrahim, R.; Karsiti, M.N.; Hassan, S.M.; Harindran, V.R. *Fractional-Order Systems and PID Controllers*; Springer: London, UK, 2020.
8. Bouzoualegh, S.; Guechi, E.H.; Kelaiaia, R. Model predictive control of a differential-drive mobile robot. *Acta Univ. Sapientiae, Electr. Mech. Eng.* **2018**, *10*, 20–41. [CrossRef]
9. Machado, J.T.; Azenha, A. Fractional-order hybrid control of robot manipulators. In Proceedings of the SMC'98 Conference Proceedings. 1998 IEEE International Conference on Systems, Man, and Cybernetics (Cat. No. 98CH36218), San Diego, CA, USA, 11–14 October 1998; Volume 1, pp. 788–793.
10. Duarte, F.; Machado, J. Chaotic phenomena and fractional-order dynamics in the trajectory control of redundant manipulators. *Nonlinear Dyn.* **2002**, *29*, 315–342. [CrossRef]

11. Monje, C.; Ramos, F.; Feliu, V.; Vinagre, B. Tip position control of a lightweight flexible manipulator using a fractional order controller. *IET Control. Theory Appl.* **2007**, *1*, 1451–1460. [CrossRef]
12. Efe, M.Ö. Fractional fuzzy adaptive sliding-mode control of a 2-DOF direct-drive robot arm. *IEEE Trans. Syst. Man. Cybern. Part (Cybern.)* **2008**, *38*, 1561–1570. [CrossRef]
13. Ferreira, N.F.; Machado, J.T.; Tar, J.K. Two cooperating manipulators with fractional controllers. *Int. J. Adv. Robot. Syst.* **2009**, *6*, 31. [CrossRef]
14. Delavari, H.; Ghaderi, R.; Ranjbar, N.; HosseinNia, S.H.; Momani, S. Adaptive fractional PID controller for robot manipulator. *arXiv* **2012**, arXiv:1206.2027.
15. Delavari, H.; Ghaderi, R.; Ranjbar, A.; Momani, S. Fuzzy fractional order sliding mode controller for nonlinear systems. *Commun. Nonlinear Sci. Numer. Simul.* **2010**, *15*, 963–978. [CrossRef]
16. Fayazi, A.; Rafsanjani, H.N. Fractional order fuzzy sliding mode controller for robotic flexible joint manipulators. In Proceedings of the 2011 9th IEEE International Conference on Control and Automation (ICCA), Santiago, Chile, 19–21 December 2011; pp. 1244–1249.
17. Bingul, Z.; Karahan, O. Tuning of fractional PID controllers using PSO algorithm for robot trajectory control. In Proceedings of the 2011 IEEE International Conference on Mechatronics, Istanbul, Turkey, 13–15 April 2011; pp. 955–960.
18. Machado, J.T. The effect of fractional order in variable structure control. *Comput. Math. Appl.* **2012**, *64*, 3340–3350. [CrossRef]
19. Bingül, Z.; KARAHAN, O. Fractional PID controllers tuned by evolutionary algorithms for robot trajectory control. *Turk. J. Electr. Eng. Comput. Sci.* **2012**, *20*, 1123–1136. [CrossRef]
20. Copot, C.; Burlacu, A.; Ionescu, C.M.; Lazar, C.; De Keyser, R. A fractional order control strategy for visual servoing systems. *Mechatronics* **2013**, *23*, 848–855. [CrossRef]
21. Dumlu, A.; Erenturk, K. Trajectory tracking control for a 3-dof parallel manipulator using fractional-order $\text{PI}^\lambda \text{D}^\mu$ control. *IEEE Trans. Ind. Electron.* **2013**, *61*, 3417–3426. [CrossRef]
22. Delavari, H.; Lanusse, P.; Sabatier, J. Fractional order controller design for a flexible link manipulator robot. *Asian J. Control.* **2013**, *15*, 783–795. [CrossRef]
23. Moreno, A.R.; Sandoval, V.J. Fractional order PD and PID position control of an angular manipulator of 3DOF. In Proceedings of the 2013 Latin American Robotics Symposium and Competition, Arequipa, Peru, 21–27 October 2013; pp. 89–94.
24. Copot, C.; Ionescu, C.M.; Lazar, C.; De Keyser, R. Fractional order PD^μ control of a visual servoing manipulator system. In Proceedings of the 2013 European Control Conference (ECC), Zurich, Switzerland, 17–19 July 2013; pp. 4015–4020.
25. Mujumdar, A.; Tamhane, B.; Kurode, S. Fractional order modeling and control of a flexible manipulator using sliding modes. In Proceedings of the 2014 American Control Conference, Portland, OR, USA, 4–6 June 2014; pp. 2011–2016.
26. Wang, Y.; Luo, G.; Gu, L.; Li, X. Fractional-order nonsingular terminal sliding mode control of hydraulic manipulators using time delay estimation. *J. Vib. Control* **2016**, *22*, 3998–4011. [CrossRef]
27. Mujumdar, A.; Tamhane, B.; Kurode, S. Observer-based sliding mode control for a class of noncommensurate fractional-order systems. *IEEE/ASME Trans. Mechatronics* **2015**, *20*, 2504–2512. [CrossRef]
28. Nojavanzadeh, D.; Badamchizadeh, M. Adaptive fractional-order non-singular fast terminal sliding mode control for robot manipulators. *IET Control. Theory Appl.* **2016**, *10*, 1565–1572. [CrossRef]
29. Fani, D.; Shahraki, E. Two-link robot manipulator using fractional order PID controllers optimized by evolutionary algorithms. *Biosci. Biotechnol. Res. Asia* **2016**, *13*, 589–598. [CrossRef]
30. Mohammed, R.H.; Bendary, F.; Elserafi, K. Trajectory tracking control for robot manipulator using fractional order-fuzzy-PID controller. *Int. J. Comput. Appl.* **2016**, *134*, 22–29.
31. Sharma, R.; Gaur, P.; Mittal, A. Design of two-layered fractional order fuzzy logic controllers applied to robotic manipulator with variable payload. *Appl. Soft Comput.* **2016**, *47*, 565–576. [CrossRef]
32. Nikdel, N.; Badamchizadeh, M.; Azimirad, V.; Nazari, M.A. Fractional-order adaptive backstepping control of robotic manipulators in the presence of model uncertainties and external disturbances. *IEEE Trans. Ind. Electron.* **2016**, *63*, 6249–6256. [CrossRef]
33. Łegowski, A.; Niezabitowski, M. Manipulator path control with variable order fractional calculus. In Proceedings of the 2016 21st International Conference on Methods and Models in Automation and Robotics (MMAR), Miedzyzdroje, Poland, 29 August–1 September 2016; pp. 1127–1132.
34. Rahmani, M.; Ghanbari, A.; Ettefagh, M.M. Hybrid neural network fraction integral terminal sliding mode control of an Inchworm robot manipulator. *Mech. Syst. Signal Process.* **2016**, *80*, 117–136. [CrossRef]
35. Ghasemi, I.; Ranjbar Noei, A.; Sadati, J. Sliding mode based fractional-order iterative learning control for a nonlinear robot manipulator with bounded disturbance. *Trans. Inst. Meas. Control.* **2018**, *40*, 49–60. [CrossRef]
36. Wang, Y.; Gu, L.; Xu, Y.; Cao, X. Practical tracking control of robot manipulators with continuous fractional-order nonsingular terminal sliding mode. *IEEE Trans. Ind. Electron.* **2016**, *63*, 6194–6204. [CrossRef]
37. Rahmani, M.; Ghanbari, A.; Ettefagh, M.M. Robust adaptive control of a bio-inspired robot manipulator using bat algorithm. *Expert Syst. Appl.* **2016**, *56*, 164–176. [CrossRef]
38. Aghababa, M.P. Optimal design of fractional-order PID controller for five bar linkage robot using a new particle swarm optimization algorithm. *Soft Comput.* **2016**, *20*, 4055–4067. [CrossRef]

39. Kumar, A.; Kumar, V. A novel interval type-2 fractional order fuzzy PID controller: Design, performance evaluation, and its optimal time domain tuning. *ISA Trans.* **2017**, *68*, 251–275. [CrossRef]

40. Feliu-Talegon, D.; Feliu-Batlle, V. A fractional-order controller for single-link flexible robots robust to sensor disturbances. *IFAC-PapersOnLine* **2017**, *50*, 6043–6048. [CrossRef]

41. Machado, J.T.; Lopes, A.M. A fractional perspective on the trajectory control of redundant and hyper-redundant robot manipulators. *Appl. Math. Model.* **2017**, *46*, 716–726. [CrossRef]

42. Guo, Y.; Ma, B.L. Global sliding mode with fractional operators and application to control robot manipulators. *Int. J. Control.* **2019**, *92*, 1497–1510. [CrossRef]

43. Kumar, A.; Kumar, V. Hybridized ABC-GA optimized fractional order fuzzy pre-compensated FOPID control design for 2-DOF robot manipulator. *AEU-Int. J. Electron. Commun.* **2017**, *79*, 219–233. [CrossRef]

44. Kumar, V.; Rana, K. Nonlinear adaptive fractional order fuzzy PID control of a 2-link planar rigid manipulator with payload. *J. Frankl. Inst.* **2017**, *354*, 993–1022. [CrossRef]

45. De la Fuente, M.S.L. Trajectory Tracking Error Using Fractional Order PID Control Law for Two-Link Robot Manipulator via Fractional Adaptive Neural Networks. In *Robotics—Legal, Ethical and Socioeconomic Impacts*; IntechOpen: London, UK, 2017; p. 35. [CrossRef]

46. Kumar, V.; Rana, K. Comparative study on fractional order PID and PID controllers on noise suppression for manipulator trajectory control. In *Fractional Order Control and Synchronization of Chaotic Systems*; Springer: New York, NY, USA, 2017; pp. 3–28.

47. Al-Saggaf, U.M.; Mehedi, I.M.; Mansouri, R.; Bettayeb, M. Rotary flexible joint control by fractional order controllers. *Int. J. Control. Autom. Syst.* **2017**, *15*, 2561–2569. [CrossRef]

48. Kumar, J.; Kumar, V.; Rana, K. A fractional order fuzzy PD+I controller for three-link electrically driven rigid robotic manipulator system. *J. Intell. Fuzzy Syst.* **2018**, *35*, 5287–5299. [CrossRef]

49. Bensafia, Y.; Ladaci, S.; Khettab, K.; Chemori, A. Fractional order model reference adaptive control for SCARA robot trajectory tracking. *Int. J. Ind. Syst. Eng.* **2018**, *30*, 138–156. [CrossRef]

50. Ahmed, S.; Wang, H.; Tian, Y. Fault tolerant control using fractional-order terminal sliding mode control for robotic manipulators. *Stud. Inform. Control.* **2018**, *27*, 55–64. [CrossRef]

51. Azar, A.T.; Kumar, J.; Kumar, V.; Rana, K. Control of a two link planar electrically-driven rigid robotic manipulator using fractional order SOFC. In *Proceedings of the International Conference on Advanced Intelligent Systems and Informatics, Cairo, Egypt, 9–11 September 2017*; Springer: Cham, Switzerland, 2017; pp. 57–68.

52. Fareh, R.; Bettayeb, M.; Rahman, M. Control of serial link manipulator using a fractional order controller. *Int. Rev. Autom. Control.* **2018**, *11*, 1–6. [CrossRef]

53. Kumar, A.; Kumar, V.; Gaidhane, P.J. Optimal design of fuzzy fractional order $PI\lambda D\mu$ controller for redundant robot. *Procedia Comput. Sci.* **2018**, *125*, 442–448. [CrossRef]

54. Ataşlar-Ayyıldız, B.; Karahan, O. Tuning of fractional order PID controller using cs algorithm for trajectory tracking control. In Proceedings of the 2018 6th International Conference on Control Engineering & Information Technology (CEIT), Istanbul, Turkey, 25–27 October 2018; pp. 1–6.

55. Yin, C.; Xue, J.; Cheng, Y.; Zhang, B.; Zhou, J. Fractional order nonsingular fast terminal sliding mode control technique for 6-DOF robotic manipulator. In Proceedings of the 2018 37th Chinese Control Conference (CCC), Wuhan, China, 25–27 July 2018; pp. 10186–10190.

56. Kathuria, T.; Kumar, V.; Rana, K.; Azar, A.T. Control of a three-link manipulator using fractional-order pid controller. In *Fractional Order Systems*; Elsevier: Berkeley, CA, USA, 2018; pp. 477–510.

57. Kumar, J.; Kumar, V.; Rana, K. Design of robust fractional order fuzzy sliding mode PID controller for two link robotic manipulator system. *J. Intell. Fuzzy Syst.* **2018**, *35*, 5301–5315. [CrossRef]

58. Kumar, J.; Azar, A.T.; Kumar, V.; Rana, K.P.S. Design of fractional order fuzzy sliding mode controller for nonlinear complex systems. In *Mathematical Techniques of Fractional Order Systems*; Elsevier: Amsterdam, The Netherlands, 2018; pp. 249–282.

59. Angel, L.; Viola, J. Fractional order PID for tracking control of a parallel robotic manipulator type delta. *ISA Trans.* **2018**, *79*, 172–188. [CrossRef]

60. Coronel-Escamilla, A.; Torres, F.; Gómez-Aguilar, J.; Escobar-Jiménez, R.; Guerrero-Ramírez, G. On the trajectory tracking control for an SCARA robot manipulator in a fractional model driven by induction motors with PSO tuning. *Multibody Syst. Dyn.* **2018**, *43*, 257–277. [CrossRef]

61. Yin, C.; Zhou, J.; Xue, J.; Zhang, B.; Huang, X.; Cheng, Y. Design of the fractional-order adaptive nonsingular terminal sliding mode controller for 6-DOF robotic manipulator. In Proceedings of the 2018 Chinese Control And Decision Conference (CCDC), Shenyang, China, 9–11 June 2018; pp. 5954–5959.

62. Muñoz-Vázquez, A.J.; Gaxiola, F.; Martínez-Reyes, F.; Manzo-Martínez, A. A fuzzy fractional-order control of robotic manipulators with PID error manifolds. *Appl. Soft Comput.* **2019**, *83*, 105646. [CrossRef]

63. Sharma, R.; Bhasin, S.; Gaur, P.; Joshi, D. A switching-based collaborative fractional order fuzzy logic controllers for robotic manipulators. *Appl. Math. Model.* **2019**, *73*, 228–246. [CrossRef]

64. Mohan, V.; Chhabra, H.; Rani, A.; Singh, V. An expert 2DOF fractional order fuzzy PID controller for nonlinear systems. *Neural Comput. Appl.* **2019**, *31*, 4253–4270. [CrossRef]

65. Kumar, V.; Rana, K.; Kler, D. Efficient control of a 3-link planar rigid manipulator using self-regulated fractional-order fuzzy PID controller. *Appl. Soft Comput.* **2019**, *82*, 105531. [CrossRef]
66. Fareh, R. Sliding mode fractional order control for a single flexible link manipulator. *Int. J. Mech. Eng. Robot. Res.* **2019**, *8*, 228–232. [CrossRef]
67. Ardeshiri, R.R.; Kashani, H.N.; Reza-Ahrabi, A. Design and simulation of self-tuning fractional order fuzzy PID controller for robotic manipulator. *Int. J. Autom. Control* **2019**, *13*, 595–618. [CrossRef]
68. Hamzeh Nejad, F.; Fayazi, A.; Ghayoumi Zadeh, H.; Fatehi Marj, H.; HosseinNia, S.H. Precise tip-positioning control of a single-link flexible arm using a fractional-order sliding mode controller. *J. Vib. Control.* **2020**, *26*, 1683–1696. [CrossRef]
69. Wang, Y.; Li, B.; Yan, F.; Chen, B. Practical adaptive fractional-order nonsingular terminal sliding mode control for a cable-driven manipulator. *Int. J. Robust Nonlinear Control* **2019**, *29*, 1396–1417. [CrossRef]
70. Singh, A.P.; Deb, D.; Agarwal, H. On selection of improved fractional model and control of different systems with experimental validation. *Commun. Nonlinear Sci. Numer. Simul.* **2019**, *79*, 104902. [CrossRef]
71. Ahmed, T.M.; Gaber, A.N.A.; Hamdy, R.; Abdel-Khalik, A.S. Position Control of Arm Manipulator Within Fractional Order PID Utilizing Particle Swarm Optimization Algorithm. In Proceedings of the 2019 IEEE Conference on Power Electronics and Renewable Energy (CPERE), Aswan City, Egypt, 23–25 October 2019; pp. 135–139.
72. Ahmed, S.; Lochan, K.; Roy, B.K. Fractional-Order Adaptive Sliding Mode Control for a Two-Link Flexible Manipulator. In *Innovations in Infrastructure*; Springer: Singapore, 2019; pp. 33–53.
73. Rahmani, M.; Rahman, M.H. Adaptive neural network fast fractional sliding mode control of a 7-DOF exoskeleton robot. *Int. J. Control. Autom. Syst.* **2020**, *18*, 124–133. [CrossRef]
74. Ahmed, S.; Wang, H.; Tian, Y. Adaptive fractional high-order terminal sliding mode control for nonlinear robotic manipulator under alternating loads. *Asian J. Control* **2021**, *23*, 1900–1910. [CrossRef]
75. Lavín-Delgado, J.; Solís-Pérez, J.; Gómez-Aguilar, J.; Escobar-Jiménez, R. Trajectory tracking control based on non-singular fractional derivatives for the PUMA 560 robot arm. *Multibody Syst. Dyn.* **2020**, *50*, 259–303. [CrossRef]
76. Al-Mayyahi, A.; Aldair, A.A.; Chatwin, C. Control of a 3-RRR planar parallel robot using fractional order PID controller. *Int. J. Autom. Comput.* **2020**, *17*, 822–836. [CrossRef]
77. Tenreiro Machado, J.A.; Lopes, A.M. Fractional-order kinematic analysis of biomechanical inspired manipulators. *J. Vib. Control.* **2020**, *26*, 102–111. [CrossRef]
78. Zhou, M.; Feng, Y.; Xue, C.; Han, F. Deep convolutional neural network based fractional-order terminal sliding-mode control for robotic manipulators. *Neurocomputing* **2020**, *416*, 143–151. [CrossRef]
79. Chhabra, H.; Mohan, V.; Rani, A.; Singh, V. Robust nonlinear fractional order fuzzy PD plus fuzzy I controller applied to robotic manipulator. *Neural Comput. Appl.* **2020**, *32*, 2055–2079. [CrossRef]
80. Yousfi, N.; Almalki, H.; Derbel, N. Robust control of industrial MIMO systems based on fractional order approaches. In Proceedings of the 2020 Industrial & Systems Engineering Conference (ISEC), New Orleans, LA, USA, 30 May–2 June 2020; pp. 1–6.
81. Zhang, Y.; Yang, X.; Wei, P.; Liu, P.X. Fractional-order adaptive non-singular fast terminal sliding mode control with time delay estimation for robotic manipulators. *IET Control. Theory Appl.* **2020**, *14*, 2556–2565. [CrossRef]
82. Shi, X.; Huang, J.; Gao, F. Fractional-order active disturbance rejection controller for motion control of a novel 6-dof parallel robot. *Math. Probl. Eng.* **2020**, *2020*, 3657848. [CrossRef]
83. Singh, A.P.; Deb, D.; Agrawal, H.; Balas, V.E. *Fractional Modeling and Controller Design of Robotic Manipulators: With Hardware Validation*; Springer Nature: Cham, Switzerland, 2020; Volume 194.
84. Al-Sereihy, M.H.; Mehedi, I.M.; Al-Saggaf, U.M.; Bettayeb, M. State-feedback-based fractional-order control approximation for a rotary flexible joint system. *Mechatron. Syst. Control.* **2020**, *48*. [CrossRef]
85. Anjum, Z.; Guo, Y. Finite time fractional-order adaptive backstepping fault tolerant control of robotic manipulator. *Int. J. Control Autom. Syst.* **2021**, *19*, 301–310. [CrossRef]
86. Su, L.; Guo, X.; Ji, Y.; Tian, Y. Tracking control of cable-driven manipulator with adaptive fractional-order nonsingular fast terminal sliding mode control. *J. Vib. Control* **2021**, *27*, 2482–2493. [CrossRef]
87. Singh, A.P.; Deb, D.; Agrawal, H.; Bingi, K.; Ozana, S. Modeling and control of robotic manipulators: A fractional calculus point of view. *Arab. J. Sci. Eng.* **2021**, *46*, 9541–9552. [CrossRef]
88. Gupta, S.; Singh, A.P.; Deb, D.; Ozana, S. Kalman Filter and Variants for Estimation in 2DOF Serial Flexible Link and Joint Using Fractional Order PID Controller. *Appl. Sci.* **2021**, *11*, 6693. [CrossRef]
89. Bingul, Z.; Karahan, O. Real-time trajectory tracking control of Stewart platform using fractional order fuzzy PID controller optimized by particle swarm algorithm. *Ind. Robot. Int. J. Robot. Res. Appl.* **2021**, *49*, 708–725. [CrossRef]
90. Anjum, Z.; Guo, Y.; Yao, W. Fault tolerant control for robotic manipulator using fractional-order backstepping fast terminal sliding mode control. *Trans. Inst. Meas. Control* **2021**, *43*, 3244–3254. [CrossRef]
91. Ding, Y.; Liu, X.; Chen, P.; Luo, X.; Luo, Y. Fractional-Order Impedance Control for Robot Manipulator. *Fractal Fract.* **2022**, *6*, 684. [CrossRef]
92. Abdulameer, H.I.; Mohamed, M.J. Fractional Order Fuzzy PID Controller Design for 2-Link Rigid Robot Manipulator. *Int. J. Intell. Eng. Syst.* **2021**, *15*, 103–117.
93. Violia, J.; Angel, L. Control Performance Assessment of Fractional-Order PID Controllers Applied to Tracking Trajectory Control of Robotic Systems. *WSEAS Trans. Syst. Control.* **2022**, *17*, 62–73.

94. Mishra, M.K.; Samantaray, A.K.; Chakraborty, G. Fractional-order Bouc-wen hysteresis model for pneumatically actuated continuum manipulator. *Mech. Mach. Theory* **2022**, *173*, 104841. [CrossRef]
95. Gaidhane, P.J.; Adam, S. The Enhanced Robotic Trajectory Tracking by Optimized Fractional-Order Fuzzy Controller Using GWO-ABC Algorithm. In *Soft Computing: Theories and Applications*; Springer: Singapore, 2022; pp. 611–620.
96. Azar, A.T.; Serrano, F.E.; Kamal, N.A.; Kumar, S.; Ibraheem, I.K.; Humaidi, A.J.; Gorripotu, T.S.; Pilla, R. Fractional-Order Euler–Lagrange Dynamic Formulation and Control of Asynchronous Switched Robotic Systems. In *Proceedings of the Third International Conference on Sustainable Computing*; Springer: Singapore, 2022; pp. 479–490.
97. Bruzzone, L.; Polloni, A. Fractional Order KDHD Impedance Control of the Stewart Platform. *Machines* **2022**, *10*, 604. [CrossRef]
98. Bruzzone, L.; Fanghella, P.; Basso, D. Application of the Half-Order Derivative to Impedance Control of the 3-PUU Parallel Robot. *Actuators* **2022**, *11*, 45. [CrossRef]
99. Feliu-Talegon, D.; Feliu-Batlle, V. Improving the position control of a two degrees of freedom robotic sensing antenna using fractional-order controllers. *Int. J. Control.* **2017**, *90*, 1256–1281. [CrossRef]
100. Feliu-Talegon, D.; Feliu-Batlle, V.; Tejado, I.; Vinagre, B.M.; HosseinNia, S.H. Stable force control and contact transition of a single link flexible robot using a fractional-order controller. *ISA Trans.* **2019**, *89*, 139–157. [CrossRef]
101. Ventura, A.; Tejado, I.; Valério, D.; Martins, J. Fractional Control of a 7-DOF Robot to Behave Like a Human Arm. *Prog. Fract. Differ. Appl.* **2019**, *5*, 99–110. [CrossRef]
102. Feliu-Talegon, D.; Feliu-Batlle, V. Control of very lightweight 2-DOF single-link flexible robots robust to strain gauge sensor disturbances: A fractional-order approach. *IEEE Trans. Control Syst. Technol.* **2021**, *30*, 14–29. [CrossRef]
103. Craig, J.J. *Introduction to Robotics: Mechanics and Control*; Pearson Educacion: London, UK, 2005.
104. Mishra, N.; Singh, S. Dynamic modelling and control of flexible link manipulators: Methods and scope-Part-1. *Indian J. Sci. Technol.* **2021**, *14*, 3210–3226. [CrossRef]
105. Sahu, V.S.D.M.; Samal, P.; Panigrahi, C.K. Modelling, and control techniques of robotic manipulators: A review. *Mater. Today Proc.* **2021**, *56*, 2758–2766. [CrossRef]
106. Lee, C.Y.; Lee, J.J. Adaptive control of robot manipulators using multiple neural networks. In Proceedings of the 2003 IEEE International Conference on Robotics and Automation (Cat. No. 03CH37422), Taipei, Taiwan, 14–19 September 2003; Volume 1, pp. 1074–1079.
107. Lin, F. *Robust Control Design: An Optimal Control Approach*; John Wiley & Sons: New York, NY, USA, 2007.
108. Devan, P.A.M.; Hussin, F.A.; Ibrahim, R.; Bingi, K.; Abdulrab, H. Fractional-order predictive PI controller for process plants with deadtime. In Proceedings of the 2020 IEEE 8th R10 Humanitarian Technology Conference (R10-HTC), Kuching, Sarawak, Malaysia, 1–3 December 2020; pp. 1–6.
109. Bingi, K.; Ibrahim, R.; Karsiti, M.N.; Hassan, S.M.; Harindran, V.R. Real-time control of pressure plant using 2DOF fractional-order PID controller. *Arab. J. Sci. Eng.* **2019**, *44*, 2091–2102. [CrossRef]
110. Bingi, K.; Ibrahim, R.; Karsiti, M.N.; Hassan, S.M. Fractional order set-point weighted PID controller for pH neutralization process using accelerated PSO algorithm. *Arab. J. Sci. Eng.* **2018**, *43*, 2687–2701. [CrossRef]
111. Abdelhedi, F.; Bouteraa, Y.; Chemori, A.; Derbel, N. Nonlinear PID and feedforward control of robotic manipulators. In Proceedings of the 2014 15th International Conference on Sciences and Techniques of Automatic Control and Computer Engineering (STA), Hammamet, Tunisia, 21–23 December 2014; pp. 349–354.

fractal and fractional

Article

Fractional-Order Impedance Control for Robot Manipulator

Yixiao Ding [1], Xiaolian Liu [1], Pengchong Chen [2], Xin Luo [1] and Ying Luo [1,*]

1. School of Mechanical Science and Engineering, Huazhong University of Science and Technology, Wuhan 430074, China
2. School of Electrical Engineering, Zhengzhou University, Zhengzhou 450001, China
* Correspondence: ying.luo@hust.edu.cn

Abstract: Impedance control is an important method in robot–environment interaction. In traditional impedance control, the damping force is regarded as a linear viscoelastic model, which limits the description of the dynamic model of the impedance system to a certain extent. For the robot manipulator, the optimal impedance parameters of the impedance controller are the key to improve the performance. In this paper, the damping force is described more accurately by fractional calculus than the traditional viscoelastic model, and a fractional-order impedance controller for the robot manipulator is proposed. A practical and systematic tuning procedure based on the frequency design method is developed for the proposed fractional-order impedance controller. The fairness of comparison between the fractional-order impedance controller and the integer-order impedance controller is addressed under the same specifications. Fair comparisons of the two controllers via the simulation and experiment tests show that, in the step response, the fractional-order impedance controller has a better integral time square error (ITSE) result, smaller overshoot and less settling time than the integer-order impedance controller. In terms of anti-disturbance, the fractional-order impedance controller can achieve stability with less recovering time and better ITSE index than integer order impedance controller.

Keywords: impedance control; fractional-order control; robot manipulator

Citation: Ding, Y.; Liu, X.; Chen, P.; Luo, X.; Luo, Y. Fractional-Order Impedance Control for Robot Manipulator. *Fractal Fract.* **2022**, *6*, 684. https://doi.org/10.3390/fractalfract6110684

Academic Editors: Kishore Bingi and Abhaya Pal Singh

Received: 6 October 2022
Accepted: 8 November 2022
Published: 18 November 2022

Publisher's Note: MDPI stays neutral with regard to jurisdictional claims in published maps and institutional affiliations.

1. Introduction

Robot manipulators have been playing an important role in industry, medical treatment and service industries. The manipulator is closely related to the environment in most working circumstances, which puts forward higher and higher requirements for the dynamic interaction between the robot manipulator and the environment [1–3]. The single trajectory control method may cause too much interaction force and result in damage, or the force may be too small to complete the task [4,5]. In order to expand the application range of robot manipulator and improve the system performance, controlling the contact force between the robot manipulator and the environment has become one of its hot research areas [6,7]. Active compliance control is one of the main ways to realize the force control, which adjusts the interaction force depending on the force feedback information from the joint or the force sensor installed at the end of the robot manipulator. The impedance control algorithm is a general strategy for the robot manipulator to realize active compliance control [8] which adopts the structure of inner loop position control and outer loop force control (also called admittance control) [9]. A user-defined dynamic relationship between the reference trajectory of the end effector and the interaction contact force can be built by the impedance control model.

Robot–environment interaction in an uncertain environment brings challenges to impedance control, such as cell injection, rehabilitation applications and complex work-piece curved surface processing. It is difficult to obtain the performance of accurate force tracking and the system robustness due to various unknown features. In an impedance control framework, choosing proper impedance parameters is the key to realize the desired impedance dynamics and guarantee the stability [10–12]. Otherwise, the compliance of robot–environment interaction would be severely affected. Accordingly, conventional impedance may not be suitable for these applications, and large errors of position and force might be produced [13].

Intelligent and advanced algorithms have been proposed to improve the performance of impedance controller in a complex uncertain environment. The impedance parameters matching the interaction environment properly are generated to cope with the uncertain environment. Zhang et al. [14] presented a variable impedance method to acquire the impedance parameters in real time, using the offline-trained fuzzy neural network system. In [15], the natural gradient actor-critic reinforcement learning algorithm was proposed to optimize the impedance parameters online. However, none of these works addresses the issue of sampling efficiency. A huge amount of training data is required in the learning methods, which is infeasible for the physical interacting system. Dong et al. [16] proposed a speed-based variable impedance adaptive interaction control method, and the damping parameter of the impedance controller was adaptively adjusted according to the interactive force-tracking error. In [12], an adaptive variable impedance control method was applied to track the desired dynamic force and compensate for uncertainties in the environment. However, the overshoot of the contact force, and the trade-off between force-tracking accuracy and system robustness were not addressed.

The traditional impedance control model is equivalent to a mass-spring-damper system, and the damping force is usually described by a linear viscoelastic model. However, the description accuracy of damping force is limited by the traditional linear model [17,18], which could affect the performance of the controller to a certain extent. Fractional calculus theory, as the extension of integer calculus, can describe physical objects more accurately [19,20]. Fractional-order control has been proven to achieve outstanding tracking performance and robustness [21,22]. Reference [23] used a fractional-order derivative to describe the damping force in the visco-elastic-dampers, which was more accurate and even required fewer parameters in comparison with other models. Fractional calculus provided flexibility in designing appropriate visco-elastic-dampers with a large variety of practicable values for parameters [23]. Wang et al. [24] extended the classical skyhook damping control strategy to the fractional-order one. A fractional-order skyhook damping control design method for full-car suspension was given, which obtained a more ideal control effect over the classical control method. The impedance model for the uncertain environment with nonlinear factors is a typical fractional-order object [9]. In this paper, fractional calculus theory is applied to optimize the performance of impedance control. The damping force in the impedance model is described more accurately using the fractional-order derivative than the traditional integer-order one. Based on the fractional-order damping force model, the traditional integer-order impedance (IO-impedance) control is extended to the fractional-order impedance (FO-impedance) control, which is applied to the compliance control of the robot manipulator. A fair comparison between IO-impedance control and FO-impedance control is addressed under the same design specifications. The simulation and experimental test show that the proposed FO-impedance controller outperforms the IO-impedance controller.

The main contributions of this paper include the following: (1) A FO-impedance controller for robot manipulator is proposed based on a proposed fractional order damping force model. (2) A systematic tuning method for the FO-impedance controller design is proposed with detailed procedure. The designed control system can meet the user given frequency domain specifications. (3) The simulation and experimental demonstration on the robot manipulator system are presented to verify the feasibility and advantages of the proposed FO-impedance controller compared with the optimal IO-impedance controller.

The remaining of this paper is organized as follows: Section 2 describes the impedance model and the FO-impedance controller optimal design method; the simulation illustration and experimental verification are shown in Sections 3 and 4, respectively. The step response and anti-disturbance robustness performance of FO-impedance controller is studied and compared with the IO-impedance controller. The conclusion is given in Section 5.

2. FO-Impedance Controller Design

2.1. Impedance Control Methodology

The dynamic model of the robot manipulator impedance control mechanical system can be described as a mass-spring-damper system. The structure is shown in Figure 1, and the system dynamic equation is as follows:

$$M_d\ddot{x}(t) + B_d\dot{x}(t) + K_dx(t) = F_{ext},\tag{1}$$

where x is the position, M_d is the mass, B_d is the damping, K_d is the stiffness, and F_{ext} is the contact force between the robot manipulator and the external environment.

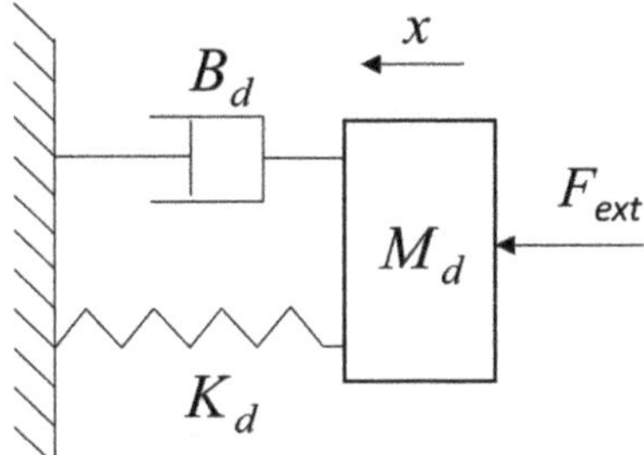

Figure 1. Dynamic model of impedance-control mechanical system.

According to Equation (1), one can get,

$$\frac{X(s)}{F(s)} = \frac{1}{M_ds^2 + B_ds + K_d}.\tag{2}$$

For impedance control, the control system adopts position control as the inner loop and force control as the outer loop. For the robot manipulator, a force sensor is usually installed at the end of the robot manipulator to sense its contact force with the environment. Through the impedance control algorithm, the position information which needs to be corrected is generated according to the force error and input into the inner loop of position control, and then the contact force with the environment is adjusted.

The robot manipulator impedance-control-system block diagram is shown in Figure 2. The reference contact force between the end of the robot manipulator and the environment is set as F_{ref}. The real contact force F_{real} between the robot manipulator and the external environment is collected by the force sensor. ΔF is the difference between the real contact force and the reference contact force. The position variation ΔX of the robot manipulator end effector is calculated according to the impedance-control algorithm. Then the position control command X_{cmd} is obtained as the target of the position control loop with position reference X_{ref} and position variation ΔX. X_{real} is the actual position of the robot manipulator end effector. K_s is the external environmental stiffness in contact with the end of the robot manipulator.

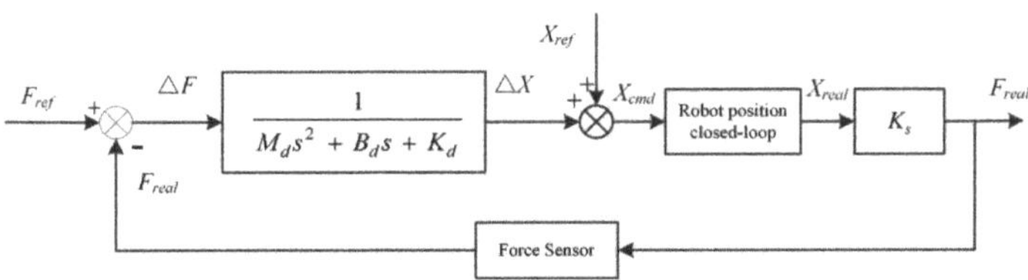

Figure 2. Robot manipulator impedance-control system diagram.

The reference contact force F_{ref} as the system input and the actual contact force F_{real} as the system output are presented in Figure 3. The closed loop of the robot position control is replaced as 1 due to the high control bandwidth and high tracking performance of the robot manipulator position servo control system.

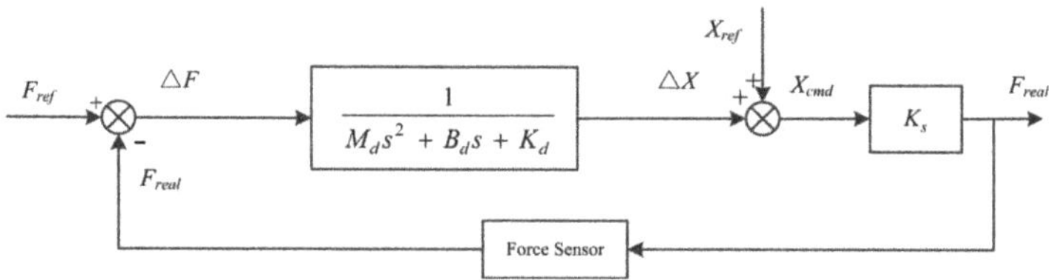

Figure 3. Simplified robot manipulator impedance-control system diagram.

The open-loop transfer function of the feedback control system in Figure 3 can be expressed as,

$$G_0(s) = \frac{K_s}{M_d s^2 + B_d s + K_d}.$$

(3)

The closed-loop transfer function is,

$$G_1(s) = \frac{K_s}{M_d s^2 + B_d s + K_d + K_s}.$$

(4)

2.2. Controllers Design

In this section, the methodologies of the IO-impedance controller and FO-impedance controller design are presented, respectively. In the position-based impedance control, virtual stiffness can lead to steady-state force-tracking errors [9,25]. Many control strategies have been proposed to attenuate the force tracking error [25–27]. Canceling the stiffness parameter in the impedance model is a simple and effective way to solve this problem [25]. According to Equation (4), the system output can stabilize at $\frac{K_s F_{ref}}{K_d + K_s}$. The FO-impedance controller design for robot contact force control is mainly studied in this paper. In order to stabilize the system output at the given reference force, the method in reference [25] is applied, and the stiffness parameter K_d is set as 0.

The open-loop transfer function Equation (3) becomes

$$G_0(s) = \frac{K_s}{M_d s^2 + B_d s}.$$

(5)

Substituting jw for s in Equation (5) yields

$$G_0(jw) = \frac{K_s}{M_d(jw)^2 + B_d(jw)}.$$

(6)

2.2.1. Design Specifications

The frequency-domain design method is applied in this paper, which constrains the gain crossover frequency and phase margin [28]. In order to ensure a fair comparison, both the gain crossover frequency and phase margin specifications are introduced for the IO-impedance controller and FO-impedance controller design in this paper, which are given as follows:

(1) Gain crossover frequency specification

At the gain crossover frequency, the amplitude of the open-loop transfer function should be 1,

$$|G_0(jw_c)|_{db} = 1,$$

(7)

where w_c is the gain crossover frequency.

(2) Phase margin specification

$$Arg[G_0(jw_c)] = -\pi + \varphi_m,\tag{8}$$

where φ_m is the phase margin required.

2.2.2. IO-Impedance Controller Design

The phase and gain of $G_0(s)$ in frequency domain can be given as follows:

$$|G_0(jw)|_{db} = \frac{K_s}{\sqrt{M_d^2 w^4 + B_d^2 w^2}},\tag{9}$$

$$\varphi(w) = atan\frac{B_d}{M_d w}.\tag{10}$$

Given the gain crossover frequency w_c and the desired phase margin φ_m, from Equations (7) and (8), one can obtain

$$\frac{K_s}{\sqrt{M_d^2 w_c^4 + B_d^2 w_c^2}} = 1,\tag{11}$$

$$\varphi_m = \pi + atan\frac{B_d}{M_d w_c}.\tag{12}$$

According to Equations (11) and (12), M_d and B_d can be obtained in the following form:

$$M_d = \frac{K_s}{\sqrt{w_c^4 + w_c^4 tan^2(\varphi_m - \pi)}},\tag{13}$$

$$B_d = tan(\varphi_m - \pi)M_d w_c.\tag{14}$$

Clearly, according to the given crossover frequency and phase margin, we can solve Equations (13) and (14) to obtain M_d and B_d.

2.2.3. FO-Impedance Controller Design

Fractional order can describe the damping characteristics more accurately [23,24]. In order to improve the control performance of the system, a FO-impedance controller is proposed with a fractional order damping model as the controller structure shown in Figure 4.

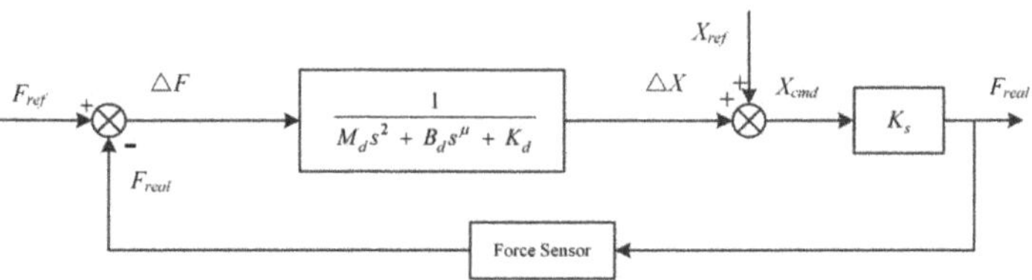

Figure 4. FO-impedance controller.

The fractional-order dynamic differential equation corresponding to the traditional integer-order dynamics in Equation (1) is as follows:

$$M_d \ddot{x}(t) + B_d x^\mu(t) + K_d x(t) = F_{ext},\tag{15}$$

where μ is the fractional order.

The stiffness parameter K_d is set as 0 to attenuate the force-tracking error [25]. The open-loop transfer function Equation (5) can be written as

$$G_{0f}(s) = \frac{K_s}{M_d s^2 + B_d s^\mu}.$$ (16)

Substituting jw for s in Equation (16) yields

$$G_{0f}(jw) = \frac{K_s}{M_d(jw)^2 + B_d p_1 + B_d p_2 j},$$ (17)

where

$$p_1 = w^u \cos(\frac{\pi}{2}\mu),$$ (18)

$$p_2 = w^u \sin(\frac{\pi}{2}\mu).$$ (19)

The gain of $G_{0f}(s)$ in frequency domain can be given as

$$\left| G_{0f}(jw) \right|_{db} = \frac{K_s}{\sqrt{[M_d(jw)^2 + (B_d p_1)]^2 + (B_d p_2)^2}}.$$ (20)

The phase of $G_{0f}(s)$ can be written as

$$\varphi(w) = -atan \frac{B_d p_2}{M_d(jw)^2 + B_d p_1}.$$ (21)

Given the fixed gain crossover frequency w_c and the desired phase margin φ_m, from Equations (7) and (8), we can obtain

$$\frac{a}{\sqrt{[(jw_c)^2 + bp_1]^2 + (bp_2)^2}} = 1,$$ (22)

$$\varphi_m = \pi - atan \frac{bp_2}{(jw_c)^2 + bp_1},$$ (23)

where

$$a = \frac{K_s}{M_d},$$ (24)

$$b = \frac{M_d}{B_d}.$$ (25)

According to Equations (24) and (25), we can establish the following equations:

$$b = \frac{tan(\pi - \varphi_m)(jw_c)^2}{p_2 - p_1 tan(\pi - \varphi_m)},$$ (26)

$$a = \sqrt{[(jw_c)^2 + bp_1]^2 + (bp_2)^2}.$$ (27)

In the fractional-order impedance controller, given the crossover frequency and phase margin, there are three unknown parameters, M_d, B_d and fractional order μ. A time-domain specification, integral time square error (ITSE), is applied to design a FO-impedance controller systematically. Sweeping all of the value range of $\mu \in (0,2)$, all of the FO-impedance controllers satisfying the pre-specified gain crossover frequency and phase margin can be obtained by Equations (24)–(27). Then, the step response simulation for all the FO-impedance controllers above can be implemented, and the corresponding ITSE value J_{ITSE} can be calculated using Equation (28). Select the parameters corresponding to the smallest J_{ITSE} as the final designed FO-impedance controller:

$$J_{ITSE} = \int_0^{tf} t[e(t)]^2 dt,$$ (28)

where t is the time, $e(t)$ is the error between the actual value and the reference value, and tf is the control time.

2.2.4. Design Procedure Illustration with an Example

The parameter setting rules of the FO-impedance controller are summarized with an example as follows with a flow chart shown in Figure 5:

(1) Given the gain crossover frequency $w_c = 10$ rad/s, the desired phase margin $\varphi_m = 40°$ and step response signal $F_{ref} = 20$ N.

(2) Sweeping all the $\mu \in (0,2)$, all of the FO-impedance controllers satisfying the pre-specified gain crossover frequency w_c and phase margin φ_m can be obtained by Equations (24)–(27) as shown in Figure 6.

(3) Implement the step response simulations and calculate the J_{ITSE} for all the FO-impedance controllers above. The correspondence diagram between μ and J_{ITSE} is shown in Figure 7. The smallest J_{ITSE} for which $\mu = 0.88$ is marked as a red star in Figure 7.

(4) Select the parameters corresponding to the smallest J_{ITSE} as the final designed FO-impedance controller, with $M_d = 11.4978$ kg, and $B_d = 111.6108$ N·s/m.

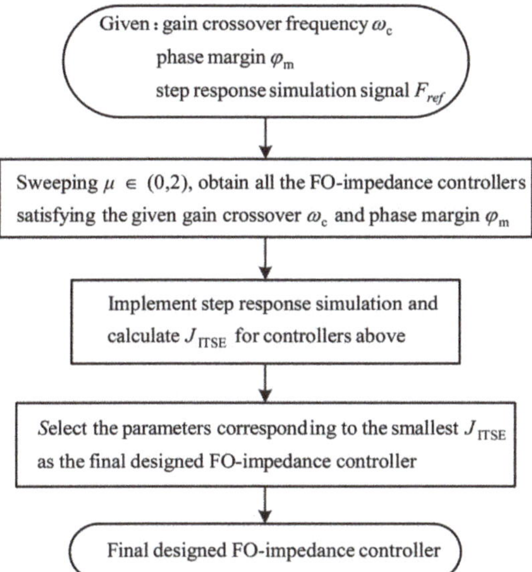

Figure 5. FO-impedance design procedure flow chart.

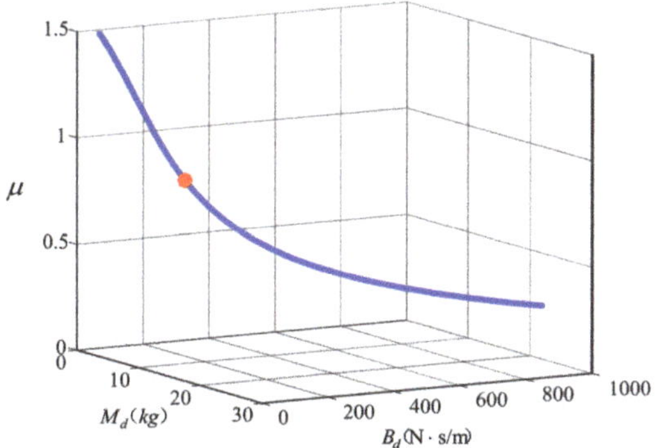

Figure 6. All the parameters satisfying the pre-specified gain crossover frequency and phase margin.

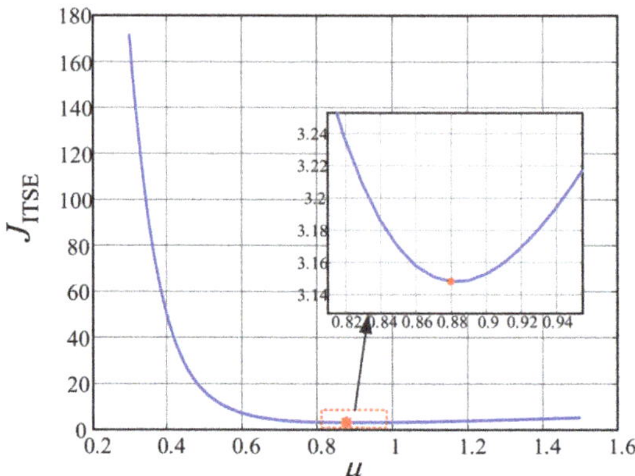

Figure 7. J_{ITSE} corresponding to the μ.

3. Simulation

The designed impedance controllers are applied to control the contact force between the end of robot manipulator and the environment in Z-axis direction. Given the frequency domain specifications, gain crossover frequency $w_c = 10\,\text{rad/s}$ and phase margin $\varphi_m = 40°$. Set the reference contact force F_{ref} as 20 N, the system control sampling period as 0.0005 s, and the stiffness coefficient K_s as 1293.83 N/m (the stiffness coefficient of the real spring in the experiment). According to the detailed process in Section 2, the IO-impedance and FO-impedance controllers can be designed and calculated. According to the given crossover frequency and phase margin, the parameters of the IO-impedance controller are obtained as follows: $M_d = 9.9113$ kg, $B_d = 83.1658$ N·s/m. For the FO-impedance controller, sweeping fractional order $\mu \in (0, 2)$, all the FO-impedance controllers which satisfy two specifications w_c and phase margin φ_m can be obtained. Then, J_{ITSE} corresponding to the above FO-impedance controllers can be calculated through step response simulation with the step signal 20 N and simulation time 3.5 s. The smallest J_{ITSE} is 3.1480, and we select the parameters corresponding to the smallest J_{ITSE} as the final designed FO-impedance controller. The final selected FO-impedance controller parameters are as follows: $\mu = 0.88$, $M_d = 11.4978$ kg, $B_d = 111.6108$ N·s/m.

3.1. Fractional-Order Operator Implementation

The fractional-order operators s^μ are implemented by the impulse response invariant discretization method [29]. The order of the approximate transfer function is 7, and the sampling frequency is 2 KHz. The discretized transfer function of the fractional order operator is shown as Equation (29). The comparison of the approximated bode diagram and true bode diagram are shown in Figure 8. Moreover, the discretized open-loop Bode plot of the FO-impedance controller is shown in Figure 9, and it is can be seen that the control system satisfies the given crossover frequency and phase margin specifications.

$$s^\mu = s^{0.88} = \frac{Nums}{Dens}, \tag{29}$$

where

$$Nums = z^7 - 4.2055z^6 + 7.1551z^5 - 6.2782z^4 + 2.9892z^3 - 0.7385z^2 + 0.0802z - 0.0023,$$
$$Dens = 0.3953z^7 - 1.6113z^6 + 2.6413z^5 - 2.2135z^4 + 0.9929z^3 - 0.2255z^2 + 0.0213z - 0.0004.$$

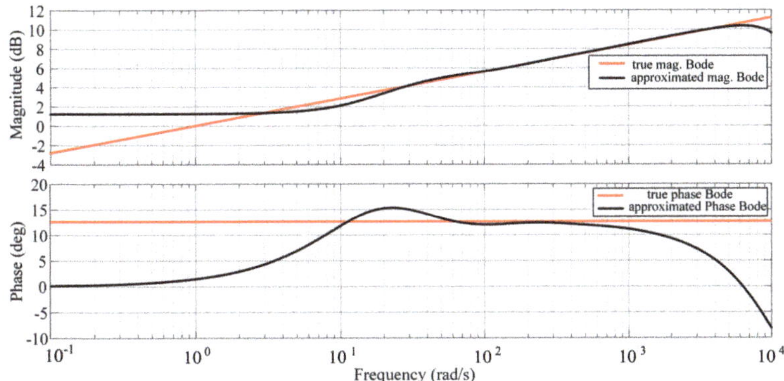

Figure 8. Comparison of approximated bode diagram and true bode diagram.

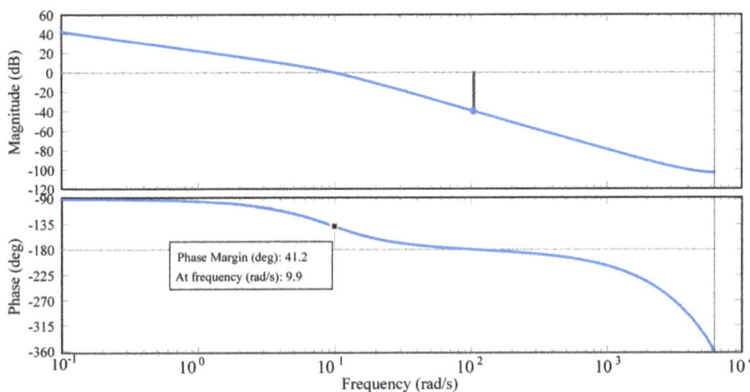

Figure 9. Bode diagram of fractional-order impedance controller.

3.2. Step Response and Anti-Disturbance Simulation

To verify the force-tracking step response performance and anti-disturbance performance, the force control simulation is performed. The simulation results are shown in Figure 10, and the control signals are shown in Figure 11. Set the reference contact force F_{ref} between the end of the robot manipulator and the environment in the Z-axis direction as 20 N.

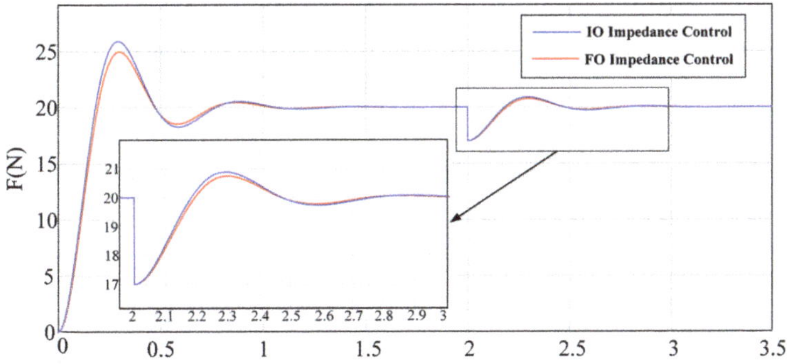

Figure 10. Simulation comparison of FO/IO-impedance controller.

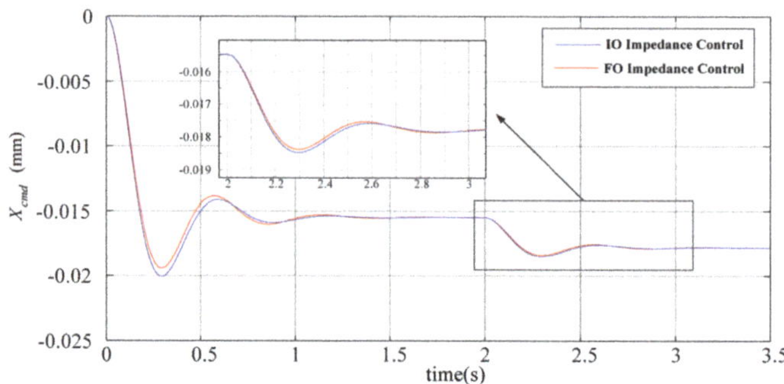

Figure 11. Control signal comparison of FO/IO-impedance controller in simulation.

The contact force step response simulations are performed, using the FO-impedance controller and IO-impedance controller. According to Figure 10, the overshoot of the IO-impedance controller is 29.4820%, the settling time is 0.9480 s, and the ITSE is 3.3350. The FO-impedance controller shows the desired force tracking performance; the overshoot is 25.3930%, the settling time is 0.9425 s and the ITSE is 3.1480. Moreover, in terms of the anti-disturbance test, F_{dis} is added to the IO/FO-impedance control systems, which is -3 N. The stabilization time using the IO-impedance controller is 0.4070 s, and using the FO-impedance controller, it is 0.3875 s. The ITSE using the IO-impedance controller is 1.7426, and that using the FO-impedance controller is 1.7411. The comparison results are shown in Table 1. Therefore, the designed FO-impedance controller achieves much better contact force-tracking and anti-disturbance performance compared to the IO-impedance controller.

Table 1. Comparison of simulation performance between FO/IO-impedance controller.

	Step Response Test			Anti-Disturbance Test	
	Overshoot (%)	Settling Time (s)	ITSE	Stabilization Time (s)	ITSE
IO-impedance	29.4820	0.9480	3.3350	0.4070	1.7426
FO-impedance	25.3930	0.9425	3.1480	0.3875	1.7411
Performance improvement	4.0890%	0.5802%	5.6072%	4.7912%	0.0861%

The robustness of the proposed FO-impedance controller is studied as follows. The designed FO-impedance controller with $K_s = 1293.83$ N/m above is tested under the different spring coefficients, with $K_s = 1500$ N/m, $K_s = 1000$ N/m and $K_s = 800$ N/m, as shown in Figure 12. The force step response results show that the designed FO-impedance controller is robust to the uncertain environment model.

In order to show the benefit of the proposed FO-impedance controller, more numerical examples are given as follows. The results are shown in Table 2 with different frequency domain specifications for the FO-impedance controller compared with the IO-impedance controller.

Table 2. Performance of FO-impedance controller compared with that of IO-impedance controller under different gain crossover w_c and phase margin φ_m.

	$w_c = 10$ rad/s $\varphi_m = 40°$	$w_c = 10$ rad/s $\varphi_m = 45°$	$w_c = 15$ rad/s $\varphi_m = 40°$	$w_c = 15$ rad/s $\varphi_m = 45°$
delta	14.6897%	10.4499%	7.4661%	6.3878%
ts	2.6912%	1.3616%	0.1577%	0.8511%
ITSE	3.8581%	0.9018%	1.9599%	1.0003 %

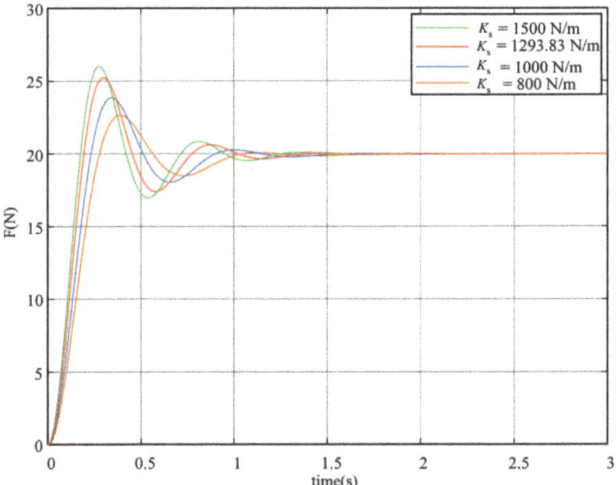

Figure 12. Robustness test for FO-impedance controller.

4. Experimental Verification

4.1. Experimental Setup

The experimental platform is mainly composed of an industrial computer, a robot manipulator, a force sensor, a spring, etc, as shown in Figure 13. The specifications of the experimental platform are given in Table 3. The robot controller software is developed in the industrial computer, which includes a real-time operating system, an Igh EtherCAT master station, and a user interface. The specific D-H parameters of the robot manipulator mechanical body are shown in Table 4. The servo drive communicates with the robot controller software as an EtherCAT slave station. The force sensor is installed at the end of the robot manipulator. The stiffness of the spring which is in contact with the end of the robot manipulator is 1293.83 N/m.

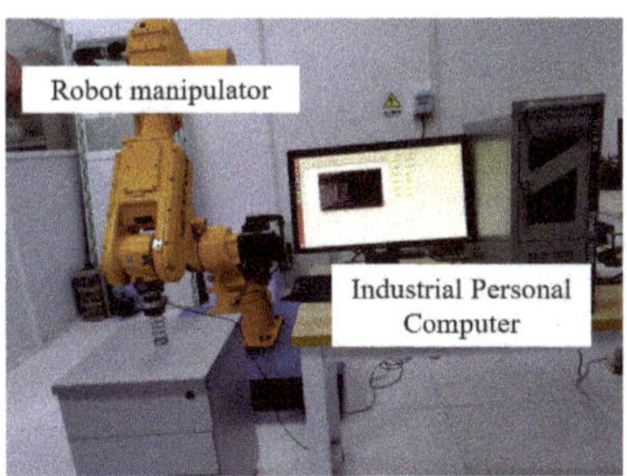

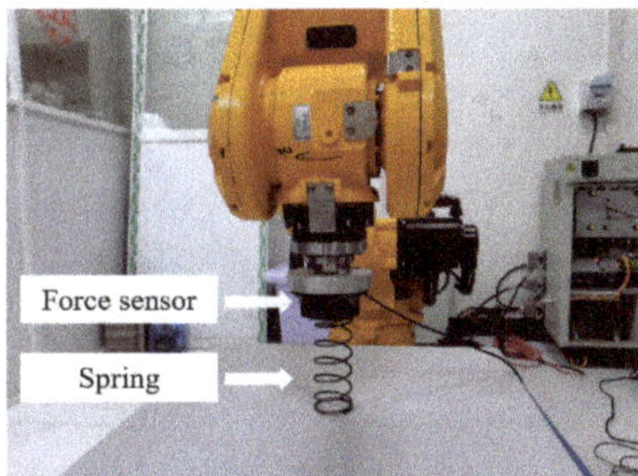

Figure 13. The experimental platform.

4.2. Step Response and Anti-Disturbance Test

The contact force experimental demonstration is performed to verify the force-tracking step response and anti-disturbance performance on the robot manipulator experimental platform. The initial pose of the robot manipulator is set as the force sensor pre-contacting with the spring. Set the reference contact force F_{ref} between the end of the robot manipulator and the environment in the Z-axis direction as 20 N.

Table 3. Model and description of the experimental platform.

Items	Brand and Model	Description
Robot manipulator mechanical body	EFFORT-ERC20C-C10	Degree-of-freedom: 6 Maximum load: 20 kg
Industrial computer	ADVANTECH	Main board: advantech AIMB-785 Processor: Intel Core™ i7-7700/3.6 GHz
Servo drive	TSINO DYNATRON CoolDrive R6	Maximum EtherCAT communication frequency: 4 KHz
Force sensor	HPS-FT060E	Range in Z-axis: ±1000 N Measurement accuracy: 0.4 N Maximum EtherCAT communication frequency: 2 KHz
Spring		Stiffness: 1293.83 N/m

Table 4. The D-H parameters of ER20C-C10.

Link i	Link Length (a_{i-1}) (mm)	Link Twist (α_{i-1}) (degree)	Joint Offset (d_i) (mm)	Joint Angle (θ_i) (degree)
1	168.46	90	504	θ_1
2	781.55	0	0	θ_2+90
3	140.34	90	−0.3	θ_3
4	0	−90	760.39	θ_4
5	0	90	0	θ_5
6	0	0	125	θ_6

The contact force step response experiment is performed. The force responses are shown in Figure 14, and the control signals are shown in Figure 15 using the designed FO-impedance controller and IO-impedance controller presented in Section 2. As shown in Figure 14, the overshoot of the IO-impedance controller is 33.1100%, the settling time is 1.8550 s, and ITSE is 3.9286. The overshoot with the designed FO-impedance controller is 29.3050%, the settling time is 1.7030 s, and ITSE is 3.8681. Moreover, in terms of the anti-disturbance test, F_{dis} is added to the IO/FO-impedance control systems, which is −3 N. The stabilization time of the IO-impedance controller is 1.0080 s and that of the FO-impedance controller is 0.9640 s. ITSE using the IO-impedance controller is 1.8998, and the using the FO-impedance controller is 1.8601. The comparison results are shown in Table 5. Thus, it is verified that the designed FO-impedance controller achieves better contact force tracking and anti-disturbance performance compared to the IO-impedance controller.

Table 5. Comparison of experimental performance between FO/IO-impedance controller.

	Step Response Test			Anti-Disturbance Test	
	Overshoot (%)	Settling Time (s)	ITSE	Stabilization Time (s)	ITSE
IO-impedance	33.1100	1.8550	3.9286	1.0080	1.8998
FO-impedance	29.3050	1.7030	3.8681	0.9640	1.8601
Performance improvement	11.4920%	8.1941%	1.5400%	4.3651%	2.0897%

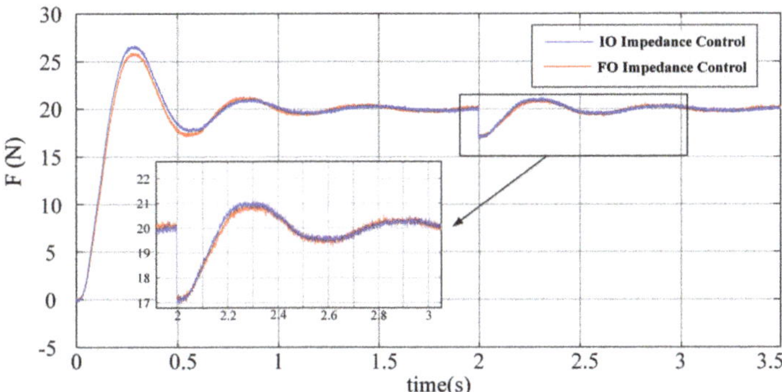

Figure 14. Experimental comparison of FO/IO-impedance controller.

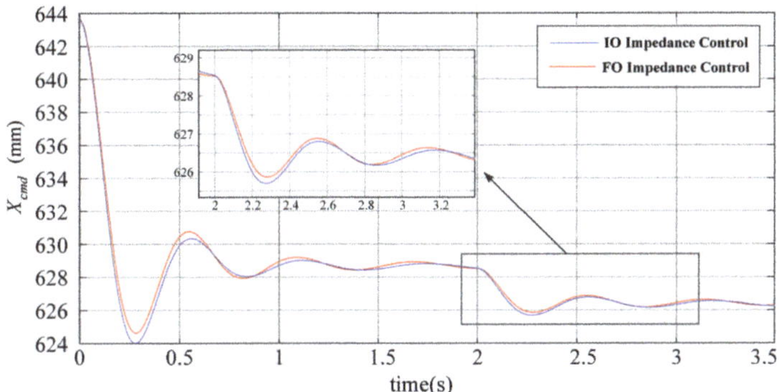

Figure 15. Control signal comparison of FO/IO-impedance controller in experiment.

5. Conclusions

A fractional-order (FO) impedance controller is proposed in this paper. A systematic parameter-tuning method based on frequency-domain specifications is presented with a summarized procedure in details. The fair comparison between the FO-impedance controller and IO-impedance controller is addressed under the same design specifications via the simulation and robot manipulator experimental demonstration. The FO-impedance controller, with the optimized impedance modeling accuracy and more flexibility for control, outperforms the IO-impedance controller in step response performance and anti-disturbance robustness. The future research may be carried out from the direction of rejecting the dynamics disturbances of the robot manipulator and further improving the control performance of the FO-impedance controller.

Author Contributions: Conceptualization, Y.L.; Data curation, Y.D.; Funding acquisition, Y.L.; Investigation, X.L. (Xiaolian Liu) and P.C.; Methodology, Y.L.; Writing—original draft, Y.D.; Writing—review & editing, X.L. (Xin Luo) and Y.L. All authors have read and agreed to the published version of the manuscript.

Funding: This work was supported by the National Natural Science Foundation of China [51975234].

Data Availability Statement: Not applicable.

Conflicts of Interest: The authors declare no conflict of interest.

References

1. Zhang, H.; Li, L.; Zhao, J.; Zhao, J. The hybrid force/position anti-disturbance control strategy for robot abrasive belt grinding of aviation blade base on fuzzy PID control. *Int. J. Adv. Manuf. Technol.* **2021**, *114*, 3645–3656. [CrossRef]
2. Zhang, W.; Li, H.; Cui, L.; Li, H.; Zhang, X.; Fang, S.; Zhang, Q. Research progress and development trend of surgical robot and surgical instrument arm. *Int. J. Med. Robot. Comput. Assist. Surg.* **2021**, *17*, e2309. [CrossRef]
3. Junge, K.; Hughes, J.; Thuruthel, T.G.; Iida, F. Improving robotic cooking using batch Bayesian optimization. *IEEE Robot. Autom. Lett.* **2020**, *5*, 760–765. [CrossRef]
4. Zhang, J.; Liao, W.; Bu, Y.; Tian, W.; Hu, J. Stiffness properties analysis and enhancement in robotic drilling application. *Int. J. Adv. Manuf. Technol.* **2020**, *106*, 5539–5558. [CrossRef]
5. Lu, H.; Zhao, X.; Tao, B.; Yin, Z. Online process monitoring based on vibration-surface quality map for robotic grinding. *IEEE/ASME Trans. Mechatronics* **2020**, *25*, 2882–2892. [CrossRef]
6. Abdi, E.; Kulić, D.; Croft, E. Haptics in Teleoperated Medical Interventions: Force Measurement, Haptic Interfaces and Their Influence on User's Performance. *IEEE Trans. Biomed. Eng.* **2020**, *67*, 3438–3451. [CrossRef] [PubMed]
7. Yang, C.; Xie, Y.; Liu, S.; Sun, D. Force Modeling, Identification, and Feedback Control of Robot-Assisted Needle Insertion: A Survey of the Literature. *Sensors* **2018**, *18*, 561. [CrossRef]
8. Hogan, N. Impedance Control: An Approach to Manipulation: Part II—Implementation. *J. Dyn. Syst. Meas. Control.* **1985**, *107*, 8–16. [CrossRef]
9. Al-Shuka, H.F.; Leonhardt, S.; Zhu, W.H.; Song, R.; Ding, C.; Li, Y. Active impedance control of bioinspired motion robotic manipulators: An overview. *Appl. Bionics Biomech.* **2018**, *2018*, 8203054. [CrossRef] [PubMed]
10. Tsumugiwa, T.; Yura, M.; Kamiyoshi, A.; Yokogawa, R. Development of mechanical-impedance-varying mechanism in admittance control. *J. Robot. Mechatronics* **2018**, *30*, 863–872. [CrossRef]
11. Li, Z.; Xu, C.; Wei, Q.; Shi, C.; Su, C.Y. Human-inspired control of dual-arm exoskeleton robots with force and impedance adaptation. *IEEE Trans. Syst. Man Cybern. Syst.* **2018**, *50*, 5296–5305. [CrossRef]
12. Duan, J.; Gan, Y.; Chen, M.; Dai, X. Adaptive variable impedance control for dynamic contact force tracking in uncertain environment. *Robot. Auton. Syst.* **2018**, *102*, 54–65. [CrossRef]
13. Abu-Dakka, F.J.; Saveriano, M. Variable Impedance Control and Learning—A Review. *Front. Robot. AI* **2020**, *7*, 590681. [CrossRef] [PubMed]
14. Zhang, F.; Lin, L.; Yang, L.; Fu, Y. Variable impedance control of finger exoskeleton for hand rehabilitation following stroke. *Ind. Robot. Int. J. Robot. Res. Appl.* **2019**, *47*, 23–32. [CrossRef]
15. Liang, L.; Chen, Y.; Liao, L.; Sun, H.; Liu, Y. A novel impedance control method of rubber unstacking robot dealing with unpredictable and time-variable adhesion force. *Robot. Comput. Integr. Manuf.* **2021**, *67*, 102038. [CrossRef]
16. Dong, J.; Xu, J.; Zhou, Q.; Hu, S. Physical human–robot interaction force control method based on adaptive variable impedance. *J. Frankl. Inst.* **2020**, *357*, 7864–7878. [CrossRef]
17. Schmidt, A.; Gaul, L. On a critique of a numerical scheme for the calculation of fractionally damped dynamical systems. *Mech. Res. Commun.* **2006**, *33*, 99–107. [CrossRef]
18. Kobayashi, Y.; Onishi, A.; Hoshi, T.; Kawamura, K.; Hashizume, M.; Fujie, M.G. Validation of viscoelastic and nonlinear liver model for needle insertion from in vivo experiments. In *International Workshop on Medical Imaging and Virtual Reality*; Springer: Berlin/Heidelberg, Germany, 2008; pp. 50–59.
19. Sun, H.; Zhang, Y.; Baleanu, D.; Chen, W.; Chen, Y. A new collection of real world applications of fractional calculus in science and engineering. *Commun. Nonlinear Sci. Numer. Simul.* **2018**, *64*, 213–231. [CrossRef]
20. Niu, H.; Chen, Y.; West, B.J. Why Do Big Data and Machine Learning Entail the Fractional Dynamics? *Entropy* **2021**, *23*, 297. [CrossRef]
21. Chen, P.; Luo, Y. A Two-Degree-of-Freedom Controller Design Satisfying Separation Principle with Fractional Order PD and Generalized ESO. *IEEE/ASME Trans. Mechatron.* **2021**, *27*, 137–148. [CrossRef]
22. Luo, Y.; Zhang, T.; Lee, B.; Kang, C.; Chen, Y. Fractional-order proportional derivative controller synthesis and implementation for hard-disk-drive servo system. *IEEE Trans. Control Syst. Technol.* **2013**, *22*, 281–289. [CrossRef]
23. Dadkhah Khiabani, E.; Ghaffarzadeh, H.; Shiri, B.; Katebi, J. Spline collocation methods for seismic analysis of multiple degree of freedom systems with visco-elastic dampers using fractional models. *J. Vib. Control* **2020**, *26*, 1445–1462. [CrossRef]
24. Wang, P.; Wang, Q.; Xu, X.; Chen, N. Fractional critical damping theory and its application in active suspension control. *Shock Vib.* **2017**, *2017*, 2738976. [CrossRef]
25. Jung, S.; Hsia, T.C. Stability and convergence analysis of robust adaptive force tracking impedance control of robot manipulators. In Proceedings of the Proceedings 1999 IEEE/RSJ International Conference on Intelligent Robots and Systems. Human and Environment Friendly Robots with High Intelligence and Emotional Quotients, Kyongju, Republic of Korea, 17–21 October 1999; Volume 2, pp. 635–640.
26. Lee, K.; Buss, M. Force tracking impedance control with variable target stiffness. *IFAC Proc. Vol.* **2008**, *41*, 6751–6756. [CrossRef]
27. Kim, T.; Kim, H.S.; Kim, J. Position-based impedance control for force tracking of a wall-cleaning unit. *Int. J. Precis. Eng. Manuf.* **2016**, *17*, 323–329. [CrossRef]

28. Luo, Y.; Chen, Y. Fractional order [proportional derivative] controller for a class of fractional order systems. *Automatica* **2009**, *45*, 2446–2450. [CrossRef]
29. Impulse Response Invariant Discretization of Fractional Order Integrators/Dierentiators. Available online: http://www.mathworks.com/matlabcentral/fileexchange/21342-impulse-response-invariant-discretization-of-fractional-orderintegrators-dierentiators (accessed on 1 September 2020).

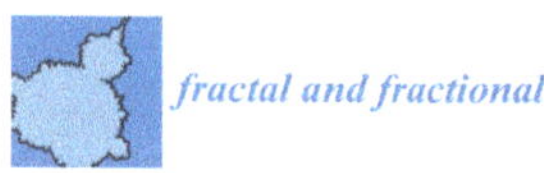

fractal and fractional

Article

Using a Fully Fractional Generalised Maxwell Model for Describing the Time Dependent Sinusoidal Creep of a Dielectric Elastomer Actuator

Timi Karner *, Rok Belšak and Janez Gotlih

Laboratory for Robotisation, Faculty of Mechanical Engineering, University of Maribor, 2000 Maribor, Slovenia
* Correspondence: timi.karner@um.si

Abstract: Actuators made of dielectric elastomers are used in soft robotics for a variety of applications. However, due to their mechanical properties, they exhibit viscoelastic behaviour, especially in the initial phase of their performance, which can be observed in the first cycles of dynamic excitation. A fully fractional generalised Maxwell model was derived and used for the first time to capture the viscoelastic effect of dielectric elastomer actuators. The Laplace transform was used to derive the fully fractional generalised Maxwell model. The Laplace transform has proven to be very useful and practical in deriving fractional viscoelastic constitutive models. Using the global optimisation procedure called Pattern Search, the optimal parameters, as well as the number of branches of the fully fractional generalised Maxwell model, were derived from the experimental results. For the fully fractional generalised Maxwell model, the optimal number of branches was determined considering the derivation order of each element of the branch. The derived model can readily be implemented in the simulation of a dielectric elastomer actuator control, and can also easily be used for different viscoelastic materials.

Keywords: dielectric elastomer actuator; fully fractional generalised Maxwell model; fractional derivation; viscoelasticity; creep; optimisation

Citation: Karner, T.; Belšak, R.; Gotlih, J. Using a Fully Fractional Generalised Maxwell Model for Describing the Time Dependent Sinusoidal Creep of a Dielectric Elastomer Actuator. *Fractal Fract.* **2022**, *6*, 720. https://doi.org/10.3390/fractalfract6120720

Academic Editors: Kishore Bingi and Abhaya Pal Singh

Received: 26 September 2022
Accepted: 2 December 2022
Published: 4 December 2022

Publisher's Note: MDPI stays neutral with regard to jurisdictional claims in published maps and institutional affiliations.

1. Introduction

Dielectric elastomer actuators, also called DEAs, can have different structures. The first DEAs had parallel or cylindrical structures [1,2]. From then on, many different structures were developed: spring roll actuators, helical actuators, and stacked actuators where the actuators are stacked on top of each other [3–5]. All these actuators have the same basic structure and activation principle. Thus, regardless of whether the DEAs are parallel or cylindrical, their activation principle is the same. Their structure is like that of a capacitor. It has an upper and a lower electrode and an elastic dielectric. Some special rules apply to both the electrodes and the dielectric. The dielectric must be elastic. The electrodes must be conductive, and their mechanical structure must be the same or similar to that of the dielectric [6]. If their mechanical properties are different, the DEA may lose its mobility and efficiency. Usually, conductive pastes of carbon are chosen for the electrodes.

The material properties of the selected elastomer VHB 4910 are viscoelastic, which means that the material has both viscous and elastic properties. When the DEA is subjected to a sinusoidal load, it exhibits sinusoidal creep behaviour. When a high voltage is applied, coupling occurs between the electrical and mechanical forces. Suo et al. were the first researchers to describe the electromechanical coupling in detail in a DEA [7]. In the work of Gu et al. [8], the viscoelastic behaviour was captured using continuum mechanics and nonequilibrium thermodynamics, where the Helmholtz free energy along with the Gent model was used to derive the constitutive equations. The work shows a good match between the experimental and simulation results. However, extensive knowledge in the above areas is required, and the model cannot be integrated easily into the control simulations. The Prandtl–Ishlinskii model and the modified Prandtl–Ishlinskii model are used in the work of Zuo et al. [9,10]. The work is based on the fourth order polynomial to describe the asymmetric behaviour, and fixed weights with thresholds of the play operators are used to describe the rate dependence, while the second order derivative of the input voltage is introduced into the fourth order polynomial to describe the peak-to-peak shifts that depend on the frequencies. This is a complex process that results in a good match between the experimental data and the calculated responses. The Prandtl–Ishlinskii model can be used as a feedforward compensator. Wissler et al. [11] used the Prony series theory to model the time dependent viscoelastic behaviour of the elastomer used in soft actuators. The drawback of this method is the inability to use it in the control simulations. A standard linear solid rheological model which satisfies the thermodynamic consistency was also used to capture rate-dependant behaviour of the soft materials in the work of Zhao et al. [12]. This model could capture the rate dependent mechanical behaviour of soft materials. However, the introduction of the internal variables in the modelling process significantly increases the complexity of the proposed constitutive model and the number of governing equations.

Fractional calculus has proved to be a powerful mathematical tool in the approach used to modelling the time dependent mechanical behaviour of viscoelastic materials. Its advantages lie in the remarkable reduction in model parameters where fractional orders of derivatives are used. This can be especially seen in the field of fractional viscoelasticity, including soils [13], polymers [14,15] and construction materials [16]. Xu et al. [17] used fractional constitutive models of a fractional Kelvin–Voigt model, a fractional Maxwell model and a fractional Poynting–Thomson model, which is a springpot connected in series with two parallel springpots. In the work of Barretta et al. [18], hereditariness and nonlocality of bending problems were presented with the help of fractional operators. Time-dependent hereditary behaviour, which is typical of viscoelastic materials, has been modelled with the help of a springpot, a fractional Kelvin–Voigt and a fractional Maxwell model.

To capture the sinusoidal creep behaviour, the fully fractional generalised Maxwell model is derived in Section 2. The generalised fractional Maxwell model has already been used in the work of Luo et al. [19] to determine the storage and loss modulus, but not with the Laplace transform, nor to derive the governing equation of the DEA. Our model is the first to describe the material behaviour of the DEA using the Laplace transform. The fully fractional generalised Maxwell model was used because it does not require complex knowledge of continuum mechanics and thermodynamics. It can readily be derived using the Laplace transform to obtain the transfer function of the system. Using the *lsim* function in the Matlab software, the excitation voltage with three different frequencies can be implemented in the transfer function of the system, and the response of the DEA can be calculated easily. An optimisation procedure using the Pattern Search global optimisation solver was used to derive the material parameters based on the experimental results. Section 3 shows the results of the optimisation procedures. Section 4 summarises the conclusions.

2. Materials and Methods

2.1. Principles of the DEA

The structure of the DEA is similar to that of a capacitor. It has an upper and lower conductive electrode and an elastic dielectric. The structure of the DEA can be seen in Figure 1. Usually, elastic materials are chosen for dielectrics, such as VHB tapes. VHB 4910 tape was used in this study [20]. A high voltage DC is required to activate the DEAs. When the voltage is applied, the electrical force also known as the Maxwell force is generated between the upper and lower conductive electrodes, causing the DEA to contract vertically and expand longitudinally, as elastomer is known to be incompressible.

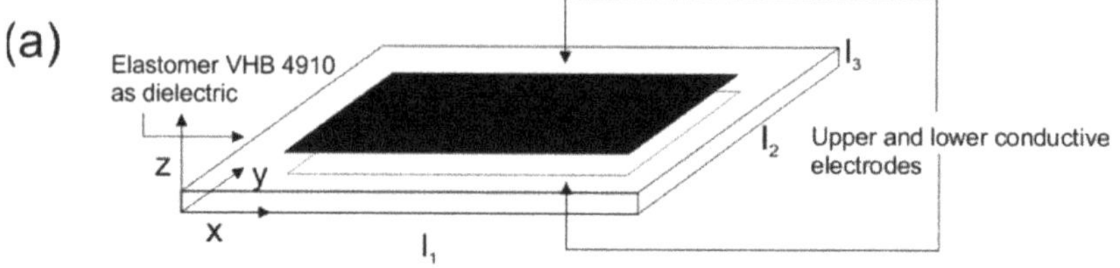

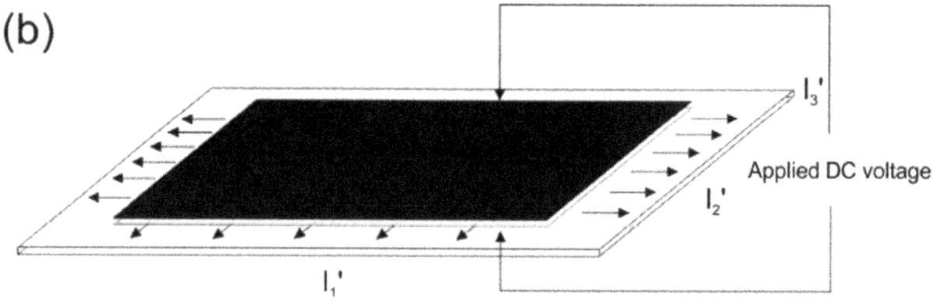

Figure 1. Structure of the parallel DEA: (**a**) initial state and (**b**) activated state.

The Maxwell force can be calculated as

$$F_{Max} = F_{el} = \varepsilon_0 \varepsilon_r E^2 l_1 l_2 = \varepsilon_0 \varepsilon_r \left(\frac{V}{l_3} \right)^2 l_1 l_2 [\text{N}] \tag{1}$$

$\varepsilon_0 -$ absolute permittivity
$\varepsilon_r = 4.7 -$ relative permittivity
$E -$ electric field
$V -$ voltage
$l_1, l_2, l_3 -$ dimensions of the DEA

The relative permittivity was chosen as 4.7 according to [8]. A conductive paste was chosen from BareConductive® [21].

2.2. Derivation of the Fully Fractional Generalised Maxwell Model

To derive the fully fractional generalised Maxwell model, one needs to use fractional derivatives. Fractional derivatives are derivatives that are not restricted to positive integers, but can be any real number. There are three definitions of fractional derivatives, namely, Riemann–Liouville, Caputo, and Gruenwald–Letnikov [22]. The Gruenwald–Letnikov definition is used, namely

$$_aD_t^p f(t) = \lim_{h \to 0} h^{-p} \sum_{r=0}^{m} (-1)^r \binom{p}{r} f(t - rh) \tag{2}$$

$a, t -$ time limits
$m -$ integer order of derivation/integration
$p- = \begin{cases} p > 0, \text{ derivation} \\ p < 0, \text{ integration} \end{cases}$

Because it can handle the fractional derivation and integration easily. It is also suitable for the numerical calculation of fractional derivatives.

The generalised Maxwell model is used to describe constitutive models with viscoelastic effects. It consists of a spring connected in parallel to the branches of the Maxwell elements. The Maxwell element consists of a spring connected in series with a dashpot [15,16]. To convert this into a fully fractional generalised Maxwell model, all elements are replaced by the so-called springpot element [23,24]. The springpot element has two parameters, one of which represents the material properties and the other the derivative order. The derivative order is bounded between 0 and 1, since there is no known physical interpretation above 1 [24]. Figure 2 shows both models.

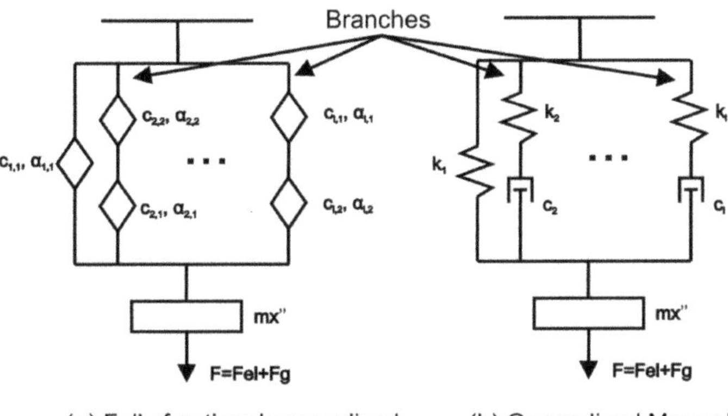

Figure 2. Two types of models used: (**a**) Fully fractional model and (**b**) Generalised model.

$c_{i,1}, c_{i,2}$ — material properties of springpot
$\alpha_{i,1}, \alpha_{i,2}$ — order of derivation
k_i — spring constants
m — mass of weight
n — number of branches with the Maxwell element
Fel — electrical force applied
Fg — weight of the mass

The Laplace transform with Laplace operator is used to derive the governing equation of motion for the DEA using the fully fractional generalised Maxwell model. The Laplace transform turns the derivative into a multiplication and the integration into a division. The fully fractional generalised Maxwell model describes the material behaviour. The electric force is calculated using Equation (1). Since the weight is used to hold the DEA in the stretched position, its inertia must be included in the governing equation of motion. All branches of the fully fractional generalised Maxwell model are subject to the same displacement, as shown in Figure 3.

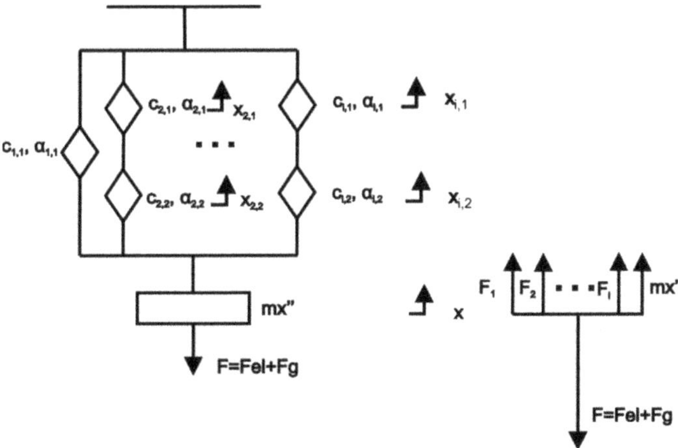

Figure 3. Displacement of the elements.

Each branch has two displacements because it has two springpots. The electric force and the force generated by the weight are distributed jointly between the branches of the fully fractional generalised Maxwell model. The governing equation of motion is calculated as

$$F = F_1 + \sum_{i=2}^{n+1} F_i + m\ddot{x}. \tag{3}$$

Displacement of each branch with the fractional Maxwell element is calculated as

$$
\begin{aligned}
x &= x_{2,1} + x_{2,2} \\
x &= x_{3,1} + x_{3,2} \\
&\;\;\vdots \\
x &= x_{i,1} + x_{i,2} \\
n \cdot x &= \sum_{i=2}^{n+1} x_{i,1} + \sum_{i=2}^{n+1} x_{i,2}
\end{aligned}
\tag{4}
$$

The force in the first branch is calculated as

$$F_1 = c_{1,1} \frac{d^{\alpha_{1,1}} x}{dt^{\alpha_{1,1}}}. \tag{5}$$

The force in each branch containing two springpots is the same in each springpot, which is calculated as

$$
\begin{aligned}
F_2 &= c_{2,1} \frac{d^{\alpha_{2,1}} x_{2,1}}{dt^{\alpha_{2,1}}} = c_{2,2} \frac{d^{\alpha_{2,2}} x_{2,2}}{dt^{\alpha_{2,2}}} \\
F_3 &= c_{3,1} \frac{d^{\alpha_{3,1}} x_{3,1}}{dt^{\alpha_{3,1}}} = c_{3,2} \frac{d^{\alpha_{3,2}} x_{3,2}}{dt^{\alpha_{3,2}}} \\
&\;\;\vdots \\
\sum_{i=2}^{n+1} F_i &= \sum_{i=2}^{n+1} c_{i,1} \frac{d^{\alpha_{i,1}} x_{i,1}}{dt^{\alpha_{i,1}}} = \sum_{i=2}^{n+1} c_{i,2} \frac{d^{\alpha_{i,2}} x_{i,2}}{dt^{\alpha_{i,2}}}
\end{aligned}
\tag{6}
$$

Expressing $x_{i,1}$ and $x_{i,2}$ from Equation (6) with the fractional integration, one gets

$$
\begin{aligned}
\sum_{i=2}^{n+1} F_i &= \sum_{i=2}^{n+1} c_{i,1} \frac{d^{\alpha_{i,1}} x_{i,1}}{dt^{\alpha_{i,1}}} \bigg/ \frac{d^{-\alpha_{i,1}}}{dt^{-\alpha_{i,1}}} \\
\sum_{i=2}^{n+1} \frac{d^{-\alpha_{i,1}} F_i}{dt^{-\alpha_{i,1}}} &= \sum_{i=2}^{n+1} c_{i,1} x_{i,1} \\
x_{i,1} &= \sum_{i=2}^{n+1} \frac{1}{c_{i,1}} \frac{d^{-\alpha_{i,1}} F_i}{dt^{-\alpha_{i,1}}} ; x_{i,2} = \sum_{i=2}^{n+1} \frac{1}{c_{i,2}} \frac{d^{-\alpha_{i,2}} F_i}{dt^{-\alpha_{i,2}}} ;
\end{aligned}
\tag{7}
$$

Inserting the results from Equation (7) into Equation (4), one gets

$$n \cdot x = \sum_{i=2}^{n+1} \frac{1}{c_{i,1}} \frac{d^{-\alpha_{i,1}} F_i}{dt^{-\alpha_{i,1}}} + \sum_{i=2}^{n+1} \frac{1}{c_{i,2}} \frac{d^{-\alpha_{i,2}} F_i}{dt^{-\alpha_{i,2}}}. \tag{8}$$

Force F_i should be expressed from Equation (8). Laplace transformation is used, since force F_i is part of the fractional integration with different orders of integration. Performing Laplace transformation on Equation (8) and expressing F_i, one gets

$$\sum_{i=2}^{n+1} F_i(s) = \frac{n \cdot x(s)}{\sum_{i=2}^{n+1} \left(\frac{1}{c_{i,1} s^{\alpha_{i,1}}} + \frac{1}{c_{i,2} s^{\alpha_{i,2}}} \right)}. \tag{9}$$

Performing the Laplace transform on Equation (3) and inserting Equation (9), one gets

$$F(s) = c_1 s^{\alpha_1} x(s) + \frac{n \cdot x(s)}{\sum\limits_{i=2}^{n+1}\left(\dfrac{1}{c_{i,1}s^{\alpha_{i,1}}} + \dfrac{1}{c_{i,2}s^{\alpha_{i,2}}}\right)} + ms^2 x(s). \tag{10}$$

Rearranging Equation (10) to get the transfer function, one gets

$$\frac{x(s)}{F(s)} = \frac{1}{ms^2 + c_{1,1}s^{\alpha_{1,1}} + \dfrac{n}{\sum\limits_{i=2}^{n+1}\left(\dfrac{1}{c_{i,1}s^{\alpha_{i,1}}} + \dfrac{1}{c_{i,2}s^{\alpha_{i,2}}}\right)}}. \tag{11}$$

Equation (11) represents the transfer function of the governing equations of motion for the fully fractional generalised Maxwell model with n branches, where the influence of the electric force and the force of the weight of the mass 316 g are combined in the equation. The add-on FOMCON is required to implement Equation (11) in Matlab [25]. Equation (11) is written in the relation force–displacement. To convert Equation (11) into the relation stress–strain, one needs to use

$$E = \frac{\sigma}{\varepsilon} = \frac{F \cdot L}{A \cdot \Delta l} = \frac{k \cdot L}{A} \rightarrow k = \frac{E \cdot A}{L} \tag{12}$$

and

$$\sigma = \eta \cdot \frac{d^\alpha \varepsilon}{dt^\alpha} \rightarrow \eta = \frac{\sigma}{\frac{d^\alpha \varepsilon}{dt^\alpha}} = \frac{F}{A\frac{d^\alpha\left(\frac{\Delta l}{L}\right)}{dt^\alpha}} = \frac{F}{A\frac{1}{L}\frac{d^\alpha(\Delta l)}{dt^\alpha}} = \frac{c \cdot L}{A} \rightarrow c = \frac{\eta \cdot A}{L}. \tag{13}$$

2.3. Experiments

The elastomer VHB 4910 was used to set up the experiment. The elastomer was cut to the initial dimensions of 49 mm × 50 mm × 1 mm, with only a 10 mm wide area used to expand the elastomer to the initial dimension, while the rest was used for clamping. The initial dimensions for the sinusoidal force excitation were 49 mm × 60 mm × 0.16 mm. The active area to which the conductive paste was applied was 40 mm × 60 mm.

Three different frequencies with an amplitude of 6 kV DC were used for the sinusoidal voltage excitation. The three frequencies were $F_1 = \frac{1}{13}$Hz, $F_2 = \frac{1}{7}$Hz, and $F_3 = \frac{1}{5}$Hz. These frequencies were chosen so that they were not multiples of each other. In this way, the frequencies are not associated with a common factor. The experiment lasted 155 s to capture the sinusoidal creep behaviour of the DEA. The displacement of the DEA was measured using a Wenglor laser sensor [26]. The experimental setup is shown in Figure 4.

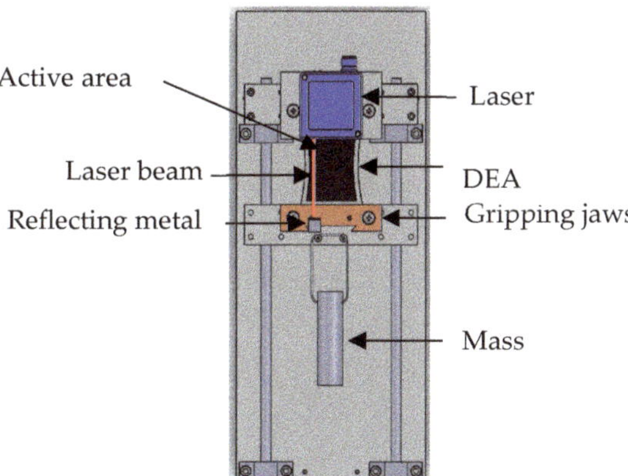

Figure 4. Experimental set up of the DEA.

2.4. Optimisation

To obtain the material parameters of the fully fractional generalised Maxwell model, the experimental results were optimised with the model. For the optimisation procedure, the global optimisation solver Pattern Search was chosen, which is also known as a direct search method in Matlab software. The flow of the optimisation procedure is shown in Figure 5. First, the user must specify the number of branches for the fully fractional generalised Maxwell model and the initial parameters for the model.

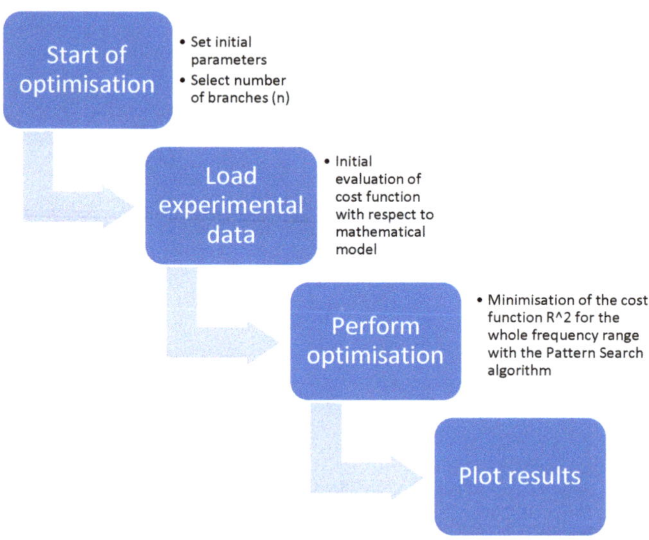

Figure 5. Optimisation procedure.

The electric and mechanical forces are calculated as the sum of the Maxwell force given by Equation (1) and the weight of the dead mass. In Equation (1), V is replaced with the calculation of the appropriate voltage regarding the frequency in use. Using the Matlab function *lsim*, the response of the transfer function of the fully fractional generalised Maxwell model to the electric and mechanical forces can be calculated easily. The responses for all frequencies were calculated and compared with the experimental results using the least squares method. The efficiency of the fully fractional generalised Maxwell model with a different number of branches was calculated using the method of the coefficient of determination, also known as the R^2 method, calculated as

$$R^2 = 1 - \frac{SS_{res}}{SS_{tot}}.$$ (14)

$$SS_{tot} = \sum_i (y_i - \overline{y})^2 \qquad \text{Total sum of squares.}$$
$$SS_{res} = \sum_i (y_i - f_i)^2 \qquad \text{Residual sum of squares.}$$

y_i Measured data.
$\overline{y}$ Average data.
f_i Calculated data.

After the optimisation procedure was completed, the best results were recorded and the final value of R^2 was calculated for each frequency. At the end, the average value of R^2 was calculated as R^2_{mean} for the whole frequency range. The number of branches was chosen to be between 1 and 5. Finally, each frequency was optimised individually for the model, to compare the results when only one frequency was optimised to the results where the whole frequency range was optimised. It was examined which the essential parameters of the model were for each frequency. Table 1 shows all symbols used in the research.

Table 1. Nomenclature table with SI units.

Symbol	Unit	Meaning
A	m^2	Area.
$\alpha_{i,1},\ \alpha_{i,2}$	/	Order or fractional derivation of the springpots.
a, t	s	Time limits of the fractional derivation.
$c_{i,1},\ c_{i,2}$	$N(s^{\alpha_{i,1}}/m),\ N(s^{\alpha_{i,2}}/m)$	Material properties of the springpots.
E	$kg/(ms^2)$	Modul of elasticity.
ε_o	F/m	Absolute permittivity.
ε_r	/	Relative permittivity.
ε	1	Strain.
F_{Max}	N	Maxwell force.
F_{el}	N	Electrical force.
F_i	N	Force in individual branch.
f_i	mm	Calculated data.
i	/	Current number of fractional Maxwell element.
k	N/m	Spring constant.
l_1, l_2, l_3	m	Dimensions of the DEA.
L	m	Initial length.
l	m	Displacement.
m	/	Integer order of derivation by the definition.
m	kg	Mass of weight.
n	/	Number of fractional Maxwell elements.
η	Ns/m^2	Viscosity.
p	/	Fractional order of derivation by the definition.
R^2	/	Coefficient of determination.
R^2_{mean}	/	Mean value of coefficient of determination.
s	/	Laplace operator.
SS_{tot}	/	Total sum of squares.
SS_{res}	/	Residual sum of squares.
σ	N/m^2	Stress.
V	V	Voltage.
x_i	m	Displacement of individual branch.
y_i	mm	Measured data.
$\overline{y}$	mm	Averaged measured data.

3. Results

After the optimisation procedure was set up using Matlab software and the Pattern Search global optimisation algorithm, the R^2_{mean} values shown in Table 2 were obtained. As can be seen, the number of branches increased the accuracy of the model, but not significantly. It can also be seen that adding more than three branches did not affect the efficiency of the model. The initial parameters used in the optimisation, as well as the optimal parameters obtained from the optimisation, are shown in Table 3.

Table 2. R^2_{mean} value of the fully fractional generalised Maxwell model for one to five branches.

Fully Fractional Generalised Maxwell Model Number of Branches	$R^2{}_{mean}$
$n = 1$	0.5456
$n = 2$	**0.5489**
$n = 3$	0.5456
$n = 4$	0.5456
$n = 5$	0.5456

Table 3. Initial and optimised parameters of the fully fractional generalised Maxwell model for one to five branches.

Param.	$n=1$	$\alpha_{1,1}$	$\alpha_{2,1}$	$\alpha_{2,2}$							
Initial		0.2	1	1							
Optimised		0.2	0.002	1							
Parameters		$c_{1,1}$	$c_{2,1}$	$c_{2,2}$							
Initial		500	500	500							
Optimised		62.952	0.036	0.142							
Param.	$n=2$	$\alpha_{1,1}$	$\alpha_{2,1}$	$\alpha_{2,2}$	$\alpha_{3,1}$	$\alpha_{3,2}$					
Initial		0.2	1	1	1	1					
Optimised		0.2	1	1	0.523	0.046					
Param.		$c_{1,1}$	$c_{2,1}$	$c_{2,2}$	$c_{3,1}$	$c_{3,2}$					
Initial		500	500	500	500	500					
Optimised		62.740	47.958	173.21	430.381	0.215					
Param.	$n=3$	$\alpha_{1,1}$	$\alpha_{2,1}$	$\alpha_{2,2}$	$\alpha_{3,1}$	$\alpha_{3,2}$	$\alpha_{4,1}$	$\alpha_{4,2}$			
Initial		0.2	1	1	1	1	1	1			
Optimised		0.2	1	1	1	0.002	1	1			
Param.		$c_{1,1}$	$c_{2,1}$	$c_{2,2}$	$c_{3,1}$	$c_{3,2}$	$c_{4,1}$	$c_{4,2}$			
Initial		500	500	500	500	500	500	500			
Optimised		62.950	0.267	0.464	0.140	0.088	0.036	0.237			
Param.	$n=4$	$\alpha_{1,1}$	$\alpha_{2,1}$	$\alpha_{2,2}$	$\alpha_{3,1}$	$\alpha_{3,2}$	$\alpha_{4,1}$	$\alpha_{4,2}$	$\alpha_{5,1}$	$\alpha_{5,2}$	
Initial		0.2	1	1	1	1	1	1	1	1	
Optimised		0.2	0.002	1	0.002	1	1	0.002	1	0.002	
Param.		$c_{1,1}$	$c_{2,1}$	$c_{2,2}$	$c_{3,1}$	$c_{3,2}$	$c_{4,1}$	$c_{4,2}$	$c_{5,1}$	$c_{5,2}$	
Initial		500	500	500	500	500	500	500	500	500	
Optimised		62.950	0.103	0.036	0.321	0.036	0.094	0.157	0.036	0.225	
Param.	$n=5$	$\alpha_{1,1}$	$\alpha_{2,1}$	$\alpha_{2,2}$	$\alpha_{3,1}$	$\alpha_{3,2}$	$\alpha_{4,1}$	$\alpha_{4,2}$	$\alpha_{5,1}$	$\alpha_{5,2}$	$\alpha_{6,1}$ $\alpha_{6,2}$
Initial		0.2	1	1	1	1	1	1	1	1	1 1
Optimised		0.2	0.002	1	1	1	1	0.002	1	0.002	1 1
Param.		$c_{1,1}$	$c_{2,1}$	$c_{2,2}$	$c_{3,1}$	$c_{3,2}$	$c_{4,1}$	$c_{4,2}$	$c_{5,1}$	$c_{5,2}$	$c_{6,1}$ $c_{6,2}$
Initial		500	500	500	500	500	500	500	500	500	500 500
Optimised		62.950	0.097	0.315	0.036	0.356	0.036	0.097	0.036	0.095	0.036 0.285

From Table 3, in some branches the order of derivation does not change from the initial values. The derivation orders of 1 represent dashpot elements, and if both elements of the branch have the order of 1, they can be combined into one dashpot element. The same applies for spring elements if the order of derivation equals 0.

Figure 6 shows the experimental and calculated results from the optimised parameters of the fully fractional generalised Maxwell model for one to five branches. Adding more than two branches did not improve the efficiency of the model. The best matching between the data is seen for the middle frequency of 1/7 Hz, where the matching was 88%. For the frequency of 1/5 Hz, the worst matching was achieved of only 12.9%.

The optimisation of the model for an individual frequency was performed after the optimisation for the whole frequency range was carried out. From Table 2, it is seen that two additional branches of the fully fractional Maxwell elements are the most optimised. In Table 4, the initial and optimised parameters of only two additional branches are shown for each individual frequency for the fully fractional generalised Maxwell model. Each frequency demands its own material parameters, as well as different topology of the fully fractional generalised Maxwell model. Figure 7 shows matching between the experimental and calculated results if each frequency was optimised individually for the model.

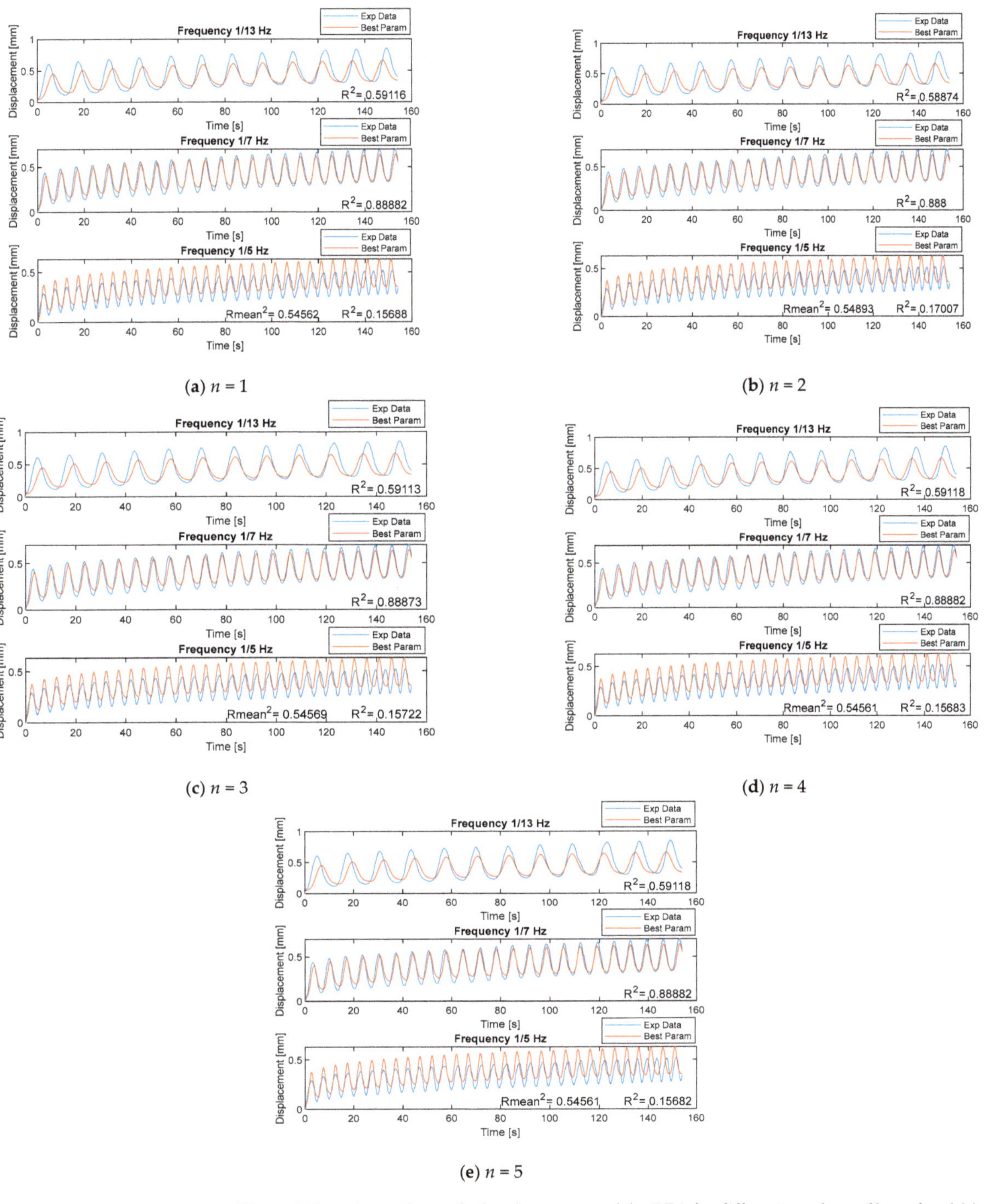

Figure 6. Experimental vs. calculated responses of the DEA for different numbers of branches (*n*) in the fully fractional generalised Maxwell model.

Table 4. Initial and optimised parameters of the fully fractional generalised Maxwell model for each individual frequency.

Parameters	F = 1/13 Hz, n = 3	$\alpha_{1,1}$	$\alpha_{2,1}$	$\alpha_{2,2}$	$\alpha_{3,1}$	$\alpha_{3,2}$	
Initial		0.2	1	1	1	1	
Optimised		0.1795	1	1	1	1	
Parameters		$c_{1,1}$	$c_{2,1}$	$c_{2,2}$	$c_{3,1}$	$c_{3,2}$	R^2
Initial		500	500	500	500	500	0.658
Optimised		51.773	0.002	0.002	0.002	0.002	
Parameters	F = 1/7 Hz, n = 3	$\alpha_{1,1}$	$\alpha_{2,1}$	$\alpha_{2,2}$	$\alpha_{3,1}$	$\alpha_{3,2}$	
Initial		0.2	1	1	1	1	
Optimised		0.188	0.255	0.046	0.225	0.880	
Parameters		$c_{1,1}$	$c_{2,1}$	$c_{2,2}$	$c_{3,1}$	$c_{3,2}$	R^2
Initial		500	500	500	500	500	0.907
Optimised		57.696	569.346	0.003	406.744	0.479	
Parameters	F = 1/5 Hz, n = 3	$\alpha_{1,1}$	$\alpha_{2,1}$	$\alpha_{2,2}$	$\alpha_{3,1}$	$\alpha_{3,2}$	
Initial		0.2	1	1	1	1	
Optimised		0.2	1	1	1	0.002	
Parameters		$c_{1,1}$	$c_{2,1}$	$c_{2,2}$	$c_{3,1}$	$c_{3,2}$	R^2
Initial		500	500	500	500	500	0.665
Optimised		73.133	19.199	188.074	305.285	191.367	
						R^2_{mean}	0.743

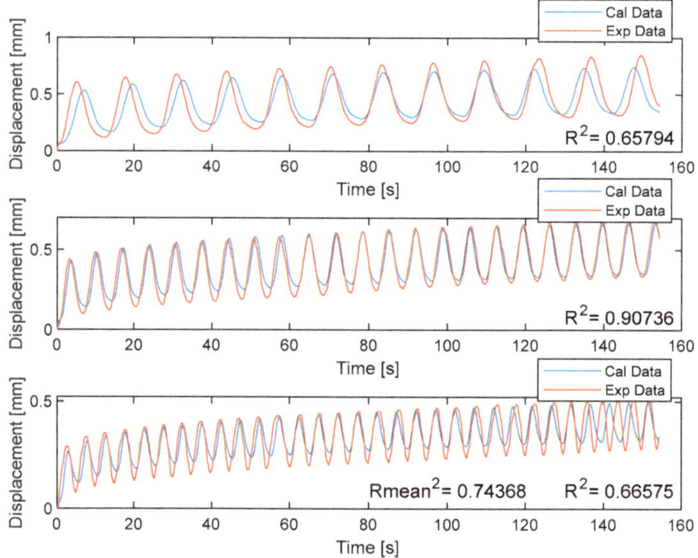

Figure 7. Responses of individually optimised frequencies with the generalised fractional Maxwell model.

4. Discussion

The fully fractional generalised Maxwell model was used for the first time to describe the material behaviour of a DEA. Table 2 shows that the number of branches contributes to the effectiveness of the model. However, the contribution of each additional branch was small. Increasing the number of branches up to two increased the effectiveness. Surprisingly, adding more branches did not improve the effectiveness of the model. It can be seen from Table 3 that when adding more than two branches, the optimised parameters were chosen such that the model could be reduced to the fully fractional generalised Maxwell model with only two branches. This was possible because the two α parameters within a branch were 1, which represents dashpots. Dashpots in series can be reduced to one dashpot. The reduced and rearranged model can be seen in Figure 8.

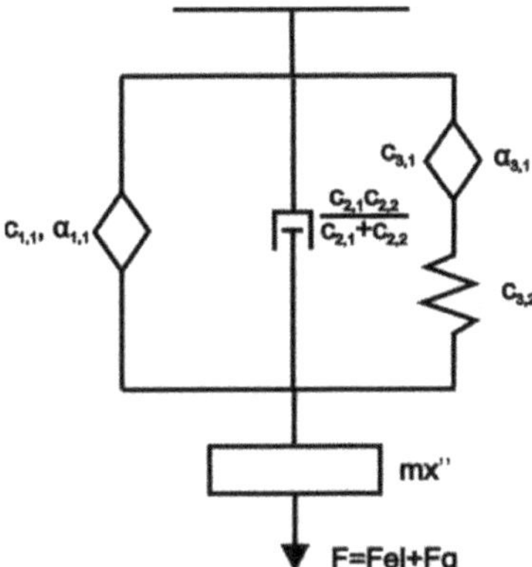

Figure 8. Reduced and rearranged optimised fully fractional generalised Maxwell model with two additional branches.

Concluding points can be itemized as follow:

- The number of fully fractional Maxwell elements slightly affected the effectiveness of the model.
- Adding more than two branches did not increase the effectiveness of the model.
- The fully fractional Maxwell model was reduced to the model seen in Figure 8.
- The middle frequency of 1/7 Hz had the best agreement of 0.88 between data.
- Optimising each frequency individually drastically improved the overall agreement between data to 0.745.
- Optimising each frequency individually has a drawback since each frequency requires its own material parameters.
- Topology optimisation cannot be included into the Pattern Search algorithm.

The reduced and rearranged model has two springpots. The first springpot has the order of derivative equal to 0.2 which means it has 80% characteristics of a spring and only 20% characteristics of a damper. The other springpot has the order of derivative equal to 0.52 which means that nearly half of its characteristics are those of a spring, and half those of a damper.

The fully fractional generalised Maxwell model best describes the frequency of 1/7 Hz, where the match between the experimental and modelled response was 0.88, which was a good match. However, at the highest frequency of 1/5 Hz, the match between the data was the lowest and was only 0.12. The average match between the data over all three frequency ranges was 0.533, which is less than the methods used in the work of Gu et al. [8] and Zuo et al. [9] where agreement between data was more than 0.9. If only data from frequency of 1/7 Hz were compared to the data from the work of Gu and Zuo, then our method proves very efficient since it is much easier to implement it and use it in the control.

The initial and optimised parameters for the optimisation of the individual frequency with the fully fractional generalised Maxwell model are listed in Table 3. The match between the experimental and calculated responses was improved. The average match increased to 0.745, which is a good match and can be easily compared with the work of Gu and Zuo. However, the material parameters of the fully fractional generalised Maxwell model were different for each frequency, which was a drawback, since one would like to have universal material parameters for the entire frequency range.

Finding the optimal number of branches for the model cannot be included in the optimisation, because changing the number of branches changed the number of lower and upper bounds on the model. This is something that cannot be included in the optimisation solver for the Pattern Search. This can only be done by observing the material parameters of the individual branches within the model manually.

5. Conclusions

The match between experimental and calculated results was lower for the whole frequency range than in the work of others. Observing only the middle frequency, the method would be easily compared with the work of others. However, the fully fractional generalised Maxwell model can be derived and implemented easily, and the responses of the model can be determined quickly. Implementation of the model in simulation control is straightforward. The proposed method can easily be used on other materials with viscoelastic behaviour. In future work, topology optimisation could be included into the optimisation procedure.

Author Contributions: Conceptualisation, methodology, software, validation, visualisation, writing—review and editing, T.K., R.B. and J.G. All authors have read and agreed to the published version of the manuscript.

Funding: Research was co-financed by the Slovenian Research Agency under the Research Programme no. P2-0157, which operates within the framework of the University of Maribor, Faculty of Mechanical Engineering.

Data Availability Statement: The data that support the findings of this study are available upon reasonable request from the authors.

Conflicts of Interest: The authors declare no conflict of interest.

References

1. Youn, J.-H.; Jeong, S.M.; Hwang, G.; Kim, H.; Hyeon, K.; Park, J.; Kyung, K.-U. Dielectric Elastomer Actuator for Soft Robotics Applications and Challenges. *Appl. Sci.* **2020**, *10*, 640. [CrossRef]
2. Bar-Cohen, Y. *Electroactive Polymer (EAP) Actuators as Artificial Muscles: Reality, Potential, and Challenges*; SPIE Press: Bellingham, WA, USA, 2001.
3. Pei, Q.; Rosenthal, M.; Stanford, S.; Prahlad, H.; Pelrine, R. Multiple-degrees-of-freedom electroelastomer roll actuators. *Smart Mater. Struct.* **2004**, *13*, N86–N92. [CrossRef]
4. Carpi, F.; Migliore, A.; Serra, G.; De Rossi, D. Helical dielectric elastomer actuators. *Smart Mater. Struct.* **2005**, *14*, 1210–1216. [CrossRef]
5. Rui, Z.; Patrick, L.; Andreas, K.; Gabor, M.K. Spring roll dielectric elastomer actuators for a portable force feedback glove. In Proceedings of Smart Structures and Materials 2006: Electroactive Polymer Actuators and Devices (EAPAD), San Diego, CA, USA, 26 February–2 March 2006.
6. Rosset, S.; Shea, H.R. Flexible and stretchable electrodes for dielectric elastomer actuators. *Appl. Phys. A* **2013**, *110*, 281–307. [CrossRef]
7. Suo, Z.; Zhao, X.; Greene, W.H. A nonlinear field theory of deformable dielectrics. *J. Mech. Phys. Solids* **2008**, *56*, 467–486. [CrossRef]
8. Gu, G.G.; Zhu, J.; Zhu, L.; Zhu, X. Modeling of Viscoelastic Electromechanical Behavior in a Soft Dielectric Elastomer Actuator. *IEEE Trans. Robot.* **2017**, *133*, 1263–1271. [CrossRef]
9. Zou, J.; Gu, G. Modeling the Viscoelastic Hysteresis of Dielectric Elastomer Actuators with a Modified Rate-Dependent Prandtl–Ishlinskii Model. *Polymers* **2018**, *10*, 525. [CrossRef] [PubMed]
10. Zou, J.; Gu, G. Dynamic modeling of dielectric elastomer actuators with a minimum energy structure. *Smart Mater. Struct.* **2019**, *28*, 085039. [CrossRef]
11. Wissler, M.; Mazza, E. Mechanical behavior of an acrylic elastomer used in dielectric elastomer actuators. *Sens. Actuators A Phys.* **2007**, *134*, 494–504.
12. Zhao, X.; Koh, S.; Suo, Z. Nonequilibrium Thermodynamics of Dielectric Elastomers. *Int. J. Appl. Mech.* **2011**, *3*, 203–217. [CrossRef]
13. Xiang, G.; Yin, D.; Cao, C.; Gao, Y. Fractional description of creep behavior for fiber reinforced concrete: Simulation and parameter study. *Constr. Build. Mater.* **2022**, *318*, 126101. [CrossRef]

14. Meng, R.; Yin, D.; Drapaca, C.S. Variable-order fractional description of compression deformation of amorphous glassy polymers. *Comput. Mech.* **2019**, *64*, 163–171. [CrossRef]
15. Gao, Y.; Zhao, B.; Yin, D.; Yuan, L. A general fractional model of creep response for polymer materials: Simulation and model comparison. *J. Appl. Polym. Sci.* **2022**, *139*, 51577. [CrossRef]
16. Su, X.; Chen, W.; Xu, W.; Liang, Y. Non-local structural derivative Maxwell model for characterizing ultra-slow rheology in concrete. *Constr. Build. Mater.* **2018**, *190*, 342–348.
17. Xu, H.; Jiang, X. Creep constitutive models for viscoelastic materials based on fractional derivatives. *Comput. Math. Appl.* **2017**, *73*, 1377–1384. [CrossRef]
18. Barretta, R.; Marotti de Sciarra, F.; Pinnola, F.P.; Vaccaro, M.S. On the nonlocal bending problem with fractional hereditariness. *Meccanica* **2022**, *57*, 807–820. [CrossRef]
19. Luo, D.; Chen, H.-S. A new generalized fractional Maxwell model of dielectric relaxation. *Chin. J. Phys.* **2017**, *55*, 1998–2004. [CrossRef]
20. 3M™. 3M™ VHB™ Tape Speciality Tapes. Available online: https://multimedia.3m.com/mws/media/986695O/3m-vhb-tape-specialty-tapes.pdf (accessed on 12 January 2021).
21. Conductive, B. Electric Paint. Available online: https://www.bareconductive.com/shop/electric-paint-50ml/ (accessed on 11 July 2020).
22. Podlubny, I. *Fractional Differential Equations: An Introduction to Fractional Derivatives, Fractional Differential Equations, to Methods of Their Solution and Some of Their Applications*; Elsevier Science: Amsterdam, The Netherlands, 1998.
23. Mainardi, F. *Fractional Calculus and Waves in Linear Viscoelasticity: An Introduction to Mathematical Models*; Imperial College Press: London, UK, 2010; pp. 51–63.
24. Mainardi, F.; Spada, G. Creep, relaxation and viscosity properties for basic fractional models in rheology. *Eur. Phys. J. Spec. Top.* **2011**, *193*, 133–160. [CrossRef]
25. FOMCON. FOMCON Toolbox for Matlab. Available online: http://fomcon.net/ (accessed on 10 December 2020).
26. Wenglor. YP05MGV80. Available online: https://www.google.com/url?sa=t&rct=j&q=&esrc=s&source=web&cd=1&ved=2ahUKEwihnaWa0bbjAhXiB50JHWepBx0QFjAAegQIBhAC&url=http%3A%2F%2Fwww.shintron.com.tw%2Fproimages%2Ftakex%2Fwenglor%2FYP05MGV80.pdf&usg=AOvVaw0hZdH3GNiqDLpm9BclkSVw (accessed on 15 July 2020).

fractal and fractional

MDPI

Article

Cascade Control for Two-Axis Position Mechatronic Systems

Dora Morar [†], Vlad Mihaly [*,†], Mircea Şuşcă and Petru Dobra

Department of Automation, Technical University of Cluj-Napoca, Str. G. Bariţiu nr. 26-28, 400027 Cluj-Napoca, Romania
* Correspondence: vlad.mihaly@aut.utcluj.ro
† These authors contributed equally to this work.

Abstract: The current paper proposes an extension for two controller design procedures for a two-axis positioning mechatronic system, followed by a comparison between them. As such, the first method consists in formulating an optimization problem in terms of linear matrix inequalities (LMIs) in order to impose the location of the closed-loop poles, considering an uncertain model of such a system. The uncertain model is treated using various forms of linear differential inclusions (LDIs), namely, polytopic LDI (PLDI) and diagonal norm-bound LDI (DNLDI). Additionally, the problem regarding the command signal constraints is characterized in terms of LMIs. The imposed structure of the controller is a cascade one, with a PI controller for the position loop and a P controller for the velocity loop, having an additional feedforward term. On the other hand, the second method consists in designing a cascade controller with an inner P controller, as in the previous method, the outer controller being a fractional-order $I^{\lambda_I} D^{\lambda_D}$ (FO–ID) controller. In terms of degrees of freedom, both methods present four degrees of freedom for each axis. The presented controller design procedures will be applied for a numerical example of such a positioning system, and a comparison of the obtained performance metrics will be performed.

Keywords: LMIs; $\mathcal{D}$-region; fractional-order controller; postion control; mechatronic system

Citation: Morar, D.; Mihaly, V.; Şuşcă, M.; Dobra, P. Cascade Control for Two-Axis Position Mechatronic Systems. *Fractal Fract.* **2023**, *7*, 122. https://doi.org/10.3390/fractalfract7020122

Academic Editors: Abhaya Pal Singh and Kishore Bingi

Received: 28 December 2022
Revised: 23 January 2023
Accepted: 25 January 2023
Published: 28 January 2023

1. Introduction

1.1. Literature Review

Recent years have been marked by an increase in the popularity of position-based mechatronic systems, especially due to the technological level they have reached, in terms of speed, performance and versatility [1]. In order to have the resulting performance at the highest possible level, it is necessary for the control system to ensure a good reference tracking in a short time, without overshoot. Moreover, one of the main problems which a control system must deal with for such an equipment consists in parametric uncertainties. To take all these into consideration, rather than classical control techniques, other methods are proposed in recent studies, based on robust, fractional or adaptive control approaches.

Starting from the classical control methods based on the idea of pole placement, the authors of [2] present the principle according to which, to ensure a certain transient response, it is not necessary to place the poles in an exact location, but in a certain region in the complex plane. These regions are called $\mathcal{D}$-regions, being convex and symmetrical to the real axis. Thus, by solving a set of linear matrix inequalities (LMIs), the state feedback matrix which ensures $\mathcal{D}$-stability (i.e., all poles of the resulting closed-loop system are placed into a certain $\mathcal{D}$-region) is obtained. The possibility to include and overimpose many design requirements, like transient response performances, command signal saturation constraints and model uncertainties, illustrates the great advantage of LMIs [3].

Due to its powerful advantage, the LMI-based control approach was proposed for various processes and applications. In [4], the authors proposed an LMI approach to design a H_∞ fuzzy controller with pole location constraints. Moreover, Ref. [5] presents an LMI-based control design that deals with the stabilization of a bilinear system with a guaranteed stability region. Parametric uncertainties are included in the control design problem using LMIs for a wind turbine, in [6], proving that using this approach better performances are obtained compared with those obtained using a classical designed controller. The journal paper [7] proposes an LMI-based H_∞ output-feedback controller, taking into consideration the input and output delays from sensors and actuators for a vehicle roll stability problem. In the conference paper [8] regarding LMI-based controllers, the authors present an energy-based control strategy with nonlinear state feedback controller for a quadrotor used for transporting payload.

In the case of PWM-driven voltage converters, an LMI-based approach used for robust linear-quadratic-regulator (LQR) design is presented in the paper [9], alongside the work of [10] in which a similar approach is considered for robust structured controller design. An LMI use case is also compared to a linear-quadratic-Gaussian (LQR) control in the paper [11], which proved that the proposed approach presented clear improvements for a MIMO helicopter process modelled at variable operating points. In the case of a pressure control process on wet clutches, as studied in the work [12], favorable results have been obtained regarding both robustness to sensor noise and modeling errors, alongside performance indices in the transient response of the closed-loop system.

Fractional-order PID controllers generalize classical PID controllers by adding extra degrees of freedom which proved to be more robust for many benchmark problems. In [13], the authors proposed a fractional-order PID (FO-PID) controller designed in order to meet both robustness and performance specifications for a DC motor used in a mechatronic system. Better performance indices have also been found in DC motor control with uncertainties and nonlinear dynamics using FO-PI regulators in [14]. Improved performances by also obtaining a reduced-order regulator through intelligent $\mathcal{H}_\infty$ techniques have been reported in [15], while a general-purpose quasi-experimental method which generalizes the well-established Kessler's symmetrical optimum principle was proposed and implemented in [16]. Moreover, a simple tuning method based in the limit cycle oscillations is proposed in [17], with application on a DC servo motor. Such a FO-PID regulator for a two-axis CNC machine is proposed in [18], designed using μ-synthesis and the minimization problem is solved using a metaheuristic Artificial Bee Colony (ABC) algorithm. In a similar manner, a different application for a highly-nonlinear twin rotor aerodynamic system with its inherent difficulties, has been studied in [19]. The integration of the fractional-order design into the robust control framework gathers the advantages of both control domains, as presented in [20]. Extensions of FO regulators directly designed for fractional-order plants have been proposed in [21,22], where the first paper presents a comprehensive theorem and corresponding algorithm for the robust stabilization problem, while the latter proposes and discusses a novel graphical tuning method for such control systems.

Position-based mechatronic systems are widely used in industry and mainly in production lines and manufacturing. Because of this, it is necessary for the control system to ensure high precision and response speed. In the last years, many studies have been made in order to find the best option when it comes to the design procedures and controller types for these classes of positioning systems. Although classical controllers are mainly used [23], such as P, PI or PID controllers, in cascade configuration or not, more advanced design techniques are proposed. In [24], an optimal ABC-based LQR is proposed, and in another paper [25], we propose a cascade control configuration obtained based on the state feedback gains designed by solving LMIs.

1.2. Contributions

The current paper proposes two control structures for an uncertain two-axis mechatronic positional system: one using the state-feedback approach, by extending our previous paper's results [25], and one based on the behavior of bringing the system to a limit cycle, which extends the ideas presented in [17]. Based on the literature review performed in the previous subsection and their limitations, the main contributions of the current paper are:

(i)   To include the parametric uncertainties of a two-axis mechatronic positional system into the LMI-based control problem in order to impose a $\mathcal{D}$-region where the poles are real and under a prescribed value, by converting the linear differential inclusion (LDI) into a polytopic LDI (PLDI), against the initial method proposed in [25], where the problem has been formulated for the nominal model of a single axis positional system;

(ii)   To reduce the size of the resulting LMI-based control problem by converting the PLDI into a diagonal norm-bound LDI (DNLDI), and to include the constraints given by the saturation phenomenon which appears on the command signal, which has not been considered in the previous paper even for the case of nominal system;

(iii)   To impose a specific structure on the LMI variables such that the resulting state-feedback can be converted into a cascade control structure for both axes, using a similar idea as in [25] for a single axis;

(iv)   To present an autotuning-type design procedure for a fractional-order integral-derivative controller by considering a relay-type nonlinearity to force a limit cycle to obtain the value of the gain-crossover frequency, and then to impose the desired phase margin, by extending the idea from [17];

(v)   To perform a set of numerical simulations to compare the performance obtained with the proposed methods in terms of quantifiable metrics, such as settling time, rise time and overshoot, robustness and implementability.

1.3. Paper Structure

The rest of the paper is organized as follows. Described in Section 2 is the LMI-based problem formulation to impose the $\mathcal{D}$-stability through a full state-feedback. Section 3 presents the two-axis positional mechatronic system, along with the possibility to convert the full state-feedback controller into a cascade control structure. The second controller design procedure is described in Section 4, while Section 5 presents the obtained results. A thorough discussion and a set of further research directions are given in Section 6, and the paper closes with some conclusions in Section 7.

2. State Feedback Controller

The purpose of the current section is to briefly describe the LMI approach to design a state feedback controller for a system described using a polytopic linear differential inclusion (PLDI). To impose a performance set, the $\mathcal{D}$ regions will be used. Consider a continuous-time PLDI described by:

$$(\Sigma_\delta): \quad \dot{\mathbf{x}}(t) = A(\delta)\mathbf{x}(t) + B(\delta)\mathbf{u}(t), \tag{1}$$

where $\mathbf{x} \in \mathbb{R}^{n_x}$, $\mathbf{u} \in \mathbb{R}^{n_u}$, $\delta \in U_\delta \subset \mathbb{R}^{n_\delta}$ is the uncertainty from the state and input matrices, and U_δ closed and bounded. The PLDI should be characterized by the following L-vertex convex hull:

$$\{(A(\delta), B(\delta)) | \delta \in U_\delta\} \subset \Omega \equiv \mathrm{Conv}\{(A_i, B_i), i = \overline{1, L}\}. \tag{2}$$

Assumption 1. *Each pair* $(A(\delta), B(\delta))$ *with* $\delta \in U_\delta$ *is detectable.*

Considering a full state feedback matrix $K \in \mathbb{R}^{n_u \times n_x}$ which gives the control law $\mathbf{u} = K\mathbf{x}$, the following PLDI closed-loop system results:

$$\dot{\mathbf{x}}(t) = \underbrace{\left(A(\delta) + B(\delta)K\right)}_{A_o(\delta)} \mathbf{x}(t). \tag{3}$$

To ensure the asymptotic stability of the closed-loop autonomous system from (3), a quadratic Lyapunov function $V : \mathbb{R}^{n_x} \to \mathbb{R}_+$ having the structure $V(\mathbf{x}) = \mathbf{x}^\top P \mathbf{x}$ should be constructed. To construct such a quadratic function, the following linear matrix inequalities should have a common solution $P \in \mathbb{S}_{n_x}^+$:

$$A_o(\delta)P + PA_o^\top(\delta) < 0, \quad \forall \delta \in U_\delta,$$

which is an uncountable set of LMIs. However, using the convex hull Ω from (2), the feasibility problem which guarantees the asymptotic stability of the closed-loop system is:

$$P \in \mathbb{S}_{n_x}^+ \quad \text{and} \quad (A_i + B_i K)P + P(A_i + B_i K)^\top < 0, \quad i = \overline{1, L}. \tag{4}$$

However, the problem is to design a state feedback, so the set of LMIs from (4) are bilinear matrix inequalities (BMIs), which are not convex by nature. However, using the substitution $Z = KP \in \mathbb{R}^{n_u \times n_x}$, the resulting problem can be described using LMIs:

$$\text{Find } P \in \mathbb{S}_{n_x}^+ \text{ and } Z \in \mathbb{R}^{n_u \times n_x} \text{ such that } A_i P + PA_i^\top + B_i Z + Z^\top B_i^\top < 0, \quad i = \overline{1, L}. \tag{5}$$

A feasible point from the convex cone of the solution of the problem (5) leads to a stabilizable feedback $K = ZP^{-1}$ for the PLDI. If an additional set of requirements should be imposed, the closed-loop eigenvalues must be placed into a specific region of the complex plane. Each convex region symmetrical to the real axis can be expressed using a combination of transformations given by two matrices $L, M \in \mathbb{R}^{m \times m}$, $L = L^\top$:

$$\mathcal{D}(L, M) = \left\{ \lambda \in \mathbb{C} : L + \lambda M + \lambda^* M^\top < 0 \right\}. \tag{6}$$

Definition 1 ([26]). *A matrix $A \in \mathbb{R}^{n_x \times n_x}$ is $\mathcal{D}$-stable if all its eigenvalues lie in a convex region $\mathcal{D}(L, M)$ defined in (6). This $\mathcal{D}$-stability is characterized using the LMI approach as follows:*

$$M_\mathcal{D}(A, P) = L \otimes P + M \otimes (AP) + M^\top \otimes (PA^\top) < 0, \tag{7}$$

where $\otimes$ is the symbol for the Kronecker product.

For the purpose of this paper we consider only two regions: the vertical strip and the conic sector. The vertical strip is used to impose the condition of having real part less than a prescribed value $\alpha < 0$, which imposes one important closed-loop performance regarding the settling time: $t_s \approx \frac{4}{|\alpha|}$. On the other hand, the conic sector is used to impose:

$$\frac{\lambda + \overline{\lambda}}{|\lambda - \overline{\lambda}|} \leq \frac{\pi}{\tan \theta}, \tag{8}$$

which leads to having a damping factor $\zeta \geq \cos(\theta)$, thus, implying a limitation for the overshoot of the closed-loop system. For the vertical strip the $\mathcal{D}$-region is characterized using $L = -2\alpha$ and $M = 1$, while for the conic region $L = 0$ and $M = \mathcal{R}_\theta$ are used, where $\mathcal{R}_\theta$ is the 2×2 matrix corresponding to the rotation with angle θ. We further call this region as a $\mathcal{D}(\alpha, \theta)$-region. A set of necessary and sufficient conditions to ensure $\mathcal{D}(\alpha, \theta)$-regional stability for a PLDI system (Σ_δ) is presented in the following theorem.

Theorem 1 ($\mathcal{D}$-stability controller [3]). *All eigenvalues of the closed-loop system $A_{cl}(\delta) \equiv A(\delta) + B(\delta)K$ could be set in a specific $\mathcal{D}(\alpha, \theta)$-region if and only if all pairs $(A(\delta), B(\delta))$, $\delta \in U_\delta$, are controllable and there are two matrices $P \in \mathbb{S}_{n_x}^+$ and $Z \in \mathbb{R}^{n_u \times n_x}$ such that:*

$$A_i P + B_i Z + (A_i P + B_i Z)^\top - 2\alpha P < 0, \quad i = \overline{1, L}, \tag{9a}$$

$$\begin{pmatrix} A_i P + B_i Z + (A_i P + B_i Z)^\top & \frac{1}{\tan\theta}(A_i P + B_i Z - (A_i P + B_i Z)^\top) \\ \frac{1}{\tan\theta}(A_i P + B_i Z - (A_i P + B_i Z)^\top)^\top & A_i P + B_i Z + (A_i P + B_i Z)^\top \end{pmatrix} < 0, \quad i = \overline{1, L}. \tag{9b}$$

Then the full state feedback gain is given by $K = ZP^{-1}$.

3. Position-Based Mechatronic System

3.1. Plant Model

The described plant is a general purpose two-axis positional system, each axis being operated individually by a servo motor. In this paper, a translational two-axis Computer Numerical Control machine (CNC), also used in our previous works [25,27], will be considered. The mathematical model of the system can be written as follows:

$$
G_\delta(s): \begin{pmatrix} \dot{\omega}_x(t) \\ \dot{\theta}_x(t) \\ \dot{\omega}_y(t) \\ \dot{\theta}_y(t) \\ \hline \theta_x(t) \\ \theta_y(t) \end{pmatrix} = \left(\begin{array}{cccc|cc} -\frac{1}{T_{Mx}} & 0 & 0 & 0 & \frac{K_{Mx}}{T_{Mx}} & K_{xy} \\ 1 & 0 & 0 & 0 & 0 & 0 \\ 0 & 0 & -\frac{1}{T_{My}} & 0 & K_{yx} & \frac{K_{My}}{T_{My}} \\ 0 & 0 & 1 & 0 & 0 & 0 \\ \hline 0 & 1 & 0 & 0 & 0 & 0 \\ 0 & 0 & 0 & 1 & 0 & 0 \end{array} \right) \begin{pmatrix} \omega_x(t) \\ \theta_x(t) \\ \omega_y(t) \\ \theta_y(t) \\ \hline u_x(t) \\ u_y(t) \end{pmatrix}, \tag{10}
$$

where ω_x, ω_y, θ_x and θ_y are the angular speeds and the positions of the X axis and Y axis, respectively, while $u_x, u_y \in [-1, 1]$ are the duty cycles, scaled to relative values, of the PWM command signals for each axis. The mathematical model has been obtained based on a priori knowledge of the physical process and then fine-tuned using system identification techniques. The electrical part presents a time constant which can be neglected against the time constant of the mechanical part, resulting in a first order model for the system from input command to angular speed. Moreover, the mechanical part presents a set of nonlinearities mainly as a result of Coulombian friction, leading to slightly different values of the model's parameters. As such, these parameters present uncertainties which encompass differences appearing in various equilibrium points. Such differences lead to the system G_δ, possible to be described using a PLDI.

The first method proposed in this paper implies an extension of finding a cascade structure from [25], comprised of a PI controller for the outer position loop and a P-type controller for the inner velocity loop. In our previous work, the controller gains have been computed using an optimal full state feedback gain matrix which is a solution for the nominal case of the LMI problem described in Section 2. The main improvement of this particular method considered in this paper consists in finding the optimal state-feedback which satisfies the constraints imposed through regional LMIs for both the nominal and uncertain plants, as in Theorem 1. As such, the initial plant model must be augmented with two additional states z_x and z_y having the state equations $\dot{z}_x = \theta_x$ and $\dot{z}_y = \theta_y$. This augmentation leads to an extended uncertain state-space model:

$$
\overline{G}_\delta(s): \begin{pmatrix} \dot{\omega}_x(t) \\ \dot{\theta}_x(t) \\ \dot{z}_x(t) \\ \dot{\omega}_y(t) \\ \dot{\theta}_y(t) \\ \dot{z}_y(t) \\ \hline \theta_x(t) \\ \theta_y(t) \end{pmatrix} = \left(\begin{array}{cccccc|cc} -\frac{1}{T_{Mx}} & 0 & 0 & 0 & 0 & 0 & \frac{K_{Mx}}{T_{Mx}} & K_{xy} \\ 1 & 0 & 0 & 0 & 0 & 0 & 0 & 0 \\ 0 & 1 & 0 & 0 & 0 & 0 & 0 & 0 \\ 0 & 0 & 0 & -\frac{1}{T_{My}} & 0 & 0 & K_{yx} & \frac{K_{My}}{T_{My}} \\ 0 & 0 & 0 & 1 & 0 & 0 & 0 & 0 \\ 0 & 0 & 0 & 0 & 1 & 0 & 0 & 0 \\ \hline 0 & 1 & 0 & 0 & 0 & 0 & 0 & 0 \\ 0 & 0 & 0 & 0 & 1 & 0 & 0 & 0 \end{array} \right) \begin{pmatrix} \omega_x(t) \\ \theta_x(t) \\ z_x(t) \\ \omega_y(t) \\ \theta_y(t) \\ z_y(t) \\ \hline u_x(t) \\ u_y(t) \end{pmatrix}. \tag{11}
$$

To model an uncertain parameter $c \in [\underline{c}, \overline{c}]$, the following transformation can be used:

$$
c = c^{(n)} + r_c \delta_c, \quad \text{with} \quad r_c = \overline{c} - \underline{c}, \tag{12}
$$

where $c^{(n)}$ is the nominal value of the parameter and $|\delta_c| \leq 1$ is a normalized uncertainty element. Using this transformation, the extended PLDI system can be converted into a diagonal norm-bound LDI (DNLDI) system, having the following state-space representation:

$$\begin{pmatrix} \dot{\mathbf{x}}(t) \\ \hline \mathbf{y}(t) \\ \hline \mathbf{v}(t) \end{pmatrix} = \left(\begin{array}{c|c|c} A & B_u & B_d \\ \hline C_y & D_{yu} & D_{yd} \\ \hline C_v & D_{vu} & D_{vd} \end{array} \right) \begin{pmatrix} \mathbf{x}(t) \\ \hline \mathbf{u}(t) \\ \hline \mathbf{d}(t) \end{pmatrix} \tag{13}$$

where $\mathbf{x} = \begin{pmatrix} \omega_x & \theta_x & z_x & \omega_y & \theta_y & z_y \end{pmatrix}^\top \in \mathbb{R}^{n_x}$ is the state vector, $\mathbf{u} = \begin{pmatrix} u_x & u_y \end{pmatrix}^\top \in \mathbb{R}^2$ is the command input vector, $\mathbf{d} = \begin{pmatrix} d_{T_{M_x}} & d_{T_{M_y}} & d_{K_{M_x}} & d_{K_{M_y}} & d_{K_{xy}} & d_{K_{yx}} \end{pmatrix}^\top \in \mathbb{R}^6$ is the disturbance input vector, $\mathbf{y} = \begin{pmatrix} \theta_x & \theta_y \end{pmatrix}^\top \in \mathbb{R}^2$ is the output vector, and $\mathbf{v} = \begin{pmatrix} v_{T_{M_x}} & v_{T_{M_y}} & v_{K_{M_x}} & v_{K_{M_y}} & v_{K_{xy}} & v_{K_{yx}} \end{pmatrix}^\top \in \mathbb{R}^6$ is the disturbance output vector. The matrices involved in the state-space realization are:

$$A = \begin{pmatrix} -\frac{1}{T_{M_x}^{(n)}} & 0 & 0 & 0 & 0 & 0 \\ 1 & 0 & 0 & 0 & 0 & 0 \\ 0 & 1 & 0 & 0 & 0 & 0 \\ 0 & 0 & 0 & -\frac{1}{T_{M_y}^{(n)}} & 0 & 0 \\ 0 & 0 & 0 & 1 & 0 & 0 \\ 0 & 0 & 0 & 0 & 1 & 0 \end{pmatrix} \quad B_u = \begin{pmatrix} \frac{K_{M_x}^{(n)}}{T_{M_x}^{(n)}} & K_{xy}^{(n)} \\ 0 & 0 \\ 0 & 0 \\ K_{yx}^{(n)} & \frac{K_{M_y}^{(n)}}{T_{M_y}^{(n)}} \\ 0 & 0 \\ 0 & 0 \end{pmatrix} \quad B_d = \begin{pmatrix} \frac{1}{T_{M_x}^{(n)}} & 0 & \frac{1}{T_{M_x}^{(n)}} & 0 & -\frac{1}{T_{M_x}^{(n)}} & 0 \\ 0 & 0 & 0 & 0 & 0 & 0 \\ 0 & 0 & 0 & 0 & 0 & 0 \\ 0 & \frac{1}{T_{M_y}^{(n)}} & 0 & \frac{1}{T_{M_y}^{(n)}} & 0 & -\frac{1}{T_{M_y}^{(n)}} \\ 0 & 0 & 0 & 0 & 0 & 0 \\ 0 & 0 & 0 & 0 & 0 & 0 \end{pmatrix} \tag{14a}$$

$$C_v = \begin{pmatrix} 0 & 0 & 0 & 0 & 0 & 0 \\ 0 & 0 & 0 & 0 & 0 & 0 \\ 0 & 0 & 0 & 0 & 0 & 0 \\ 0 & 0 & 0 & 0 & 0 & 0 \\ -\frac{r_{T_{M_x}}}{T_{M_x}^{(n)}} & 0 & 0 & 0 & 0 & 0 \\ 0 & 0 & 0 & -\frac{r_{T_{M_y}}}{T_{M_y}^{(n)}} & 0 & 0 \end{pmatrix} \quad D_{vu} = \begin{pmatrix} r_{K_{M_x}} & 0 \\ 0 & r_{K_{M_y}} \\ r_{K_{xy}} & 0 \\ 0 & r_{K_{yx}} \\ r_{T_{M_x}} \frac{K_{M_x}^{(n)}}{T_{M_x}^{(n)}} & r_{T_{M_x}} K_{xy}^{(n)} \\ r_{T_{M_y}} K_{yx}^{(n)} & r_{T_{M_y}} \frac{K_{M_y}^{(n)}}{T_{M_y}^{(n)}} \end{pmatrix} \quad D_{vd} = \begin{pmatrix} 0 & 0 & 0 & 0 & 0 & 0 \\ 0 & 0 & 0 & 0 & 0 & 0 \\ 0 & 0 & 0 & 0 & 0 & 0 \\ 0 & 0 & 0 & 0 & 0 & 0 \\ \frac{r_{T_{M_x}}}{T_{M_x}^{(n)}} & 0 & \frac{r_{T_{M_x}}}{T_{M_x}^{(n)}} & 0 & -\frac{r_{T_{M_x}}}{T_{M_x}^{(n)}} & 0 \\ 0 & \frac{r_{T_{M_y}}}{T_{M_y}^{(n)}} & 0 & \frac{r_{T_{M_y}}}{T_{M_y}^{(n)}} & 0 & -\frac{r_{T_{M_y}}}{T_{M_y}^{(n)}} \end{pmatrix} \tag{14b}$$

$$C_y = \begin{pmatrix} 0 & 1 & 0 & 0 & 0 & 0 \\ 0 & 0 & 0 & 0 & 1 & 0 \end{pmatrix} \quad D_{yu} = \begin{pmatrix} 0 & 0 \\ 0 & 0 \end{pmatrix} \quad D_{yd} = \begin{pmatrix} 0 & 0 & 0 & 0 & 0 & 0 \\ 0 & 0 & 0 & 0 & 0 & 0 \end{pmatrix}. \tag{14c}$$

The conversion of a PLDI into DNLDI offers the possibility to avoid using L LMIs to impose the vertical strip region, leading to the following LMI problem: find $P \in \mathbb{S}_{n_x}^+$, $T \in \mathbb{S}_{n_d}^+$, and $Z \in \mathbb{R}^{n_u \times n_x}$ such that:

$$\begin{pmatrix} AP + PA^\top + B_u Z + Z^\top B_u^\top - 2\alpha P & \star \\ TB_d^\top + C_v P + D_{yu} Z & TD_{vd} + D_{vd}^\top T - 2T \end{pmatrix} < 0. \tag{15}$$

Additionally, to impose the allowed maximum command input in a symmetrical manner $\mathbf{u} \in [\underline{\mathbf{u}} = -\overline{\mathbf{u}}, \overline{\mathbf{u}}]$ for an initial value of the state vector lying in the ellipsoid:

$$\mathcal{E}_Q = \{\mathbf{x} \in \mathbb{R}^{n_x} \mid \mathbf{x}^\top Q \mathbf{x} \leq 1\},$$

with $Q \in \mathbb{S}_{n_x}^+$, the following LMI can be used:

$$\begin{pmatrix} \mathrm{diag}(\overline{\mathbf{u}})^2 & Z \\ Z^\top & Q \end{pmatrix} \geq 0, \tag{16}$$

where $\mathrm{diag}(\overline{\mathbf{u}})$ is the diagonal matrix formed using the vector $\overline{\mathbf{u}}$.

3.2. From State Feedback to Cascade Control

To convert the state-feedback control structure to a cascade control structure in the case of a two-axis mechatronic system, the following forms of the matrices P and Z will be considered:

$$P = \begin{pmatrix} P_1 & O \\ O & P_2 \end{pmatrix} \in \mathbb{S}_6^+ \quad \text{and} \quad Z = \begin{pmatrix} Z_1 & O \\ O & Z_2 \end{pmatrix} \in \mathbb{R}^{2 \times 6}, \tag{17}$$

with $P_1, P_2 \in \mathbb{S}_3^+$ and $Z_1, Z_2 \in \mathbb{R}^{1 \times 3}$. The resulting form of the static state-feedback gain matrix K is:

$$K = \begin{pmatrix} k_1 & k_2 & k_3 & 0 & 0 & 0 \\ 0 & 0 & 0 & k_4 & k_5 & k_6 \end{pmatrix}. \tag{18}$$

For brevity, we consider the behavior on a single axis $K_1 = \begin{pmatrix} k_1 & k_2 & k_3 & 0 & 0 & 0 \end{pmatrix}$, and we analyze the equivalence between the the control law given by the state-feedback and the control law given by the cascade structure. As such, on X-axis we have the following command signal given by the state-feedback approach:

$$u_{x,K}(t) = \theta_x^\star(t) + K_1 x(t) = \theta_x^\star(t) + k_1 \omega_x(t) + k_2 \theta_x(t) + k_3 \int \theta_x(t) dt, \tag{19}$$

with $\theta^\star$ being the outer position loop reference signal. On the other hand, for a cascade control scheme having a P controller on the inner loop, along with a PI controller on the outer loop, the control law resulted in this case can be rearranged as:

$$u_{x,PI-P}(t) = P_v^{(x)} \left(-\omega_x(t) - P_p^{(x)} \theta_x(t) - I_p^{(x)} \int \theta_x(t) dt + P_p \theta_x^\star(t) + I_p^{(x)} \int \theta_x^\star(t) dt \right), \tag{20}$$

with inner loop proportional gain P_v, and the pair P_p and I_p as the outer loop proportional and integral regulator gains, respectively. In the exact same manner, the signals $u_{y,K}$ and $u_{y,PI-P}$ can be expressed. Therefore, using the relations exposed in Equations (19) and (20), the following equivalence between the aforementioned gains from the cascade control structure and those obtained using the state feedback controller can be performed:

$$P_v^{(x)} = -k_1, \quad P_p^{(x)} = \frac{k_2}{k_1}, \quad I_p^{(x)} = \frac{k_3}{k_1}, \tag{21}$$

$$P_v^{(y)} = -k_4, \quad P_p^{(y)} = \frac{k_5}{k_4}, \quad I_p^{(y)} = \frac{k_6}{k_4}. \tag{22}$$

3.3. Feedforward Component

As noticed in Equations (19) and (20), an extra integral term in $\theta_x^\star$ appears in the cascade control structure case. As such, an extra analysis should be performed to emphasize this difference. For state-feedback control, the obtained transfer function from $\theta_x^\star$ to θ_x is:

$$H_{o,K}(s) = C_{y,1}(sI_n - (A + B_{u,1}K_1))^{-1}B_{u,1} = \frac{\frac{K_{M_x}}{T_{M_x}}s}{\prod(s - \lambda_i)}, \tag{23}$$

where λ_i are the closed-loop state matrix $A + B_{u,1}K_1$ eigenvalues, while the index 1 is used to mark the first row or column of the matrices which describes the augmented plant. Similarly, in the case of using the PI–P cascade control strategy, the resulting closed-loop transfer function from $\theta_x^\star$ to θ_x becomes:

$$H_{o,PI-P}(s) = \frac{P_v^{(x)} K_{M_x}(P_p^{(x)}s + I_p^{(x)})}{\prod(s - \lambda_i)}. \tag{24}$$

Therefore, in both cases, transmission zeros appear in closed-loop, causing changes in terms of the performance indices initially imposed through the position of the closed-loop system's poles. For the state-feedback strategy for the augmented model, a transmission zero in the origin $\mathfrak{s} = 0$ is obtained. On the other hand, the cascade controller brings a transmission zero in the value $\mathfrak{s} = -\frac{I_p^{(x)}}{P_p^{(x)}}$.

To cancel the effect of the resulting transmission zero, the feedforward gain will be used as an additional degree of freedom. To cancel the closed-loop zero for the cascade control structure, the feedforward gains must be:

$$K_{ff}^{(x)} = -P_p^{(x)} \quad \text{and} \quad K_{ff}^{(y)} = -P_p^{(y)}. \tag{25}$$

Moreover, using this feedforward term, the integral action present on the reference manages to cancel the transmission zero $\mathring{s} = 0$ which shows in the state-feedback case. Therefore, the additional feedforward gains $K_{ff}^{(x)}$ and $K_{ff}^{(y)}$ lead to obtaining two closed-loop systems without transmission zeros, so the performances imposed through $\mathcal{D}$-region are kept unchanged.

4. 4DOF Fractional-Order Controller

The previous section described a 4DOF controller for each axis organized in a cascade-type structure, having an outer loop PI controller and an inner loop P controller, along with a feedforward gain. It must be pointed out that this structure is imposed by the physical stand, as mentioned in [25]. In the current section we propose an alternative control structure having four degrees-of-freedom as well. As such, for each axis:

- For the inner loop, a simple P controller is used;
- For the outer loop, a particular structure of fractional-order PID controller is proposed:

$$K_{FO-ID}(s) = K_i \cdot \frac{T_i s + 1}{s^\lambda}. \tag{26}$$

The differential equations which describe the dynamics of the velocities ω_x and ω_y in terms of command inputs u_x and u_y are:

$$\dot{\omega}_x = -\frac{1}{T_{M_x}}\omega_x + \frac{K_{M_x}}{T_{M_x}}u_x + K_{xy}u_y; \tag{27a}$$

$$\dot{\omega}_y = -\frac{1}{T_{M_y}}\omega_x + K_{yx}u_x + \frac{K_{M_y}}{T_{M_y}}u_y. \tag{27b}$$

The inner P controllers K_{ω_x} and K_{ω_y} should be designed to impose the desired speed profile. The dynamics of the resulting inner closed-loop systems are given by:

$$H_{in,X}(s) = \frac{K_{\omega_x}}{s + \frac{1}{T_{M_x}}(1 + K_{\omega_x}K_{M_x})} \quad \text{and} \quad H_{in,Y}(s) = \frac{K_{\omega_y}}{s + \frac{1}{T_{M_y}}\left(1 + K_{\omega_y}K_{M_y}\right)}. \tag{28}$$

For designing the *FO–ID* controller we impose a limit cycle oscillation. Let us consider a symmetrical bipositional relay having a hysteresis width of ε and an amplitude h, denoted by $R(\varepsilon, h)$, having the inverse describing function:

$$N_i(R(\varepsilon, h)) = -\frac{\pi\varepsilon}{4h}\left(\sqrt{\left(\frac{A}{\varepsilon}\right)^2 - 1} + j\right), \quad A > \varepsilon,$$

where A is the amplitude of the input sinusoidal signal. If such a nonlinearity is inserted into a negative feedback system next to a linear system, a stable limit cycle appears. The parameters of the resulting limit cycle can be determined using the intersection between the negative inverse description locus and the Nyquist diagram of the linear system. The limit cycle is characterized by a frequency ω_{osc} and an amplitude A_{osc}. Moreover, using the Fourier decomposition of $N_i(R(\varepsilon, h))$, the equivalent gain of the relay nonlinearity is:

$$K_{eq} = \frac{4h}{\pi A_{\text{osc}}}. \tag{29}$$

The outer open-loop systems have the following dynamics:

$$H_{ol,out,i}(s) = \frac{K_{\omega_i}}{s\left(s + \frac{1}{T_{M_i}}\left(1 + K_{\omega_i}K_{M_i}\right)\right)}, \quad i \in \{X, Y\}. \tag{30}$$

The time constant of the *FO–ID* controller is set to be in both cases:

$$T_1^{(x)} = \frac{T_{M_x}}{1 + K_{\omega_x}K_{M_x}} \quad \text{and} \quad T_1^{(y)} = \frac{T_{M_y}}{1 + K_{\omega_y}K_{M_y}}, \tag{31}$$

and we want to impose in each case the frequency ω_{osc} of the resulting limit cycle as the gain-crossover frequency ω_{gc}, while the fractional order of the integral effect is designed such that a prescribed value of the phase margin $\gamma_k^\star$ is met. The resulting equation is:

$$\gamma_k^\star = \arg(N_i(R(\varepsilon, h))) + \arctan\left(T_1^{(i)}\omega_{osc}\right) - \lambda^{(i)}\frac{\pi}{2}, \tag{32}$$

which implies:

$$\lambda^{(i)} = \frac{2}{\pi}\left(\arg(N_i(R(\varepsilon, h))) + \arctan\left(T_1^{(i)}\omega_{osc}\right) - \gamma_k^\star\right). \tag{33}$$

Additionally, to ensure $\omega_{gc} = \omega_{osc}$, the gain of the *FO–ID* controller should be:

$$K_i = \frac{4h}{\pi A_{osc}}\frac{\omega_{osc}^\lambda}{\sqrt{1 + (T_1\omega_{osc})^2}} = K_{eq} \cdot \frac{\omega_{osc}^\lambda}{\sqrt{1 + (T_1\omega_{osc})^2}}. \tag{34}$$

5. Numerical Results

The purpose of this section is to illustrate the numerical results obtained by implementing the proposed controllers for the plant described in Section 3.1. The mathematical model for the MIMO system was obtained based on the measured data in various operating points for the described two axes of the mechatronic system. The mechanical system inherent to each CNC axis is presented in Figure 1a), showing its components and interconnections, alongside the brushless DC (BLDC) motor characteristics involved in the system dynamics in Figure 1b). The actuation, measurements and control laws are managed through a Siemens CNC Sinumerik and MC206X Motion Coordinator devices. As such, considering various persistent inputs around a value u_0, i.e., by considering an additional pseudo-random binary signal over the constant component, a set of numerical values have been identified using an auto-regressive with exogenous inputs method in MATLAB. The resulting parameters for the state-space model (10) are described in Table 1, with their nominal values and the uncertainty ranges based on the previously mentioned identification steps.

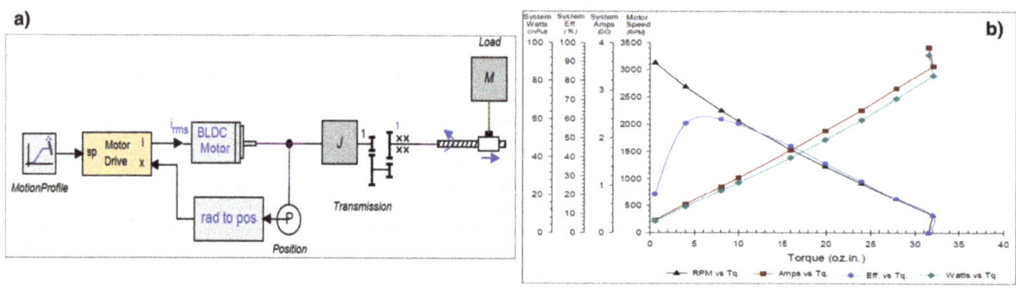

Figure 1. Single-axis mechanical model for the case study in (**a**), with the brushless DC motor characteristics in (**b**), used for system identification and control.

Table 1. Nominal values of the system's parameters, along with their uncertainty ranges.

Parameter	Nominal Value	Percentage	Parameter	Nominal Value	Percentage
T_{M_x}	0.0245	$\pm10\%$	K_{M_x}	25.8017	$\pm10\%$
T_{M_y}	0.0114	$\pm10\%$	K_{M_y}	24.9174	$\pm10\%$
K_{xy}	26.65	$\pm10\%$	K_{yx}	24.46	$\pm10\%$

Cascade Control from State-Feedback Structure

As mentioned in the theoretical part, in order to formulate the control problem of finding the PI-P cascade structure for each axis subsystem, it is necessary to augment the plant model with two extra states. The extended state-space model used is:

$$\overline{G}_\delta(s): \begin{pmatrix} \dot{\omega}_x(t) \\ \dot{\theta}_x(t) \\ \dot{z}_x(t) \\ \dot{\omega}_y(t) \\ \dot{\theta}_y(t) \\ \dot{z}_y(t) \\ \theta_x(t) \\ \theta_y(t) \end{pmatrix} = \left(\begin{array}{cccccc|cc} -40.85 & 0 & 0 & 0 & 0 & 0 & 1054 & 26.65 \\ 1 & 0 & 0 & 0 & 0 & 0 & 0 & 0 \\ 0 & 1 & 0 & 0 & 0 & 0 & 0 & 0 \\ 0 & 0 & 0 & -87.81 & 0 & 0 & 24.46 & 2188 \\ 0 & 0 & 0 & 1 & 0 & 0 & 0 & 0 \\ 0 & 0 & 0 & 0 & 1 & 0 & 0 & 0 \\ 0 & 1 & 0 & 0 & 0 & 0 & 0 & 0 \\ 0 & 0 & 0 & 0 & 1 & 0 & 0 & 0 \end{array} \right) \begin{pmatrix} \omega_x(t) \\ \theta_x(t) \\ z_x(t) \\ \omega_y(t) \\ \theta_y(t) \\ z_y(t) \\ u_x(t) \\ u_y(t) \end{pmatrix}. \quad (35)$$

Moreover, for the matrices corresponding to the DNLDI system in (14), A, B_u, B_d, C_v, D_{vu}, D_{vd}, C_y, D_{yu} and D_{yd}, the nominal values from Table 1 are used, along with the following terms for the uncertainties:

$$\begin{cases} r_{T_{M_x}} = 0.0024; \\ r_{T_{M_y}} = 0.0011; \\ r_{K_{M_x}} = 2.5802; \end{cases} \qquad \begin{cases} r_{K_{M_y}} = 2.4917; \\ r_{K_{xy}} = 2.6650; \\ r_{K_{yx}} = 2.4460. \end{cases}$$

The performances for the closed loop system imposed through the $\mathcal{D}(\alpha, \theta)$-region are:

- A settling time $t_s \approx 0.2[sec]$, imposed using the corresponding vertical strip parameter $\alpha = 20$.
- A very small overshoot, tending to zero, imposed by the conic region corresponding parameter $\theta = 0.01$.
- The command signal allowed values $u_x, u_y \in [-1, 1]$, imposed by the corresponding LMI, having the initial conditions in the ellipsoid described using $Q = \mathcal{I}_6$.

As such, one LMI using the DNLDI form of the system is necessary for the settling time constraints, while for imposing the maximum overshoot, 32 LMIs are necessary. Considering the input saturation, the final problem involves finding a common solution to the 35 LMIs. Using the LMI Solver from MATLAB's Robust Control Toolbox [28], a feasible solution has been successfully found:

$$P = 10^3 \times \begin{pmatrix} 2.4209 & -0.0835 & 0.0031 & 0 & 0 & 0 \\ -0.0835 & 0.0031 & -0.0001 & 0 & 0 & 0 \\ 0.0031 & -0.0001 & 0.00004 & 0 & 0 & 0 \\ 0 & 0 & 0 & 1.1275 & -0.0249 & 0.0007 \\ 0 & 0 & 0 & -0.0249 & 0.0007 & 0.00002 \\ 0 & 0 & 0 & 0.0007 & 0.00002 & 0.00007 \end{pmatrix},$$

$$Z = \begin{pmatrix} -0.0453 & -0.9385 & 0.0395 & 0 & 0 & 0 \\ 0 & 0 & 0 & -0.1431 & -0.4835 & 0.0149 \end{pmatrix}, \quad (36)$$

which leads to the following state-feedback matrix:

$$K = \begin{pmatrix} -0.8719 & -47.9103 & -616.4164 & 0 & 0 & 0 \\ 0 & 0 & 0 & -0.4557 & -33.7512 & -497.9665 \end{pmatrix}. \tag{37}$$

Using the equivalence between the gains obtained for the state-feedback controller and the gains needed for the PI-P cascade control structure, we have:

$$P_v^{(x)} = 0.8719, \quad P_p^{(x)} = 54.9470 \quad I_p^{(x)} = 706.9512,$$

$$P_v^{(y)} = 0.4557, \quad P_p^{(y)} = 74.0668, \quad I_p^{(y)} = 1092.8.$$

Figure 2 shows the step responses obtained for the nominal models of each axis, along with 50 uncertain samples obtained with Monte Carlo simulations. As noticed, the imposed settling time for both axes is fulfilled, along with a zero steady-state error. The overshoot appears due to the transmission zero, as proved in Section 3.2. As such, a feedforward component is necessary to mitigate the transmission zero causing the overshoot and so that the performances imposed through the $\mathcal{D}(\alpha, \theta)$-region are fully met. The values of the feedforward gains are:

$$K_{ff}^{(x)} = -54.947 \quad \text{and} \quad K_{ff}^{(y)} = -74.0668.$$

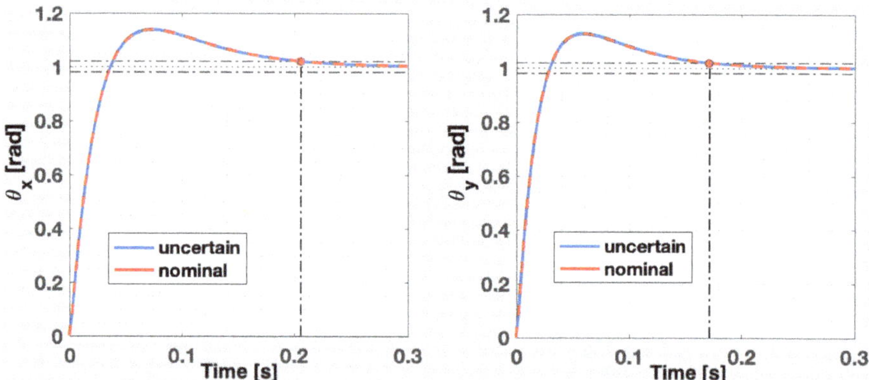

Figure 2. The step responses of the nominal (red) closed-loop system, along with 50 Monte Carlo simulations (blue) for X axis (**left**) and Y axis (**right**), without considering the feedforward component.

The final configuration of the cascade control and feedforward component, together with the mutual effect between the axes and integrating the uncertainties, is depicted in Figure 3. The two resulting closed-loop systems have no transmission zeros and the eigenvalues obtained for the nominal state-space representation become:

$$\Lambda\left(H_0^{(x)}\right) = \{-904.85, \ -33.72, \ -21.29\};$$

$$\Lambda\left(H_0^{(y)}\right) = \{-1013, \ -50.6, \ -21.3\}.$$

The step responses for both nominal and uncertain systems of each axis obtained using the final control structure from Figure 3 are presented in Figure 4. As noticed, the settling time is robustly kept at $t_s \approx 0.2$ [s], with no detectable overshoot or steady-state error.

Next, we consider several numerical simulations obtained with the second proposed control structure, described in Figure 5. The inner P controller used for the speed loop should be designed considering the desired speed profile. For the purpose of this paper, we considered the same values $K_v^{(x)} = 0.8719$ and $K_v^{(y)} = 0.4557$ as in the previous structure in order to have a fair comparison.

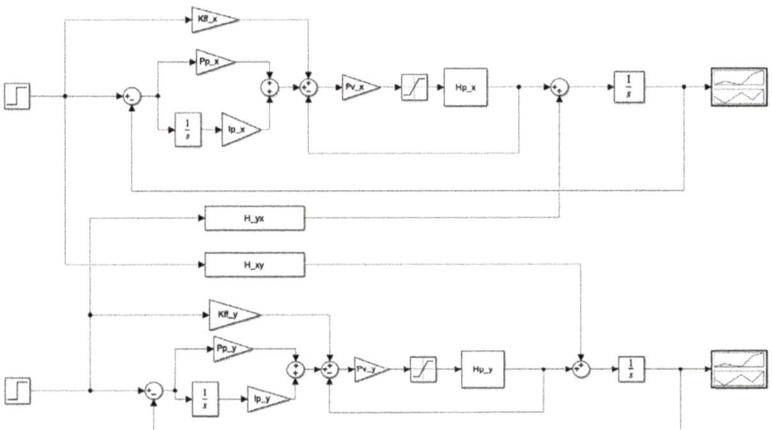

Figure 3. Cascade and feedforward configuration for the MIMO system.

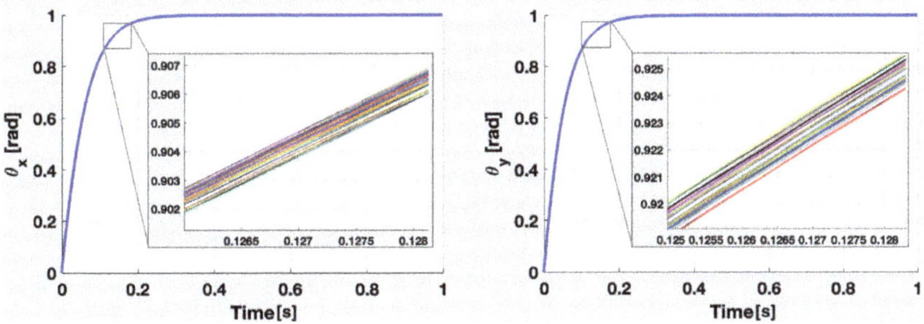

Figure 4. The step responses of the nominal closed-loop system, along with 50 Monte Carlo simulations for X axis (**left**) and Y axis (**right**), obtained by adding the feedforward component.

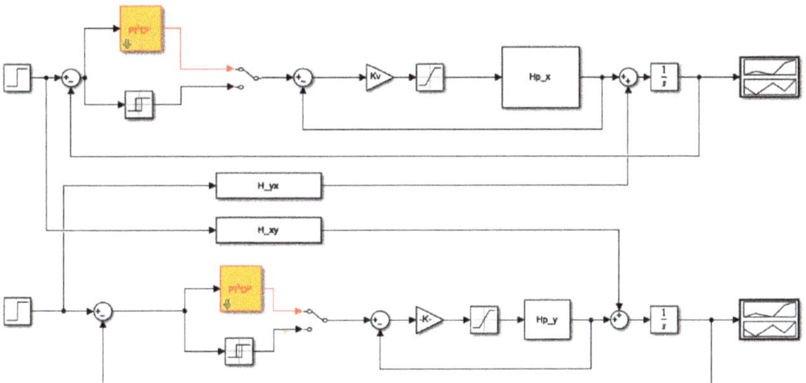

Figure 5. Closed-loop control structure for the two-axis positional mechatronic system having a P controller on the inner speed loop, and a *FO–ID* controller on the outer position loop, the switching element being used to commute between the controller (working mode) and the relay (design mode).

For both axes we considered a bipositional relay with a hysteresis $h = \pm 0.5$ and with the switching points at ± 0.05. The resulting limit cycle for the X axis has the frequency $\omega_{osc} = 71.23$ [rad/s] and the amplitude $A_{osc} = 0.01$, while the resulting Y axis limit cycle has the frequency $\omega_{osc} = 69.38$ [rad/s] and the amplitude $A_{osc} = 0.01$. Because the amplitudes of both limit cycles are the same, the equivalent gains will be equal for both axes:

$$K_{eq}^{(x)} = K_{eq}^{(y)} = 63.662. \tag{38}$$

The resulting time constants of the outer fractional-order integral-derivative controllers presented in (31) are $T_1^{(x)} = 0.01$ [s] and $T_1^{(y)} = 9.21 \times 10^{-4}$ [s]. By imposing the phase margin $\gamma^\star = \frac{13\pi}{30}$, the resulting fractional orders of both controllers are $\lambda_I^{(x)} = \lambda_I^{(y)} = 0.133$. Moreover, to ensure $\omega_{gc} = \omega_{osc}$, the controllers' gains are $K_I^{(x)} = 112.1283$ and $K_I^{(y)} = 111.8148$. As such, the outer controllers are:

$$H_{FO-ID}^{(x)}(s) = 112.1283\frac{0.01s + 1}{s^{0.133}} \quad \text{and} \quad H_{FO-ID}^{(y)}(s) = 111.8148\frac{9.21 \cdot 10^{-4}s + 1}{s^{0.133}}. \tag{39}$$

For the numerical results, the *FO–ID* controllers has been simulated using MATLAB's FOMCON toolbox [29]. The approximation order of the fractional-order elements has been set to 5, while the frequency range is $[0.1, 10^4]$ [rad/s]. The performance of the closed-loop systems is illustrated in Figure 6. As noticed, the value of the settling time is $t_s \approx 0.2$ [s], comparable to the value obtained using the previous structure, and there is no steady-state error. However, due to the lack of coupling consideration between the axes, a small overshoot $\sigma \approx 5\%$ appears, being the main limitation of the proposed method.

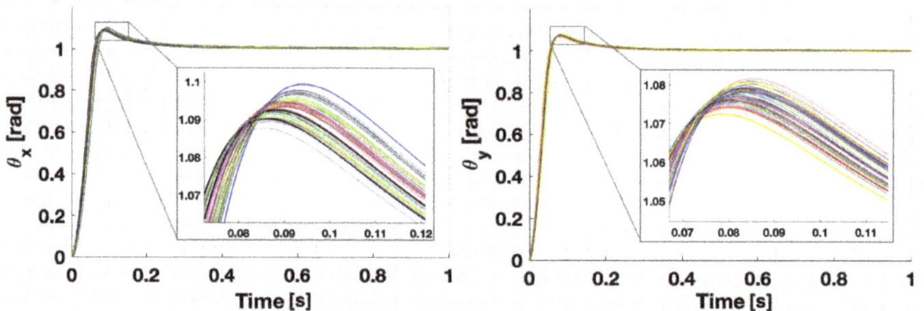

Figure 6. The step responses of the nominal closed-loop system, along with 50 Monte Carlo simulations for X axis (**left**) and Y axis (**right**).

Moreover, the fractional-order ID controller requires an approximation for the numerical implementation, which cannot be considered on the equipment described in [27] and whose model has been considered in the current paper.

6. Discussions

The comparison between the two proposed methods should be on even grounds, as both ensure zero steady-state error and comprise of four degrees of freedom. Regarding the time-domain performances, the settling time values are similar, with an average value of 0.22 [s] for both axes with the first method, having the values between 0.215 [s] and 0.225 [s], while with the second method, the settling time has an average value of 0.185 [s], varying between 0.17 [s] and 0.2 [s]. Moreover, the second method presents a smaller rise time of ≈ 0.07 [s] against to the first method where the rise time is ≈ 0.12 [s]. The overshoot has been imposed to be 0 in the first method, while using the second method the resulting overshoot is ≈ 9.5 [%] for the X axis and ≈ 7.5 [%] for the Y axis, varying between 8.5 [%] and 10 [%] for the X axis and between 7 [%] and 8 [%] for the Y axis. Regarding the implementability, the first structure manages to deal with the limitations imposed by the control structure given by the MC206X Motion Coordinator, which provides five servo gains for control that allow the design of a *PID* controller for the outer position loop, a *P* controller for the inner angular speed loop and a feedforward angular speed gain [27]. As such, regarding the implementability of the proposed control laws, the *FO–ID* regulator cannot be implemented on a generic industrial controller where the structure could be fixed, as in our case.

Comparing the proposed methods with the available results in the literature, the first method presents a set of improvements against the similar method proposed in [25], where the uncertainties and the command signal limitations have not been considered, while the coupling between axes has been studied as a disturbance and only the capacity of globally rejecting such disturbances has been analyzed. Moreover, against the second paper which has been extended here, the problem of designing a *FO–ID* controller for each axis considering the positional systems has been treated in the current paper, compared to [17], where only a single axis angular velocity control system has been treated. The overshoot presented using the second controller can be justified by the lack of consideration of the interconnections between the axes, but the approximation of the fractional-order element could also represent a problem. As noticed in other available research papers, the problem of a small overshoot appears even in the case of a single axis position system, as in [14–16]. However, the great advantage of the second method consists in being an autotuning method which manages to fulfill the vast majority of constraints imposed by such a problem: fast response with zero steady-state error.

As future work, we propose to investigate the possibility to mitigate the small overshoot presented in the case of using the *FO–ID* controller by adding an extra degree of freedom such that the coupling between axes can also be removed. Moreover, a possible extension will be to add both control structures in the μ-synthesis control framework by considering the possibility to integrate the fractional-order element into the Robust Control Framework, starting from the ideas underlined in [19,20]. Additionally, another research direction will be the possibility to develop a graphical tuning method for such a system having a fractional-order model instead of an integer-order one, as in [21,22].

7. Conclusions

The current paper presents two design techniques for a two-axis positional-based mechatronic system. The first technique converts the well-known full state-feedback control structure into a cascade structure having a P controller on the inner loop and a PI controller on the outer loop. The state-feedback matrix has been computed based on the solution of an LMI-based problem in which a set of performances have been imposed by describing a $\mathcal{D}$-region where the closed-loop poles will be placed, along with an extra condition to overcome the saturation effect which appears on both command signals. The results show that the settling time requirements are fulfilled, having an average value of 0.22 [s] for X axis and 0.175 [s] for Y axis, with no steady state-error, but with a small overshoot caused by a transmission zero. This transmission zero effect has also been canceled using a feedforward gain for each axis, leading to an average value of the settling time of $\approx$0.2 [s] for each axis, as imposed. Moreover, a great advantage of this method compared to the classical LQR scheme, as part of an extended range of optimal-control based control laws, is the ability to impose the regional location of the closed-loop system poles in order to obtain specific time-domain performances and counteract uncertainties.

The second technique presents a methodology to design a similar control structure, with two modifications: the feedforward gain has been removed, and the outer PI controller has been replaced with a *FO–ID* controller designed by bringing the system to the limit cycle using a relay-type nonlinearity. However, the numerical results reveal one issue with this approach: the overshoot appears due to the design procedure of the outer controller which does not consider the coupling between the axes. The settling time has similar values in both control structures, with no steady-state error, but the first control structure presents no overshoot, while the second structure induces a small overshoot. As such, based on our findings, the LMI-based controller is suitable for this particular problem. Regarding the robustness, as illustrated in the previous section, the uncertainties do not significantly affect the imposed performances in both cases. Moreover, the method presents the advantage of being an autotuning one.

Author Contributions: Conceptualization, D.M. and V.M.; methodology, V.M.; software, V.M. and D.M.; validation, M.Ş. and P.D.; formal analysis, D.M. and P.D.; investigation, V.M.; resources, D.M.; data curation, M.Ş. and D.M.; writing—original draft preparation, D.M., V.M. and M.Ş.; writing—review and editing, M.Ş., D.M. and P.D.; visualization, P.D.; supervision, P.D.; project administration, D.M.; funding acquisition, P.D. All authors have read and agreed to the published version of the manuscript.

Funding: This paper was financially supported by the Project "Entrepreneurial competences and excellence research in doctoral and postdoctoral programs—ANTREDOC", co-funded by the European Social Fund financing agreement no. 56437/24.07.2019.

Conflicts of Interest: The authors declare no conflict of interest.

Abbreviations

The following abbreviations are used in this manuscript:

ABC	Artificial Bee Colony
CNC	Computer Numerical Control
DNLDI	Diagonal Norm-Bound Linear Differential Inclusion
FO-ID	Fractional-Order Integral-Derivative
FO-PID	Fractional-Order Proportional-Integral-Derivative
LDI	Linear Differential Inclusion
LMI	Linear Matrix Inequality
LQG	Linear-Quadratic-Gaussian
LQR	Linear-Quadratic-Regulator
MIMO	Multiple-Inputs and Multiple-Outputs
PID	Proportional Integral Derivative
PLDI	Polytopic Linear Differential Inclusion
PWM	Pulse Width Modulation

List of Symbols

$\mathbb{S}_n^+$	The set of symmetric and positive definite matrices of order n
ω_x, ω_y	Angular speed for X and Y axes, respectively
θ_x, θ_y	Angular position for X and Y axes, respectively
u_x, u_y	Command signal for X and Y axes, respectively
$\theta_x^\star, \theta_y^\star$	Angular position's reference for X and Y axes, respectively
u_x, u_y	Command signal for X and Y axes, respectively
z_x, z_y	Additional states resulting after augmentation
T_{M_x}, T_{M_y}	Time constant of the subsystem $u \to \omega$ for X and Y axes, respectively
K_{M_x}, K_{M_y}	Gain factor of the subsystem $u \to \omega$ for X and Y axes, respectively
K_{xy}, K_{yx}	Gain factor representing the interconnection between X and Y axes
$c^{(n)}$	The nominal value of an uncertain parameter c
d_c	The disturbance input corresponding to an uncertain parameter c
v_c	The disturbance output corresponding to an uncertain parameter c
$\underline{c}, \overline{c}$	Lower and upper bound of an uncertain parameter c
$P_v^{(x)}, P_v^{(y)}$	The resulting inner loop controllers' parameters for both X and Y axes, respectively
$P_p^{(x)}, I_p^{(x)}, P_p^{(y)}, I_p^{(y)}$	The resulting outer loop controllers' parameters for both X and Y axes, respectively
$K_{ff}^{(x)}, K_{ff}^{(y)}$	The feedforward gains for both X and Y axes, respectively
$T_1^{(x)}, T_1^{(y)}$	Time constant of the outer $FO{-}ID$ controller for both X and Y axes, respectively
$\lambda_I^{(x)}, \lambda_I^{(y)}$	Fractional order of the outer $FO{-}ID$ controller for both X and Y axes, respectively
$K_I^{(x)}, K_I^{(y)}$	The gain of the outer $FO{-}ID$ controller for both X and Y axes, respectively

References

1. Lee, T.H.; Liang, W.; de Silva, C.W.; Tan, K.K. *Force and Position Control of Mechatronic Systems—Design and Applications in Medical Devices*; Springer Nature: Cham, Switzerland , 2021.
2. Chilali, M.; Gahinet, P. $\mathcal{H}_\infty$ Design with Pole Placement Constraints: An LMI Appproach. *IEEE Trans. Autom. Control* **1996**, *41*, 358–367. [CrossRef]
3. Boyd, S.; El Ghaoui, L.; Feron, E.; Balakrishnan, V. *Linear Matrix Inequalities in System and Control Theory*; SIAM Studies in Applied Mathematics: Philadelphia, PA, USA, 1994.
4. Assawinchaichote, V.; Nguang, S.K. H_∞ Fuzzy Control Design For Nonlinear Singularly Perturbed Systems with Pole Placement Constraints: An LMI Approach. *IEEE Trans. Syst. Man Cybern. Part B (Cybern.)* **2004**, *34*, 579–588. [CrossRef] [PubMed]
5. Tarbouriech, S.; Queinnec, I.; Calliero, T.R.; Peres, P.L.D. Control design for bilinear systems with a guaranteed region of stability: An LMI-based approach. In Proceedings of the 17th Mediterranean Conference on Control and Automation, Thessaloniki, Greece, 24–26 June 2009; pp. 809–814.
6. Sloth, C.; Esbensen, T.; Niss, M.O.K.; Stoustrup, J.; Odgaard, P.F. Robust LMI-based control of wind turbines with parametric uncertainties. In Proceedings of the 2009 IEEE Control Applications (CCA) & Intelligent Control (ISIC), St. Petersburg, Russia, 8–10 July 2009; pp. 776–781.
7. Redondo, J.P.; Boada, B.L.; Díaz, V. LMI-Based H∞ Controller of Vehicle Roll Stability Control Systems with Input and Output Delays. *Sensors* **2021**, *21*, 7850. [CrossRef] [PubMed]
8. Guerrero-Sánchez, M.-E.; Hernández-González, O.; Lozano, R.; García-Beltrán, C.-D.; Valencia-Palomo, G.; López-Estrada, F.-R. Energy-Based Control and LMI-Based Control for a Quadrotor Transporting a Payload. *Mathematics* **2019**, *7*, 1090. [CrossRef]
9. Olalla, C.; Leyva, R.; El Aroudi, A.; Queinnec, I. Robust LQR control for PWM converters: An LMI approach. *IEEE Trans. Ind. Electron.* **2009**, *56*, 2548–2558. [CrossRef]
10. Daher, F.A.; Stoustrup, J. Robust structured control design via LMI optimization. *IFAC Proc.* **2011**, *44*, 7933–7938.
11. da Silva, L.R.T.; de Campos, L.A.H.; Potts, A.S. Robust Control for Helicopters Performance Improvement: An LMI Approach. *J. Aerosp. Technol. Manag.* **2020**, *12*, e3620. [CrossRef]
12. Cocetti, M.; Donnarumma, S.; De Pascali, L.; Ragni, M.; Biral, F.; Panizzolo, F.; Rinaldi, P.P.; Sassaro, A.; Zaccarian, L. Hybrid nonovershooting set-point pressure regulation for a wet clutch. *IEEE/ASME Trans. Mechatron.* **2020**, *25*, 1276–1287. [CrossRef]
13. Lino, P.; Maione, G. Fractional-order controllers for mechatronics and automotive applications. In *Handbook of Fractional Calculus with Applications—Volume 6: Applications in Control*; De Gruyter: Berlin, Germany, 2019; pp. 267–292.
14. Dulf, E.-H.; Șușcă, M.; Kovács, L. Novel Optimum Magnitude Based Fractional Order Controller Design Method. *IFAC-PapersOnLine* **2018**, *51*, 912–917. [CrossRef]
15. Rahman, M.Z.U.; Leiva, V.; Martin-Barreiro, C.; Mahmood, I.; Usman, M.; Rizwan, M. Fractional Transformation-Based Intelligent H-Infinity Controller of a Direct Current Servo Motor. *Fractal Fract.* **2023**, *7*, 29. [CrossRef]
16. Dulf, E.-H. Simplified Fractional Order Controller Design Algorithm. *Mathematics* **2019**, *7*, 1166. [CrossRef]
17. Duma, R.; Dobra, P.; Trusca, M. Embedded application of fractional order control. *Electron. Lett.* **2012**, *48*, 1526–1528. [CrossRef]
18. Mihaly, V.; Șușcă, M.; Morar, D.; Stănese, M.; Dobra, P. μ-Synthesis for Fractional-Order Robust Controllers. *Mathematics* **2021**, *9*, 911. [CrossRef]
19. Mihaly, V.; Șușcă, M.; Dulf, E.H. μ-Synthesis FO-PID for Twin Rotor Aerodynamic System. *Mathematics* **2021**, *9*, 2504. [CrossRef]
20. Mihaly, V.; Șușcă, M.; Dulf, E.H.; Dobra, P. Approximating the Fractional-Order Element for the Robust Control Framework. In Proceedings of the 2022 American Control Conference, Atlanta, GA, USA, 8–10 June 2022; pp. 1151–1157.
21. Ghorbani, M.; Tepljakov, A.; Petlenkov, E. Stabilizing region of fractional-order proportional integral derivative controllers for interval fractional-order plants. *Trans. Inst. Meas. Control* **2022**, *45*, 546–556 . [CrossRef]
22. Zheng, S.; Tang, X.; Song, B. A graphical tuning method of fractional order proportional integral derivative controllers for interval fractional order plant. *J. Process. Control* **2014**, *24*, 1691–1709. [CrossRef]
23. Sławomir, M. Comparison of automatically tuned cascade control systems of servo-drives for numerically controlled machine tools. *Elektron. Elektrotechnika* **2014**, *20*, 16–23.
24. Morar D.; Dobra, P. Optimal LQR weight matrices selection for a CNC machine controller. In Proceedings of the 2021 23rd International Conference on Control Systems and Computer Science (CSCS), Bucharest, Romania, 26–28 May 2021.
25. Morar, D.; Mihaly, V.; Șușcă, M.; Dobra, M. LMI Conditions for CNC Cascade Controller Design - A State Feedback Approach. In Proceedings of the 2022 IEEE International Conference on Automation, Quality and Testing, Robotics (AQTR), Cluj-Napoca, Romania, 19–21 May 2022; pp. 1–6.
26. Shaul, G.; Jury, E. A general theory for matrix root-clustering in subregions of the complex plane. *IEEE Trans. Autom. Control* **1981**, *26*, 853–863.
27. Morar, D. Advanced Control Techniques for CNC Machines. Ph.D. Thesis, Technical University of Cluj-Napoca, Cluj-Napoca, Romania, 2022.

28. Balas, G.; Chiang, R.; Packard, A.; Safonov, M. *Robust Control Toolbox—User's Guide*; The MathWorks: Natick, MA, USA, 2020.
29. Tepljakov, A. *FOMCON Toolbox for MATLAB*; GitHub: San Francisco, CA, USA, 2022. Available online: https://github.com/extall/fomcon-matlab/releases/tag/v1.50.4 (accessed on 3 December 2022).

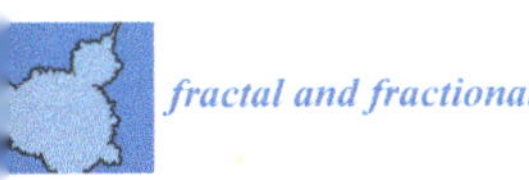

fractal and fractional

Article

Digital-Twin-Based Real-Time Optimization for a Fractional Order Controller for Industrial Robots

Xuan Liu [1], He Gan [1], Ying Luo [1,*], Yangquan Chen [2] and Liang Gao [1]

1   School of Mechanical Science and Engineering, Huazhong University of Science and Technology, Wuhan 430074, China
2   School of Engineering, University of California, 5200 N. Lake Road, Merced, CA 95343, USA
*   Correspondence: ying.luo@hust.edu.cn

Abstract: Digital twins are applied in smart manufacturing towards a smarter cyber-physical manufacturing system for effective analysis, fault diagnosis, and system optimization of a physical system. In this paper, a framework applying a digital twin to industrial robots is proposed and realizes the real-time monitoring and performance optimization of industrial robots. This framework includes multi-domain modeling, behavioral matching, control optimization, and parameter updating. The properties of the industrial robot are first modeled in a digital environment to realize the strong interactive and all-around 3D visual monitoring. Then, behavioral matching is performed to map the virtual system to the physical system in real time. Furthermore, the control performance of the system is improved by using a fractional order controller based on the improved particle swarm optimization algorithm. This framework is applied to the experimental verification of real-time control optimization on an industrial robot. The time-domain performance is improved in the simulation and experimental results, where the overshoot is promoted at least 42%, the peak time is promoted at least 32%, and the settling time is promoted at least 33%. The simulation and experimental results demonstrate the effectiveness of the proposed framework for a digital twin combined with fractional order control (FOC).

Keywords: digital twin; industrial robots; smart manufacturing; FOC

Citation: Liu, X.; Gan, H.; Luo, Y.; Chen, Y.; Gao, L. Digital-Twin-Based Real-Time Optimization for a Fractional Order Controller for Industrial Robots . *Fractal Fract.* **2023**, *7*, 167. https://doi.org/10.3390/fractalfract7020167

Academic Editors: Kishore Bingi and Abhaya Pal Singh

Received: 19 December 2022
Revised: 26 January 2023
Accepted: 29 January 2023
Published: 7 February 2023

1. Introduction

With the development and application of new information technologies, countries have proposed different manufacturing strategies [1], and smart manufacturing is a common way to improve the level of the manufacturing industry [2]. Smart manufacturing requires not only high quality standards but also enhanced robustness and autonomy to achieve production targets [3]. Most of the work is related to the physical system, and the content of the digital system only plays an auxiliary role most of the time in traditional manufacturing [4].

Compared with smart manufacturing, the innovation cycle of the traditional industrial field is much longer. Therefore, one of the key challenges is to achieve Cyber-Physical Systems (CPS) [5]. CPS realizes real-time interactions and close combinations of the network and the physical systems through computing, communication, and control [6]. To achieve this transformation, an emerging technology is urgently needed—namely, digital twins [7].

The digital twin concept was officially proposed in NASA's technical report in Mid-2010 [8]. The key of a digital twin is to create a virtual model of the physical system in a digital way and then simulate, verify, control, and predict the whole life cycle process of the physical system with the help of the digital twin data, the virtual system, and the connection between physical and virtual systems [9]. With the development of smart manufacturing, the application of digital twins in the manufacturing industry has been widely studied [10].

Tao et al. constructed a general digital twin framework for complex equipment, which is used for prognostics and health management to improve accuracy and efficiency [11]. Rodolfo et al. presented a digital twin-based optimization procedure for an ultra-precision motion system [12]. Aivaliotis et al. presented a methodology to calculate the remaining useful life (RUL) of machinery equipment using physics-based simulation models based on the digital twin concept, and the RUL of industrial robots was predicted effectively [13].

Zhang et al. proposed an optimal state control framework based on digital twins, which helps the synchronized production operational system maintain an optimal state when uncertainties effect the system [14]. Viola et al. applied a digital twin to the framework of intelligent control engineering. The framework reproduced the system behavior through a multi-domain simulation and completed the real-time interface between physical and virtual systems by adjusting the behavioral matching technology to the digital twin [3].

Gallala et al. proposed a digital twin approach for human–robot interactions (HRIs) in hybrid teams; however, this approach lacked a description of performance optimization [15]. Lei et al. presented a web-based digital twin thermal power plant and discussed the architecture, modeling, control algorithm, rule model, and physical–digital twin control, which is beneficial to study the applications of digital twinning in other fields [16].

However, implementing digital twins to industrial motion systems still lacks a thorough understanding of the concept, framework, and development methods, which hinders the progress of real digital twin application in smart manufacturing [17]. There are two major research questions that need to be solved: (1) how to match the virtual model with the actual motion state in real time to ensure the accuracy of the model in the optimization process; and (2) how to further optimize the control performance of the physical system in the proposed digital twin framework.

The greatest challenge for the first question is how to construct real-time behavioral matching based on an optimal algorithm to ensure the accuracy of the virtual model. The real-time interaction is realized through the database. The digital twin data collected from the virtual and physical system, including static and dynamic model information of the physical system, information collected by sensors during physical system operation, and information collected by virtual sensors during virtual system operation, are all stored in the database.

Digital twin data can be read at any time as a database client. Then, the real-time interaction between the physical and the virtual systems can be realized, which lays a solid foundation for the accuracy and effectiveness of the virtual model optimization. In order to achieve intelligent optimal control, the methods of introducing artificial intelligence algorithms to optimize the performance of the control system can be divided into two categories.

One is to use artificial intelligence algorithms directly for control [18], thereby, replacing the traditional controllers. The other is to combine the approach with classical control theory and use artificial intelligence algorithms for parameter tuning [19]. As for the second question, the proposed methodology in this paper is fractional order control (FOC) optimization using artificial intelligence algorithms in a digital twin framework to achieve optimal control performance.

Fractional calculus is the quantitative analysis of functions using non-integer-order integration and differentiation [20], and this has attracted a great deal of interest in system modeling and control fields [21]. Fractional order controllers have been found to obtain more control options and flexibility compared with integer order controllers [22]. Among them, the fractional order $PI^\lambda D^\mu$ was proposed by I. Podlubny [23]. Due to the fact that the fractional order $PI^\lambda D^\mu$ controller achieves better tracking performance with less overshoot and faster response [24], the fractional order $PI^\lambda D^\mu$ controller has been widely used in the control fields [22,25].

Therefore, the fractional order $PI^\lambda D^\mu$ controller design and optimization method is proposed in the industrial robot motion system digital twin framework in this paper. There exist some tuning methods of fractional order controllers, which can be divided into two categories: the frequency-domain-design method and time-domain optimization algorithms.

The frequency-domain-design method generally refers to the numerical solution of the parameters of the controller by specifying the frequency specifications, combined with the robustness criterion [26]. According to the frequency-domain-design method, the parameters of the controller can be obtained through analytical calculation, which can realize effective control of the system [27]. Monje et al. proposed a design scheme of fractional order controller with given frequency-domain indexes, which is robust to equipment uncertainty, load disturbance and high-frequency noise [28].

Chen et al. used the frequency-domain-design scheme to design the parameters of FOPID-BICO, which ensures the robustness and anti-interference of the control system [29]. However, there exist some problems, including the parameter limitation for optimization, the complexity of the algorithm, and a large amount of calculation for the real-time optimization process. Time-domain-optimization algorithms have also been developed [30], which overcome the uncertainty and cumbersomeness of manually adjusting parameters by introducing intelligent optimization algorithms.

The major contributions of this paper are as follows: (1) we introduce a framework of a digital twin for industrial motion system, which realizes real-time monitoring and optimization of the running state with experimental verification of control optimization; (2) digital twin behavioral matching based on real-time data analysis and dynamic mapping between virtual system and physical systems; and (3) FOC optimization using the intelligent algorithm in the digital twin framework to achieve optimal motion system performance.

This paper is structured as follows. Section 2 presents the framework of a digital twin for industrial robots and introduces the four phases of implementation in detail. Section 3 shows the implementation of the proposed framework using related software. Section 4 describes the specific application example, which is how to implement the digital twin framework on an industrial robot control system. Section 5 presents the simulation and experimental results to demonstrate the effectiveness and advantages of the proposed digital twin framework combined with FOC optimization. Finally, our conclusions and future work are presented in Section 6.

2. Digital Twin Approach

The idea of a digital twin first appeared in the product lifecycle management course taught by Grieves around 2003. In 2014, he further elaborated on digital twins in a white paper and proposed a general standard system on digital twin, which is a three-dimensional structure [31]. This three-dimensional architecture consists of three main objects—namely, physical products in real space, virtual products in virtual space, and the connections of data and information that tie the virtual and physical products together.

In order to improve the accuracy and efficiency of prognostics and health management for complex systems, Tao proposed an extended five-dimension digital twin architecture, which adds digital twin data and services based on the Grieves' architecture [11]. In addition, other digital twin architectures have been proposed by other researchers [32]. Referring to Tao's five-dimensional model of digital twins, this paper presents a five-dimensional architecture of digital twins applied to industrial robots as shown in Figure 1.

This framework also consists of five aspects—namely, the physical system, virtual system, digital twin data, service applications, and the connection between the above four aspects. The virtual system is built based on the physical system and digital twin data first, adjusted to realize real-time synchronization to the physical system, and then optimized according to the service applications proposed by physical systems, such as fault detection, control optimization, and three-dimensional monitoring. Finally, the monitoring and optimization of the physical system can be realized based on the digital twin data stored after the virtual system optimization.

In classical control theory, it is necessary to manually adjust the parameters based on the error signal through the control algorithm offline, which is a slow, tedious, and inefficient process. The framework in this paper focuses on the utilization of the digital twin concept to automatically adjust the controller parameters when the operating state of the robot changes. More specifically, a real-time optimization strategy for FOC is proposed in the framework, which can achieve optimal tracking control and robustness performance.

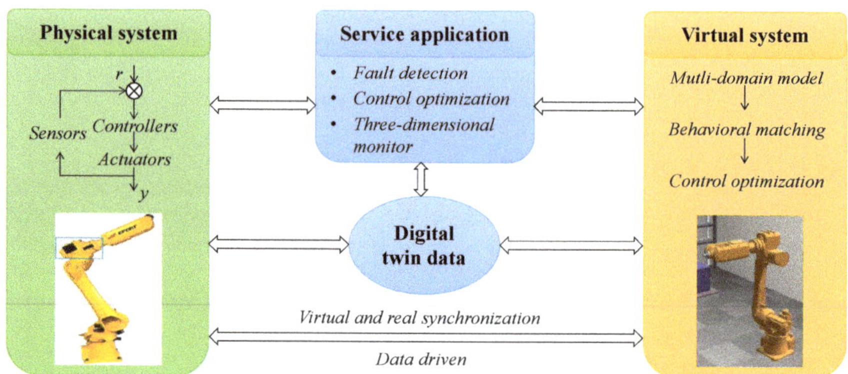

Figure 1. A five-dimensional framework of digital twins: applications to industrial robots.

Based on the service application of three-dimensional monitoring and control optimization, the framework is composed of four steps: multi-domain modeling, behavioral matching, control optimization, and parameter updating.

2.1. Multi-Domain Modeling

The purpose of the first step is to establish a virtual system representing the behavior of the physical system. As a copy of the physical system, the virtual system needs to truly reflect the state of the physical system at every moment to realize monitoring of the physical system.

The modeling process includes two parts as depicted in Figure 2: the first part is to model the position, geometric size, material, and dependency of the physical system; and the second part is to model the kinematic and dynamic characteristics of the physical system, which is also the most important part of the modeling.

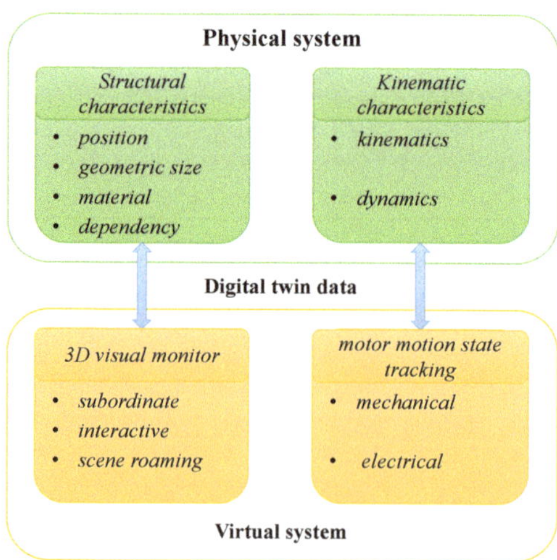

Figure 2. A multi-domain model as a virtual system based on a physical system.

The first part of modeling is to represent the model of robot and its environment in 3D. The target is to realize the strong interactive and all-around 3D visual monitoring effect of the physical entities. The second part of modeling is to achieve motor motion state tracking. The complete model of each motor consists of a number of elements, which represent the dynamic behavior of each motor's component based on the modeling of the mechanical, electrical, and other functions.

2.2. Behavioral Matching

The purpose of this step is to find the parameters of the virtual model so that the virtual model can adapt to its complete system dynamics and be consistent with the real state of the physical system. The parameters of the virtual model can be divided into two parts. The first part is all available data related to the physical system. It is worth mentioning that most of the modeled elements can use the parameters directly from collection.

The second part of the parameters will change continuously due to the operation of the machine and other external factors and cannot be defined directly using the collected data. These parameters need to be optimized online in real time by using the digital twin data to achieve the target of the behavior of the virtual system matching the behavior of the physical system. This process is called behavioral matching [3] as depicted in Figure 3.

This process is set as an optimization cycle. Intelligent optimization algorithms are used to optimize the parameters. The cost function will be computed continuously until the input and output data streams of the system are consistent with a certain tolerance rate or after a fixed number of iterations.

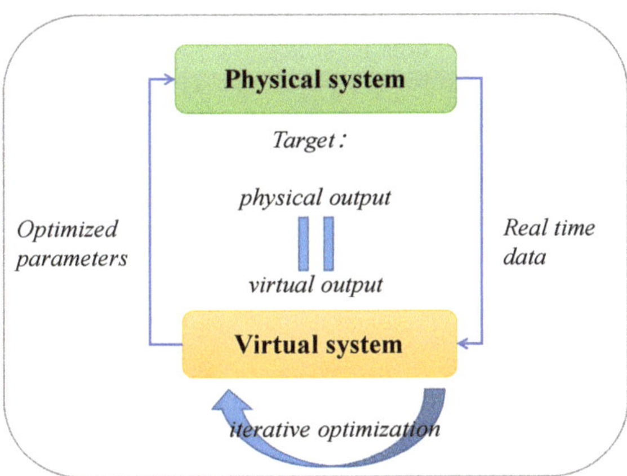

Figure 3. Behavioral matching.

2.3. Control Optimization

The purpose of this step is to optimize the control performance of the virtual model. After behavioral matching, the mapping and interactions between the physical system and virtual system are deployed. Therefore, the control performance optimization of the virtual model is of great significance to the optimization of physical system. In order to achieve better control performance, the fractional order controller is used to control and optimize based on the accurate virtual model.

There exist some tuning methods of fractional order controllers, which can be divided into two categories: frequency-domain-design method and time-domain-optimization algorithms. To overcome the uncertainty and cumbersomeness of manually adjusting parameters, the intelligent optimization algorithm is used in this paper. Using optimal control rules, we first calculating the error between the reference and feedback of the closed-loop system, and then using the optimization algorithms and tools to minimize the cost function value representing the dynamic performance index of the system, and search for the optimal parameters. The time-domain-optimization algorithm is applied in the proposed digital-twin-system framework with data collection from the physical system to the virtual system as shown in Figure 4.

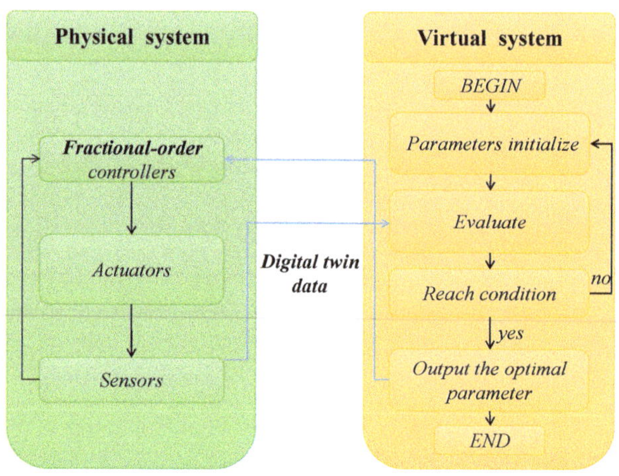

Figure 4. Digital-twin-system framework control optimization.

2.4. Parameter Updating

The purpose of this step is to update the optimal parameters obtained by the virtual system to the physical system to achieve the predetermined control targets. The iteratively optimized parameters are stored in the virtual system and are updated in the physical system through the data communication channel between the virtual and physical systems. By optimizing the controller parameters in real time, the high performance index can be maintained when the running state of the physical system changes.

3. Digital Twin Deployment

The previous section describes a framework of digital twins for the industrial motion system. The implementation of multi-domain modeling with related software and behavioral matching of the framework are presented in this section. Multi-domain modeling is divided into two parts. The first part is to model the position, geometric size, material, and dependency of the physical system, which is performed in Unity3D [33]. The 3D platform used to build the model needs to follow three rules: 3D visualization requirement, key function requirement, and cross-platform operation requirement.

Furthermore, as a popular virtual system development engine, Unity3D is used to build 2D and 3D scenes, and add scripts, shaders, and physical features to scenes. Thus, Unity3D can be used to model the virtual 3D environment. There are two modeling methods of Unity3D. One is to build the components by users in Unity3D and build the components hierarchically according to the principle of behavioral consistency; the other is to import the model directly, which is actually the most general method [34].

Then, the correct subordinate relationship should be established in virtual model based on the hierarchical structure of the physical system according to the design principles of hierarchical consistency and behavioral consistency. Through the strong interaction and scene roaming function of Unity3D, the omni-directional real-time visual monitoring of the physical system can be realized.

The second part of multi-domain modeling is to model the kinematic and dynamic characteristics of the physical system in MATLAB Simulink, which is a well-known multi-domain simulation and model-based design tool and can provide the environment for modeling, simulation, and comprehensive analysis of dynamic systems as well as an interactive graphical environment and a customizable module library for design, simulation, execution, and testing.

Thus, the behavioral matching and control optimization process can be directly performed in MATLAB. After behavioral matching, the four parameters in the electromechanical model can be obtained to make the behavior of the virtual model closer to that of the physical system. After control optimization, the parameters of the controller will be obtained to optimize the control performance of the virtual system. These parameters are all stored as digital twin data.

As multi-domain modeling, behavioral matching, and control optimization are all supported by digital twin data, including sensor data, control state data in the physical system and simulation data in the virtual system. Therefore, a database structure needs to be built to store all digital twin data generated in real time. In this paper, Mysql as a C/S architecture [35] is applied to build the database. A server is used to store and manage the database, and the client is the program that issues the operation request.

Furthermore, Mysql can store data for each period of time during the running process. After completing the behavioral matching of the virtual system, optimizing the controller parameters based on the accurate model, and sending the optimized controller parameters back to the physical system for control, it is achievable by monitoring the digital twin data to verify whether the performance of the physical system has been improved.

4. A Case Study

A case study on the industrial robot has taken place to demonstrate the functionality of the proposed framework. In order to describe the above proposed framework in more detail, each stage is described in the case study.

4.1. Multi-Domain Modeling

The robot studied in this case study is the EFORT ER20C-C10 as a six-axes robot, which has six rotating joints, and each axis is driven by a permanent magnet synchronous motor (PMSM). One of the six axes on this industrial robot is focused on in this paper to build a digital twin model.

The first part is to model the position, geometric size, material, and dependency of the physical system. The second method is adopted in this paper to construct Unity3D model, which is to directly import the model. Furthermore, the correct subordinate relationship is established in the virtual model based on the hierarchical structure of the physical system. Then, the interactive functions and scene roaming functions are added to the virtual system scene. Furthermore, through the digital twin data collected by the data interface, the function of running state reproduction can be realized. The model established in Unity3D is shown in Figure 5.

The second part is to model the kinematic and dynamic characteristics of the physical system. The PMSM model of the robot axis consists of an electromagnetic link model and mechanical link model; the former conforms to the voltage equation, and the latter conforms to the mechanical characteristic equation as follows:

$$u_q - C_e n = R i_q + L \frac{di_q}{dt} \tag{1}$$

$$T_e - T_L = B\omega + J \frac{d\omega}{dt} \tag{2}$$

where, in (1), u_q is the armature voltage, C_e is the back EMF coefficient, n is the motor speed (the unit is rpm), R is the armature resistance, i_q is the armature current, and L is the armature inductance; in (2), T_e is the electromagnetic torque, T_L is the equivalent torque of load, ω is the motor angular speed (the unit is rad/s), B is the damping coefficient, and J is the moment of inertia.

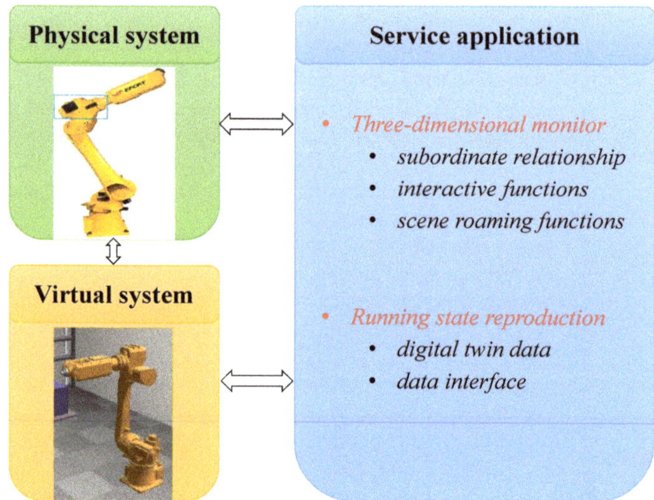

Figure 5. Model of six-axes robot in Unity3D.

Figure 6 shows the speed servo system of PMSM, where the electromagnetic part is in the red dashed block, and the mechanical part is in the green dashed block. n_r and n are the reference motor speed and actual motor speed, respectively; i_{qr} and i_q are the reference q-axis current and actual q-axis current feedback, respectively; u_q is the q-axis voltage; and E is the back EMF. $C_v(s)$ is the speed controller to be designed, and $C_i(s)$ is the current controller to be designed. In order to optimize the control performance in the physical system, a fractional order controller is designed in this control system. In the simulation and experiment, the cascaded fractional order PI^λ-PI^λ controller and the cascaded integer order PI–PI controller are designed with controller optimization for a fair control performance comparison.

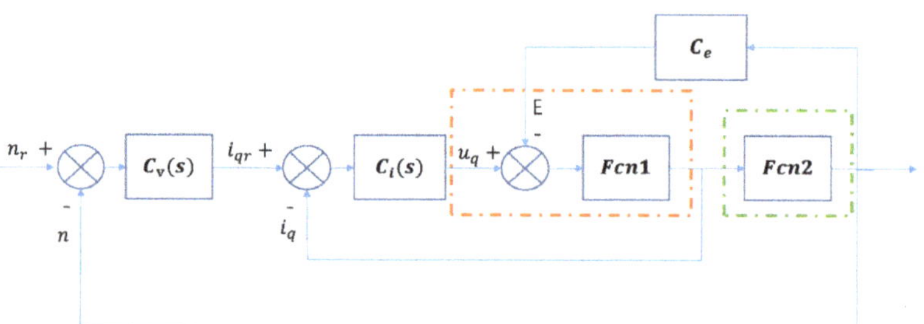

Figure 6. The permanent magnet synchronous motor (PMSM) speed closed−loop control system.

4.2. Behavioral Matching

In order to meet the requirements of real-time mapping of the physical system to the virtual system, behavioral matching is performed based on the established model with real-time adjustment of parameters. In this case, the parameters of the electromechanical model are the most relevant for building the accurate virtual model. Therefore, the four parameters in the electromechanical model, including the moment of inertia, damping coefficient, resistance, and inductance—namely, J, B, R, and L are the parameters for behavioral matching. Then, we evaluate whether the identified four parameters can match the virtual system behavior with the physical system behavior.

Based on the parameter identification method in [36], the transfer function model of each part in Figure 6 can be obtained as

$$Fcn1(s) = \frac{250}{s + 150} \tag{3}$$

$$Fcn2(s) = \frac{2563.72}{s + 2.06}. \tag{4}$$

The data used for behavioral matching is derived from the digital twin system, including data collected by sensors in the physical system and also simulated statutes in the virtual system. The input data collected from the physical system sensor is taken as the input of the virtual system model. Then, the output data collected from the physical system sensor and the output data of the virtual model are compared as the cost function for parameter optimization,

$$J = \int_0^t \left(y_{real} - y_{virtual}\right)^2 \tag{5}$$

where y_{real} is the output of the physical system, $y_{virtual}$ is the output of the virtual system, and t is the time.

The optimization algorithm used in this case is an improved particle swarm optimization algorithm. The concept of the standard PSO algorithm is simple, and it has a short-term memory function, which makes particles slide in the local optimal or global optimal position. However, the population is prone to premature convergence, and improper setting of inertial weight will lead to a local optimal solution [37]. In order to overcome the tendency of falling into useless solutions and improve its convergence, some improved particle swarm optimization algorithms were proposed [38].

The inertia weight directly affects the search speed and accuracy of the algorithm [39]. When the inertia weight is large, the global search ability of the particle is enhanced. When the inertia weight is small, the local search ability of the particle is improved, which is conducive to improving the search accuracy of the particle. Therefore, it is expected that the inertia weight of the algorithm is large in the early stage but becomes smaller in the late stage. In this paper, the method of dynamically changing the inertia weight is proposed to improve the performance of the particle swarm optimization algorithm.

This optimization algorithm is used to perform iterative optimization of the parameters by minimizing the cost function. The optimization cycle ends when the cost function reaches the allowable error range or the number of iterations is reached. In this case, the number of iterations is set as 50. The convergence process of the cost function in behavioral matching is shown in Figure 7. After behavioral matching, the parameters can be updated in the virtual model as shown in (6) and (7) and stored in the database as digital twin data.

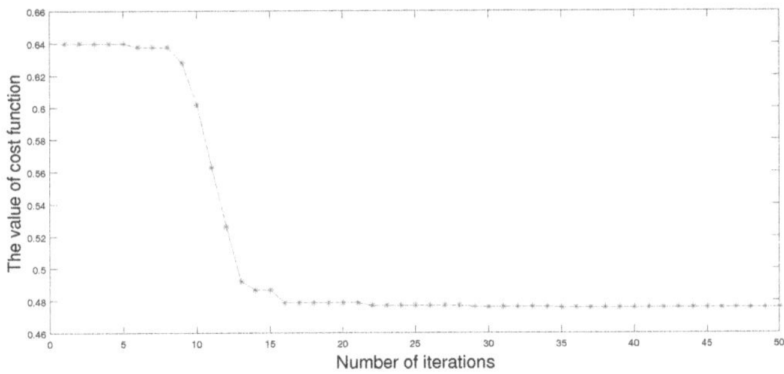

Figure 7. The convergence process of behavioral matching applying the improved particle swarm optimization algorithm.

$$Fcn1_{BM}(s) = \frac{100}{s + 97.16}$$

(6)

$$Fcn2_{BM}(s) = \frac{2436.87}{s + 1}.$$

(7)

4.3. Control Optimization

After the behavioral matching phase is done, the virtual system is called the twin of the physical system. This step is based on the accurate virtual model to optimize the control performance. The parameters of the controllers are also optimized by the improved particle swarm optimization algorithm. The optimization procedures are as follows:

4.3.1. Define the Parameter Search Scope

The four parameters of the cascaded integer order PI–PI controller and the six parameters of the cascaded fractional order PI^{λ}-PI^{λ} controller are considered as the solution set. Then, we initialize the parameters of the particle swarm optimization algorithm by randomly initializing the position and initial velocity of the particle. The size of the particle swarm is set as 100, and the maximum number of iterations is set as 50. A linearly decreasing inertia weight is used, which starts with ω_{start} set as 0.9 and ends with ω_{end} set as 0.4. The learning factors c_1 and c_2 are set as 0.9.

4.3.2. Choose a Fitness Function

The commonly used comprehensive performance evaluation standards are mainly based on the relationship between the deviation of the system w.r.t the time t. Generally, the performance index functions of the control system can be the error absolute value integral (IAE), the error square integral (ISE), the integral of timed square error (ITSE) or the absolute value of the error multiplied by the time integral (ITAE). IAE and ISE are not restricted by time, which is easy to cause the contradiction of reducing the overshoot and reducing time.

Furthermore, ITSE focuses on the error that occurs in the later stage of the transient response but seldom considers the large initial error in the response. For fast, stable, and small overshoot systems, ITAE is one of the commonly used performance indicators. The fitness function chosen in this paper is ITAE. The initialization process of the algorithm is as follows: the fitness of each particle is calculated according to the fitness function, and then the optimal individual is found in the initialized particle swarm, which is initialized to the optimal population, and the optimal fitness of the particle itself to a single particle is initialized.

4.3.3. The Iterative Optimization

At the time $t + 1$, as shown in Figure 8, the inertia weight decreases linearly during the iteration, and the particle position is updated as follows:

$$\omega^t = \omega_{start} - (\omega_{start} - \omega_{end})\frac{t}{K}$$

(8)

$$v_{id}^{t+1} = \omega^t v_{id}^t + c_1 r_1 (p_{id}^t - x_{id}^t) + c_2 r_2 (p_{gd}^t - x_{id}^t)$$

(9)

$$x_{id}^{t+1} = x_{id}^t + v_{id}^{t+1}$$

(10)

where ω^t is the weight of the inertia at time t, which balances the global search and local search; K is the maximum number of iterations; r_1 and r_2 are random numbers in the interval (0,1); v_{id}^t and v_{id}^{t+1} are the velocity of the particle at time t and $t + 1$, respectively; x_{id}^t and x_{id}^{t+1} are the positions of the particle at time t and $t + 1$, respectively; p_{id}^t is the personal best position at time t; and p_{gd}^t is the global best position at time t.

Then, the updated particle fitness is calculated and compared. Furthermore, the particle's own optimal fitness and the global optimal fitness are updated.

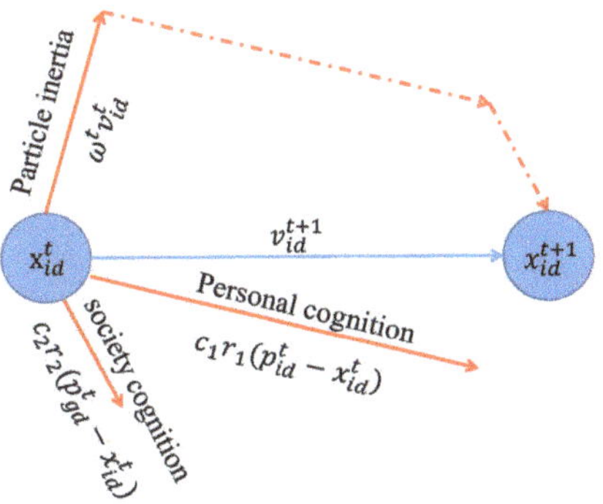

Figure 8. Iterative graph of the particle swarm optimization.

4.3.4. End Condition Judgment

Judge whether the maximum number of iterations is reached or the fitness function reaches the error tolerance range, if not, jump to (3), else, jump out of the loop. The convergence process of ITAE value in control optimization is shown in Figure 9.

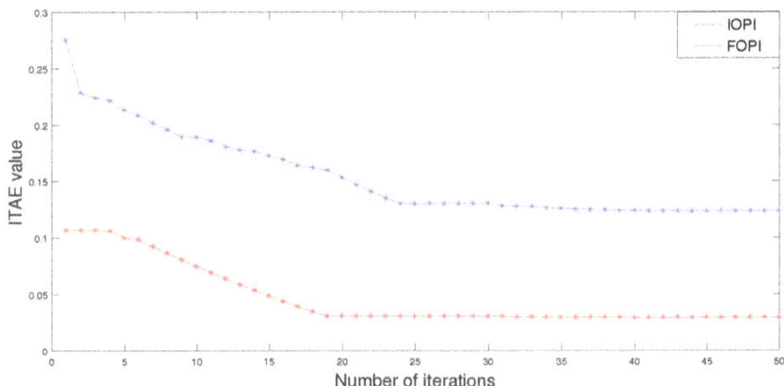

Figure 9. The convergence process of controller parameter optimization applying the improved particle swarm optimization algorithm.

4.3.5. Output Optimized Parameters

The output global optimal particles are the four parameters of the cascaded integer order PI–PI controller or the six parameters of the cascaded fractional order PI^{λ}-PI^{λ} controller.

4.4. Parameter Updating

The data transmission flow chart of the process is shown in Figure 10. First, the operating data of the industrial robot is collected by sensors and stored in MySQL as digital twin data. Then, the virtual system is modeled and optimized based on the digital twin data. The controller parameters obtained by iterative optimization in the virtual system need to be updated in the physical system to realize the control optimization of the physical system. Thus, the optimized parameters are stored in the database as the Mysql client, and Unity3D reads the digital twin data stored in the Mysql server and transmits the parameters to the controller through Socket communication.

The controller uses EtherCAT to transmit data to the driver with a transmission frequency of 500 Hz and then updates the controller parameters. After the controller parameters are updated, the running state data of the physical system operation process is collected again to verify whether the parameters obtained through virtual model simulation optimization have an optimal effect on the behavior of the physical system.

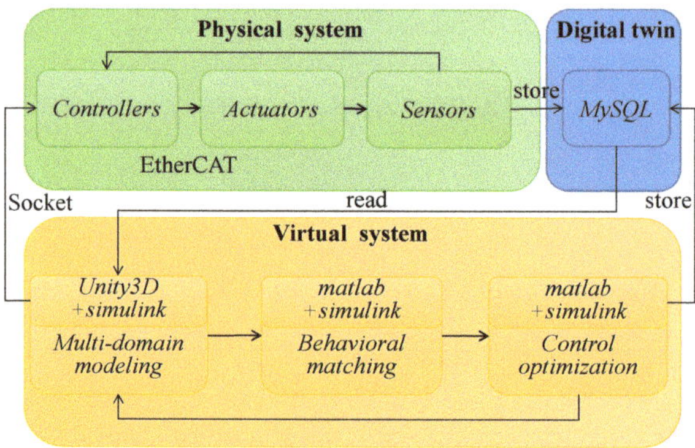

Figure 10. The data transmission flow chart of parameter updating.

4.5. Summary

According to the above case study, the whole six-axes industrial robot is modeled in unity3D, and then the PMSM of the single axis is modeled in Simulink to realize the virtual mapping of the fourth axis running state. It means that the running action of the robot can be clearly observed through unity3D, while the running state of the single axis can be well observed through Simulink. The real-time data obtained from the sensor of the physical system and the simulation of the virtual model are collected to identify the behavior of components and adjust their virtual model accordingly.

The four parameters with the greatest influence on the PMSM model (namely, J, B, R, and L) are evaluated. Based on the iterative algorithms and real-time digital twin data, the four parameters are adjusted in real time. The virtual model after behavioral matching can better match the real running state of the physical system compared with the initially identified model and can truly realize the real-time mapping between the physical system and the virtual system.

Then, the controller parameters can be optimized based on this accurate virtual model. The optimized parameters are transmitted back to the controllers of the physical system reliably in real time to achieve better performance. Finally, simulations and experiments are conducted in the next section to show the feasibility and effectiveness of the proposed framework.

5. Simulation and Experiment

The motor step-response operation reference value is 400 rpm. Figure 11 shows the results of behavioral matching, in which the blue line is the speed–output data curve collected during motor operation, the green line is the speed–output data curve obtained by simulation based on the identified model, and the red line is the speed–output data curve obtained by simulation based on the model after behavioral matching. In Figure 11a, both the physical system and the virtual system adopt the controller parameters originally set in the physical system.

In Figure 11b, both the physical system and virtual system adopt the integer order controller parameters optimized based on the accurate model after behavioral matching. In Figure 11c, both the physical system and the virtual system adopt fractional order controller parameters optimized based on the accurate model after behavioral matching. The differences in Figure 11a–c are in the different controller parameters. The simulation results show that when the controller parameters and even the controller form change, behavioral matching still plays an important role, making the virtual model much closer to the behavior of the physical system.

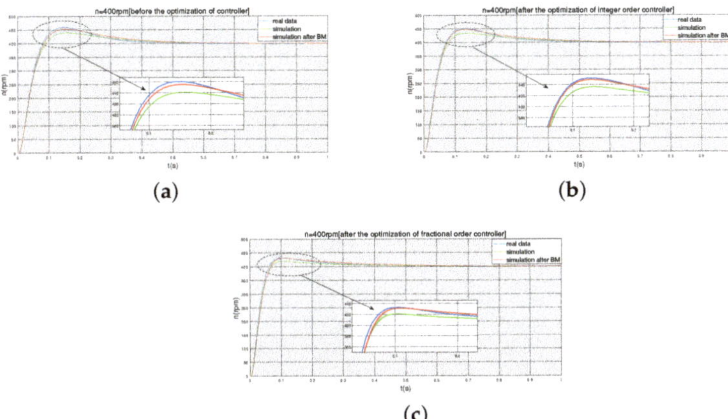

(a)　　　　　　　　(b)

(c)

Figure 11. Verify the effectiveness of behavioral matching before and after the controller parameter optimization. (**a**) Step response comparison with the initial integer order controller parameters of physical system. (**b**) Step response comparison with the optimized integer order controller parameters in the virtual system. (**c**) Step response comparison with the optimized fractional order controller parameters in the virtual system.

Figure 12 shows the simulation results of the control parameter optimization based on the accurate model after behavioral matching, in which the blue line is the simulation curve using the cascaded integer order controller parameters in the initial physical system, the green line is the simulation curve using the cascaded integer order controller parameters after optimization, and the red line is the simulation curve using the cascaded fractional order controller parameters after optimization. From the comparison of simulation results, it can be seen that the controller optimization based on the accurate model after behavioral matching improved the performance of the system, including the rise time and overshoot. The fractional order controller performed better than the integer order controller.

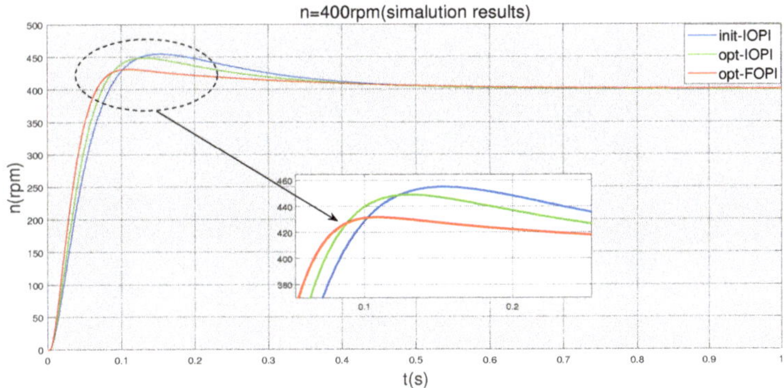

Figure 12. Simulation result comparison between the optimized fractional order controller and optimized integer order controller based on the accurate model after behavioral matching.

Figure 13 shows the experimental results of control parameter optimization, in which the blue line is the experimental curve using the cascaded integer order controller parameters in the initial physical system, the green line is the experimental curve using the cascaded integer order controller parameters after optimization, and the red line is the experimental curve using the cascaded fractional order controller parameters after optimization.

From the comparison of experimental results in Table 1, we also see the effectiveness of real-time optimization based on the proposed digital twin framework for improving physical system performance, including the rise time and overshoot. This also demonstrates that the fractional order controller outperformed the integer order controller.

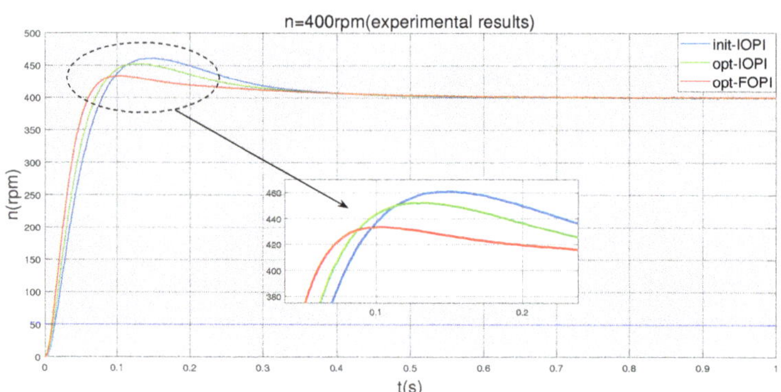

Figure 13. Experimental result comparison between the optimized fractional order controller and optimized integer order controller based on the accurate model after behavioral matching.

Table 1. Step response performances of simulation and experimental results.

	Simulation Results			Experimental Results		
	Overshoot	Peak Time	Settling Time	Overshoot	Peak Time	Settling Time
init-IOPI	13.8%	0.157 s	0.337 s	15%	0.151 s	0.296 s
opt-IOPI	12.3%	0.131 s	0.301 s	12.7%	0.132 s	0.262 s
opt-FOPI	8.0%	0.108 s	0.229 s	8.2%	0.098 s	0.198 s

6. Conclusions and Future Work

In this paper, we proposed a framework of a digital twin applied to industrial robots and applied it to a specific case. This framework employs four phases: establishing the virtual model of the physical system to reflect the characteristics of the physical system, mapping the physical system behavior in real time by using behavioral matching, optimizing the behavior of the virtual model by using the time-domain-optimization algorithm and fractional order controller, and finishing with the optimized parameters being updated in the physical system.

Using this framework, the mapping and interaction between the virtual system and the physical system can be realized, and the real-time optimization of the physical system based on a digital twin can be achieved. Moreover, by introducing the concept of fractional order into the proposed framework to design the fractional order controller, the optimization effect of the physical system can be improved. The simulation and experimental results show the feasibility and effectiveness of the proposed digital twin framework applied to industrial robots.

As a future activity, further investigation will be conducted for overall virtual modeling and real-time control of the six-axes industrial robot based on a digital twin in order to integrate the proposed framework into more general aspects. Furthermore, the algorithm used in the behavioral matching process and the control optimization process will be optimized to achieve better results.

Author Contributions: Conceptualization, Y.L.; methodology, Y.L. and X.L.; validation, X.L. and H.G.; data curation, X.L. and H.G.; writing—original draft preparation, X.L.; writing—review and editing, Y.L., Y.C. and L.G. All authors have read and agreed to the published version of the manuscript.

Funding: This work was supported by the National Natural Science Foundation of China grant number 51975234.

Data Availability Statement: Not applicable.

Conflicts of Interest: The authors declare no conflict of interest.

References

1. Tao, T.; Zhang, M. Digital Twin Shop-floor: A New Shop-floor Paradigm towards Smart Manufacturing. *IEEE Access* **2017**, *5*, 20418–20427. [CrossRef]
2. Zhang, X.; Zhu, W. Application framework of digital twin-driven product smart manufacturing system: A case study of aeroengine blade manufacturing. *Int. J. Adv. Robot. Syst.* **2019**, *16*, 1100–1108. [CrossRef]
3. Viola, J.; Chen, Y. Digital Twin Enabled Smart Control Engineering as an Industrial AI: A New Framework and Case Study. In Proceedings of the 2020 second International Conference on Industrial Artificial Intelligence (IAI), Shenyang, China, 23–25 October 2020; pp. 1–6.
4. Zhou, J.; Zhou, Y.; Wang, B.; Zang, J. Human–cyber–physical systems (HCPSs) in the context of new-generation intelligent manufacturing. *Engineering* **2019**, *5*, 624–636. [CrossRef]
5. Tao, F.; Cheng, J.; Qi, Q.; Zhang, M.; Zhang, H.; Sui, F. Digital twin-driven product design, manufacturing and service with big data. *Int. J. Adv. Manuf. Tech.* **2018**, *94*, 3563–3576. [CrossRef]
6. Liu, Y.; Peng, Y.; Wang, B.; Yao, S.; Liu, Z. Review on Cyber-physical Systems. *CAA J. Autom. Sin.* **2017**, *4*, 27–40. [CrossRef]
7. Kamel Boulos, M.N.; Zhang, P. Digital twins: From personalised medicine to precision public health. *J. Pers. Med.* **2021**, *11*, 745. [CrossRef]
8. Tuegel, E.; Ingraffea, A.; Eason, T. Reengineering aircraft structural life prediction using a digital twin. *Int. J. Aerosp. Eng.* **2011**, *2011*, 1687–5966. [CrossRef]
9. Qi, Q.; Tao, F. Digital Twin and Big Data Towards Smart Manufacturing and Industry 4.0: 360 Degree Comparison. *IEEE Access* **2018**, *6*, 3585–3593. [CrossRef]
10. Kritzinger, W.; Karner, M.; Traar, G.; Henjes, J.; Sihn, W. Digital Twin in manufacturing: A categorical literature review and classification. *IFAC-PapersOnLine* **2018**, *51*, 1016–1022. [CrossRef]
11. Tao, F.; Zhang, M.; Liu, Y.; Nee, A. Digital twin driven prognostics and health management for complex equipment. *CIRP Ann.* **2018**, *67*, 169–172. [CrossRef]
12. Guerra, R.; Quiza, R.; Villalonga, A. Digital Twin-Based Optimization for Ultraprecision Motion Systems with Backlash and Friction. *IEEE Access* **2019**, *7*, 93462–93472. [CrossRef]
13. Aivaliotis, P.; Georgoulias, K.; Chryssolouris, G. The use of Digital Twin for predictive maintenance in manufacturing. *Int. J. Comput. Integr. Manuf.* **2019**, *32*, 1067–1080. [CrossRef]
14. Zhang, K.; Qu, T.; Zhou, D.; Jiang, H.; Lin, Y.; Li, P.; Huang, G. Digital twin-based opti-state control method for a synchronized production operation system. *Robot. Comput.-Integr. Manuf.* **2019**, *63*, 101892. [CrossRef]
15. Gallala, A.; Kumar, A.; Hichri, B.; Plapper, P. Digital Twin for Human–Robot Interactions by Means of Industry 4.0 Enabling Technologies. *Sensors* **2022**, *22*, 4950. [CrossRef]
16. Lei, Z.; Zhou, H.; Hu, W.; Liu, G.; Guan, S.; Feng, X. Toward a Web-Based Digital Twin Thermal Power Plant. *IEEE Trans. Ind. Inform.* **2022**, *18*, 1716–1725. [CrossRef]
17. Aheleroff, S.; Mostashiri, N.; Xu, X.; Zhong, R. Mass Personalisation as a Service in Industry 4.0: A Resilient Response Case Study. *Adv. Eng. Inform.* **2021**, *50*, 101438. [CrossRef]
18. Wang, Y. Robot algorithm based on neural network and intelligent predictive control. *J. Amb. Intel. Hum. Comp.* **2020**, *11*, 6155–6166. [CrossRef]
19. Zhao, Z.; Liu, S.; Pan, J. A PID parameter tuning method based on the improved QUATRE algorithm. *Algorithms* **2021**, *14*, 173. [CrossRef]
20. Niu, H.; Chen, Y.; West, B. Why do big data and machine learning entail the fractional dynamics. *Entropy* **2021**, *23*, 297. [CrossRef]
21. Shi, L.; Yuan, Y.; Gao, J.; Mao, J. Compact fractional-order model of on-chip inductors with BCB on high resistivity silicon. *IEEE Trans. Compon. Packag. Manuf. Technol.* **2020**, *10*, 878–886. [CrossRef]
22. Zheng, W.; Huang, R.; Luo, Y.; Chen, Y.; Wang, X. A Look-Up Table Based Fractional Order Composite Controller Synthesis Method for the PMSM Speed Servo System. *Fractal Fract.* **2022**, *6*, 47. [CrossRef]
23. Podlubny, I. Fractional-order systems and PID controllers. *IEEE Trans. Autom. Control.* **1999**, *44*, 208–214. [CrossRef]
24. Luo, Y.; Zhang, T.; Lee, B.; Kang, C.; Yang, C. Fractional-order proportional derivative controller synthesis and implementation for hard-disk-drive servo system. *IEEE Trans. Autom. Control.* **2013**, *22*, 281–289. [CrossRef]
25. Shah, P.; Agashe, S. Review of fractional PID controller. *Mechatronics* **2016**, *38*, 29–41. [CrossRef]
26. Chen, P.; Luo, Y. A Two-Degree-of-Freedom Controller Design Satisfying Separation Principle with Fractional Order PD and Generalized ESO. *IEEE/ASME Trans. Mechatron.* **2021**, *27*, 137–148. [CrossRef]
27. Chen, P.; Luo, Y.; Peng, Y.; Chen, Y. Optimal robust fractional order PID controller synthesis for first order plus time delay systems. *ISA Trans.* **2021**, *114*, 136–149. [CrossRef]

28. Monje, C.A.; Vinagre, B.M.; Feliu, V.; Chen, Y.Q. Tuning and auto-tuning of fractional order controllers for industry applications. *Control. Eng. Pract.* **2008**, *16*, 798–812. [CrossRef]
29. Chen, P.; Luo, Y. Analytical Fractional Order PID Controller Design with Bodes Ideal Cut-Off Filter for PMSM Speed Servo System. *IEEE Trans. Ind. Electron.* **2022**, *70*, 1783–1793. [CrossRef]
30. Petras, I. The fractional order controllers: Methods for their synthesis and application. *J. Electr. Eng.* **1999**, *50*, 284–288.
31. Grieves, M. Digital Twin: Manufacturing Excellence through Virtual Factory Replication. *White Pap.* **2014**, *1*, 1–7.
32. Aheleroff, S.; Xu, X.; Zhong, R.; Lu, Y. Digital Twin as a Service (DTaaS) in Industry 4.0: An Architecture Reference Model. *Adv. Eng. Inform.* **2021**, *47*, 101225. [CrossRef]
33. Lu, G.; Xue, G.; Chen, Z. Design and implementation of virtual interactive scene based on unity 3D. In *Advanced Materials Research*; Trans Tech Publications Ltd.: Stafa-Zurich, Switzerland, 2011; Volume 317, pp. 2162–2167.
34. Kuang, Y.; Bai, X. The Research of Virtual Reality Scene Modeling Based on Unity 3D. In Proceedings of the 2018 13th International Conference on Computer Science Education (ICCSE), Colombo, Sri Lanka, 8–11 August 2018; pp. 1–3.
35. Korhonen, K.; Donadini, F.; Riisager, P.; Pesonen, L. GEOMAGIA50: An archeointensity database with PHP and MySQL. *Geochem. Geophys. Geosystems* **2008**, *9*, Q04029. [CrossRef]
36. Zheng, W.; Luo, Y.; Chen, Y.; Pi, Y. Fractional-order modeling of permanent magnet synchronous motor speed servo system. *J. Vib. Control* **2016**, *22*, 2255–2280. [CrossRef]
37. Shami, T.; El-Saleh, A.; Alswaitti, M.; Al-Tashi, Q.; Summakieh, M.; Mirjalili, S. Particle Swarm Optimization: A Comprehensive Survey. *IEEE Access* **2022**, *10*, 10031–10061. [CrossRef]
38. Yiyang, L.; Xi, J.; Hongfei, B.; Zhining, W.; Liangliang, S. A General Robot Inverse Kinematics Solution Method Based on Improved PSO Algorithm. *IEEE Access* **2021**, *9*, 32341–32350. [CrossRef]
39. Wang, F.; Li, J.; Li, Z.; Ke, D.; Du, J.; Garcia, C.; Rodriguez, J. Design of Model Predictive Control Weighting Factors for PMSM Using Gaussian Distribution-Based Particle Swarm Optimization. *IEEE Trans. Ind. Electron.* **2022**, *69*, 10935–10946. [CrossRef]

Article

Robust Trajectory Tracking Control for Serial Robotic Manipulators Using Fractional Order-Based PTID Controller

Banu Ataşlar-Ayyıldız

Department of Electronics and Communications Engineering, Kocaeli University, Kocaeli 41001, Turkey; banu.ayyildiz@kocaeli.edu.tr

Abstract: The design of advanced robust control is crucial for serial robotic manipulators under various uncertainties and disturbances in case of the forceful performance needs of industrial robotic applications. Therefore, this work has proposed the design and implementation of a fractional order proportional tilt integral derivative (FOPTID) controller in joint space for a 3-DOF serial robotic manipulator. The proposed controller has been designed based on the fractional calculus concept to fulfill trajectory tracking with high accuracy and also reduce effects from disturbances and uncertainties. The parameters of the controller have been optimized with a GWO–PSO algorithm, which is a hybrid tuning method, by considering the time integral performance criterion. The superior and contribution of the GWO–PSO-based FOPTID controller has been demonstrated by comparing the results with those offered by PID, FOPID and PTID control strategies tuned by the GWO–PSO. The examination of the results showed that the proposed controller, which is based on the GWO–PSO algorithm, demonstrates better trajectory tracking performance and increased robustness against various trajectories, external disturbances, and joint frictions as compared to other controllers under the same operating conditions. In terms of the trajectory tracking performance for robustness, the superiority of the proposed controllers tuned by GWO–PSO has been confirmed by 20.2% to 44.5% reductions in the joint tracking errors. Moreover, for assessing the energy consumption of the tuned controllers, the total energy consumption of the proposed controller for all joints has been remarkably reduced by 2.45% as compared to others. Consequently, the results illustrated that the proposed controller is robust and stable and sustains against the continuous disturbance.

Keywords: robotic manipulator; fractional order controllers; FOPTID; PTID; FOPID; PID; GWO–PSO

Citation: Ataşlar-Ayyıldız, B. Robust Trajectory Tracking Control for Serial Robotic Manipulators Using Fractional Order-Based PTID Controller. *Fractal Fract.* **2023**, *7*, 250. https://doi.org/10.3390/fractalfract7030250

Academic Editors: Kishore Bingi, Abhaya Pal Singh and Norbert Herencsar

Received: 4 January 2023
Revised: 2 March 2023
Accepted: 7 March 2023
Published: 9 March 2023

1. Introduction

Robotic manipulators are dynamically coupled and highly non-linear systems. Furthermore, in the case of various uncertainties and external or internal disturbances during their operations, effective control is needed to provide highly precise trajectory tracking and execute accurate positioning in various fields such as process industries, space applications and medical areas [1]. Due to the highly non-linear and uncertain dynamics of the robotic manipulators, accurate and robust trajectory tracking becomes even more challenging. For this reason, traditional proportional-integral-derivative (PID) controllers are generally not suitable for providing the high-performance trajectory tracking control in such operations that require high precision. In order to design a robust control strategy which is able to improve stability and performance tracking, fractional order (FO) controller design is considered using the incorporation of fractional calculus and traditional PID control approaches.

In the design of the FO controller, the orders of integral and derivative operators are indicated by non-integer values as compared to integers. Thus, in controller design, extra flexibility is provided by adding integral and derivative fractional powers to the full-order controller. The first use of FO operators in control was suggested by Oustaloup [2], who proposed a robust FO control approach called CRONE (Commande Robuste d'Ordre Non-Entier) [3,4]. The most well-known FO controller among control engineers is the fractional order PID (FOPID) controller presented by Podlubny [5]. For the controller design of the robotic manipulators in trajectory tracking control, methods based on the fractional order calculus have been widely used and cited by several authors. Bingul and Karahan [6] designed a FOPID controller optimized with Particle Swarm Optimization (PSO) and Genetic Algorithm (GA) for the trajectory tracking problem of a 2-DOF planar robotic manipulator. By employing the Matlab FMINCON function, which searches the optimal parameters of the controller, Angel and Viola [7] evaluated the FOPID controller with computer torque control strategy under external disturbances for the trajectory tracking control of a robotic manipulator type delta. The FOPID controller tuned by the Bat optimization algorithm was proposed by Al-Mayyahi et al. [8] for circular path tracking of a 3-RRR planar parallel robot platform without and with disturbance. Zhang et al. [9] studied fast spatial positioning and trajectory tracking of a 5-DOF drilling anchor manipulator by using FOPID control based on the four intelligent optimization algorithms such as Whale Algorithm (WOA), GA, PSO, and Search Algorithm (GPS) in that paper. Sharma et al. [10] presented two degree of freedom fractional order PID (2-DOF FOPID) controller tuned by Cuckoo Search (CS) algorithm for trajectory tracking task of a 2-DOF robotic manipulator with payload under model uncertainties, external disturbances, random noise and payload variations with time. Considering a 3-DOF parallel manipulator known as the Maryland manipulator, Dumlu and Erenturk [11] designed the FOPID control approach using a pattern search algorithm for improving the tracking performance of the manipulator in the case of high speed, high accuracy and high acceleration needed.

Different control strategies based on FO controllers have been designed and used in different applications in order to make an efficient control. One of them is the tilt-integral-derivative (TID) controller, which has been firstly presented by Lurie [12]. In the TID controller, which is closely related to the FOPID controller, the proportional parameter of the PID is replaced with a tilted one having a transfer function $s^{-1/n}$. By means of the resulting transfer function of the entire controller, the TID controller can achieve better disturbance rejection and reduce the effects of the system parameter changes for the closed-loop system as compared to the PID controller. Various applications of TID controller have been made in the literature, depending on a suitable choice of optimization algorithms for fine-tuning the controller parameters, for validating its superiority over PID controller in terms of improving the stability of the system and enhancing the speed of the controller response [13–19]. On the other hand, in order to improve the control performance and enhance the transient response of the TID controller, a concept of a fractional order-based TID controller has been recently indicated in the literature review. Sharma et al. [20] proposed a fractional order-based TID controller tuned by Salp Swarm Algorithm (SSA) for frequency regulation in a hybrid power system. Moreover, the results from the designed controllers based on SSA and Gray Wolf Optimization (GWO) algorithms were compared with existing controllers in terms of transient response characteristics and error indices. As a result, in that paper, the simulations prove the advisability of the fractional order TID controller in the presence of system parameter uncertainties, random load changes and different types of the system. In another study by the same authors [21], a dual-stage controller composed of fractional order-based TID and integer order proportional derivative (PD) controllers was presented for exhibiting fast and robust disturbance rejection performance of the proposed control scheme in load frequency control applications. On the other hand, a systematic tuning approach of the fractional-based robust TID controller was proposed by Lu et al. [22] for first-order plus time delay and high-order processes. In that paper, the design process of the robust TID controller and the corresponding steps were given in detail. Finally, the simulation results clearly indicated that the proposed controller achieved superior robustness and improved transient performance compared to the PID, FOPI and FOPID controllers.

For the purpose of enhancing the TID controller with more degree of freedom as compared to its integer derivative and integral terms, the fractional order integral and derivative terms are added. Thus, a hybrid controller is obtained for utilizing the features of FOPID and TID controllers. Mohamed et al. [23] designed a hybrid controller composed of TID and FOPID controllers for the load frequency control. Furthermore, the six different tunable parameters in the designed controller were optimized with a Manta Ray Foraging (MRF) algorithm by using the integral squared error criterion. In that paper, the robustness of the designed controller was examined under the variation of the system parameters and the load disturbances. By using the same control strategy, Ahmet et al. [24] proposed a modified hybrid fractional order controller, including FOPID and TID controllers for load frequency and the control of electric vehicles. Moreover, for determining the optimal parameters of the hybrid controller, the Artificial Ecosystem Optimization (AEO) algorithm was employed in that paper. It was noteworthy from the simulation results that the proposed hybrid controller demonstrates substantially superior, robust and stable performance over a wide range of fast responses during transients and parameters uncertainty. Based on this hybrid control scheme, Choudhary et al. [25] suggested a hybrid controller comprising a fractional order PI (FOPI) controller and fractional order proportional tilted integral derivative (FOPTID) controller for stabilizing the frequency and tie-line power variations in a power system. The parameters of the FOPI-FOPTID controller were adjusted by employing Global Neighborhood Algorithm (GNA) and Ant Colony Optimization (ACO) algorithms. The results revealed that the proposed FOPI-FOPTID controller provides better dynamic response and error criteria than PID, FOPID and FOPI-FOPID controllers optimized with the same optimization algorithms. Another hybrid controller based on FOPTID was proposed by Yanmaz et al. [26] for the effective control of a static compensation system. In that study, a FOPID-based model predictive controller (FOPID-MPC), TID-based MPC controller and the proposed FOPTID-based MPC (FOPTID-MPC) controller were optimized with Pathfinder Optimization Algorithm (POA) and also their transient responses and error indices were compared for showing their control performance. Consequently, the simulation results have demonstrated the effectiveness of the proposed FOPTID-MPC controller.

It is clear from the available literature that various control designs based on the TID controller using different optimization algorithms have been addressed for various applications. Furthermore, the compatibility of FOPID and TID controllers and the effect of the combination of them have not been evaluated and tackled in the literature for trajectory tracking control of the robotic manipulator. In this context, an efficient FOPTID controller tuned via GWO–PSO is demonstrated in realizing the trajectory tracking of a serial robotic manipulator in this work. The main contributions of this research article can be summarized as follows:

- To the best knowledge of the author, a FOPTID controller based on the combination of TID and FOPID controllers is firstly designed with a GWO–PSO algorithm to provide the trajectory tracking of a 3-DOF serial robotic manipulator under friction, external disturbance and different trajectories. This hybrid controller has major advantages in improving trajectory tracking control performance and enhancing robustness.
- In order to demonstrate the effectiveness of the proposed controller, PID, FOPID and PTID controllers are designed with the same optimization algorithm for carrying out trajectory-tracking tasks under the same conditions.
- By eliminating the effects of internal and external disturbances as total disturbance, the proposed FOPTID controller is more capable of dealing with the total disturbance during the reference trajectory tracking than existing controllers. Accordingly, better tracking accuracy is provided by the FOPTID controller.

The organization of the paper is as follows: In Section 2, the mathematical model of the first three links of the Staubli RX-60 manipulator is presented. The structures of the fractional order controllers are described in Section 3. The hybrid optimization algorithm GWO–PSO is given in Section 4. Furthermore, the objective function chosen for optimization studies and the proposed overall control system are also described in the same section. Simulation results are presented and discussed in Section 5, and finally, concluding remarks are presented in Section 6.

2. Dynamic Model of the Manipulator

In this study, a robust control for trajectory tracking is designed by considering the first three links of the Staubli RX-60 manipulator having the frame configuration presented in Figure 1. A brief overview of the mathematical model of the system is presented in this section.

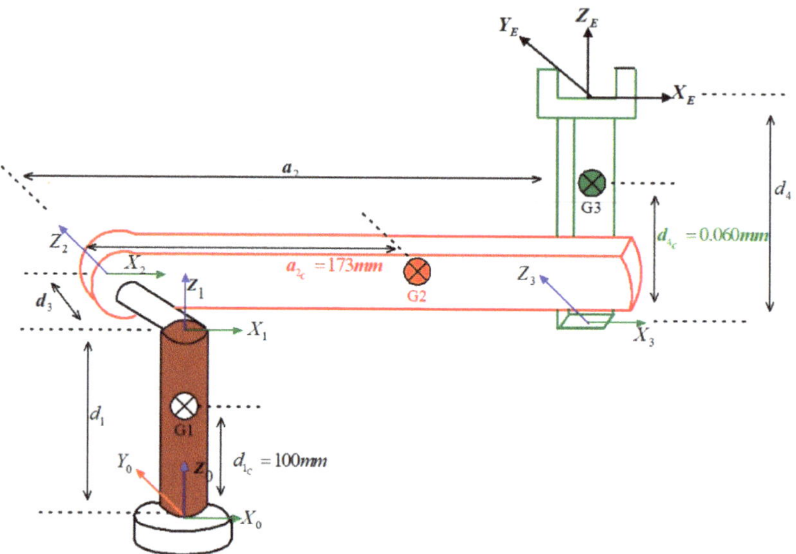

Figure 1. The model of the first three links of Staubli RX-60 robot arm [27].

The dynamics of the rigid body for robotic manipulators can be given as the following formulation:

$$\tau = D(\theta)\ddot{\theta} + C\left(\theta,\dot{\theta}\right) + G(\theta) + \tau_f \tag{1}$$

where θ, $\dot{\theta}$, and $\ddot{\theta}$ are joint angles, velocities and accelerations, respectively. $D(\theta), C\left(\theta,\dot{\theta}\right)$, and $G(\theta)$ are the inertia matrix, the coriolis/centripetal matrix and the gravity vector, respectively. τ_f is the robotic uncertainties and disturbances comprising viscous and static friction torque, and finally, τ is the control input torque.

The dynamics of the first three links of the robot can be modelled as:

$$\begin{bmatrix} \tau_1 \\ \tau_2 \\ \tau_3 \end{bmatrix} = \begin{bmatrix} d_{11} & d_{12} & d_{13} \\ d_{21} & d_{22} & d_{23} \\ d_{31} & d_{23} & d_{33} \end{bmatrix} \begin{bmatrix} \ddot{\theta}_1 \\ \ddot{\theta}_2 \\ \ddot{\theta}_3 \end{bmatrix} + \begin{bmatrix} c_1\left(\theta,\dot{\theta}\right) \\ c_2\left(\theta,\dot{\theta}\right) \\ c_3\left(\theta,\dot{\theta}\right) \end{bmatrix} + \begin{bmatrix} g_1(\theta) \\ g_2(\theta) \\ g_3(\theta) \end{bmatrix} + \begin{bmatrix} \tau_{f_1} \\ \tau_{f_2} \\ \tau_{f_3} \end{bmatrix} \tag{2}$$

The Denavit-Hartenberg (D-H) parameters of the robot and the other details about the elements of matrices $D(\theta)$, $C\left(\theta,\dot{\theta}\right)$, and $G(\theta)$ are available in [27]. The friction torque for each joint i is defined as:

$$\tau_{f_i} = F_{c_i} sign\left(\dot{\theta}_i\right) + F_{v_i}\dot{\theta}_i \tag{3}$$

where F_{c_i} and F_{v_i} are the Coulomb friction and viscous friction constants, respectively. By substituting the system parameters into Equation (2), the control input torque equation for each joint is obtained as follows [27]:

$$\begin{aligned}
\tau_1 &= [c^2\theta_2(a_2^2 m_3 + m_2 a_{2_c}^2 + B_{yy_2}) + A_{xx_2}s^2\theta_2 + B_{yy_3}c^2(\theta_2+\theta_3) - c\theta_2(-2m_3 a_2 s(\theta_2+\theta_3)d_{4_c} + 2F_2 s\theta_2) \\
&\quad -2F_3 c(\theta_2+\theta_3)s(\theta_2+\theta_3) + s^2(\theta_2+\theta_3)(m_3 d_{4_c}^2 + A_{xx_3}) + m_3 d_3^2 + C_{zz_1}]\ddot\theta_1 + [c\theta_2 D_2 + s\theta_2(E_2 + a_2 d_3 m_3) \\
&\quad + c(\theta_2+\theta_3)(D_3 - d_3 m_3 d_{4_c}) + E_3 s(\theta_2+\theta_3)]\ddot\theta_2 + [c(\theta_2+\theta_3)(D_3 - d_3 m_3 d_{4_c}) + E_3 s(\theta_2+\theta_3)]\ddot\theta_3 \\
&\quad + \big[-s(2\theta_2)(a_2^2 m_3 + m_2 a_{2_c}^2 + B_{yy_2}) + s(2\theta_2)A_{xx_2} - s(2(\theta_2+\theta_3))B_{yy_3} + 2m_3 a_2 d_{4_c}c(\theta_2+(\theta_2+\theta_3)) \\
&\quad -2F_2 c(2\theta_2) - 2F_3 c(2(\theta_2+\theta_3)) + s(2(\theta_2+\theta_3))(m_3 d_{4_c}^2 + A_{xx_3})\big]\dot\theta_2\dot\theta_1 \\
&\quad + [-s\theta_2 D_2 + c\theta_2(E_2 + a_2 d_3 m_3) - s(\theta_2+\theta_3)(D_3 - d_3 m_3 d_{4_c}) + E_3 c(\theta_2+\theta_3)]\dot\theta_2^2 \\
&\quad + 2[-s(\theta_2+\theta_3)(D_3 - d_3 m_3 d_{4_c}) + E_3 c(\theta_2+\theta_3)]\dot\theta_2\dot\theta_3 + \big[-s(2(\theta_2+\theta_3))B_{yy_3} - c\theta_2 c(\theta_2+\theta_3)(-2m_3 a_2 d_{4_c}) \\
&\quad -2F_3 c(2(\theta_2+\theta_3)) + s(2(\theta_2+\theta_3))(m_3 d_{4_c}^2 + A_{xx_3})\big]\dot\theta_3\dot\theta_1 + [-s(\theta_2+\theta_3)(D_3 - d_3 m_3 d_{4_c}) + E_3 c(\theta_2+\theta_3)]\dot\theta_3^2 \\
&\quad + F_{c_1}\,sign(\dot\theta_1) + F_{v_1}\dot\theta_1
\end{aligned} \tag{4}$$

$$\begin{aligned}
\tau_2 &= [c\theta_2 D_2 + s\theta_2(E_2 + a_2 d_3 m_3) + c(\theta_2+\theta_3)(D_3 - d_3 m_3 d_{4_c}) + E_3 s(\theta_2+\theta_3)]\ddot\theta_1 + \Big[C_{zz_2} + C_{zz_3} + m_2 a_{2_c}^2 + m_3\big(a_2^2 + d_{4_c}^2\big) \\
&\quad + 2s\theta_3 a_2 d_{4_c}m_3]\ddot\theta_2 + [m_3 d_{4_c}^2 - a_2 m_3 s\theta_3 d_{4_c} + C_{zz_3}]\ddot\theta_3 - \tfrac{1}{2}\big[s(2\theta_2)(A_{xx_2} - (a_2^2 m_3 + m_2 a_{2_c}^2 + B_{yy_2})) \\
&\quad + s(2(\theta_2+\theta_3))((m_3 d_{4_c}^2 + A_{xx_3}) - B_{yy_3}) + 2m_3 a_2 d_{4_c}c(\theta_2+(\theta_2+\theta_3)) - 2F_2 c(2\theta_2) - 2F_3 c(2(\theta_2+\theta_3))]\dot\theta_1^2 \\
&\quad + [2c\theta_3 a_2 m_3 d_{4_c}]\dot\theta_3\dot\theta_2 + [-c\theta_3 a_2 m_3 d_{4_c}]\dot\theta_3^2 + g_0 m_3 d_{4_c}s(\theta_2+\theta_3) + g_0 m_2 a_{2_c}c\theta_2 + g_0 a_2 m_3 c\theta_2 \\
&\quad + F_{c_2}\,sign(\dot\theta_2) + F_{v_2}\dot\theta_2
\end{aligned} \tag{5}$$

$$\begin{aligned}
\tau_3 &= [c(\theta_2+\theta_3)(D_3 - d_3 m_3 d_{4_c}) + E_3 s(\theta_2+\theta_3)\ddot\theta_1 + [m_3 d_{4_c}^2 - s\theta_3 a_2 m_3 d_{4_c} + C_{zz_3}]\ddot\theta_2 + [m_3 d_{4_c}^2 + C_{zz_3}]\ddot\theta_3 \\
&\quad - \tfrac{1}{2}[-s(2(\theta_2+\theta_3))B_{yy_3} - c\theta_2 c(\theta_2+\theta_3)(-2m_3 a_2 d_{4_c})] - 2F_3 c(2(\theta_2+\theta_3)) \\
&\quad + s(2(\theta_2+\theta_3))(m_3 d_{4_c}^2 + A_{xx_3})]\dot\theta_1^2 - \tfrac{1}{2}[2c\theta_3 a_2 m_3 d_{4_c}]\dot\theta_2^2 + g_0 m_3 d_{4_c}s(\theta_2+\theta_3) + F_{c_3}\,sign(\dot\theta_3) + F_{v_3}\dot\theta_3
\end{aligned} \tag{6}$$

3. Design of Controllers

In this work, the design of a fractional order proportional tilt integral derivative (FOPTID) controller for the presented 3-DOF serial robotic manipulator has been proposed and investigated. Furthermore, several controllers are applied to the same system under the same conditions in order to examine the performance of the proposed controller. The purposed FOPID, PTID and FOPTID controllers contain non-integer order integral and derivative. Therefore, fractional calculus is needed for implementation of them.

3.1. Fractional Calculus

Fractional calculus includes operations where the degree of derivative and integral is not an integer but with real or even complex values [28]. The fractional order operator $_aD_t^\alpha$ is defined as follows:

$$_aD_t^\alpha = \begin{cases} d^\alpha/dt^\alpha, & \Re(z) > 0 \\ 1, & \Re(z) = 0 \\ \int_a^t d\tau^\alpha & \Re(z) < 0 \end{cases} \tag{7}$$

where a and t are the limits of the operation, and α is the non-integer degree of the derivative and integral. Several approaches have been developed for designing the fractional order derivative and integral operators. One of the approaches is Riemann–Liouville (R–L) [17]. The definition of Riemann and Loiuville is as follows:

$$_aD_t^\alpha f(t) = \frac{1}{\Gamma(n-\alpha)}\frac{d^n}{dt^n}\int_a^t \frac{f(\tau)}{(t-\tau)^{\alpha-n+1}}d\tau \tag{8}$$

where $\Gamma(\cdot)$ is the Euler's Gamma Function:

$$\Gamma(z) = \int_0^\infty e^{-t}t^{z-1}dt, \text{ for } \Re(z) > 0 \tag{9}$$

In general, the Laplace transform is used to describe derivative and integral for simplicity. The Laplace transform can be defined as:

$$\mathcal{L}\{_aD_t^\alpha f(t)\} = s^\alpha F(s) - \sum_{k=0}^{n-1} s^k\, _aD_t^{\alpha-k-1}f(t)\Big|_{t=0} \tag{10}$$

where $\mathcal{L}\{f(t)\}$ is the Laplace Transform of the function $f(t)$. Under zero initial conditions, with the non-integer value of α, the Laplace transform of $_aD_t^\alpha f(t)$ is:

$$\mathcal{L}\{_aD_t^\alpha f(t)\} = s^\alpha F(s) \tag{11}$$

The Oustaloup Recursive Approximation Method is one of the several approximation methods proposed in the literature for the implementation of the fractional order function s^α. The Oustaloup Recursive Approximation Method uses an Nth order analog filter to approximate the fractional order function in a certain frequency range $\{\omega_b, \omega_h\}$. ω_b and ω_h are the lower and upper-frequency bounds, respectively. The approximate transfer function for s^α is expressed by integer order equivalent transfer function:

$$s^\alpha = K \prod_{k=-N}^{N} \frac{s + \omega_{z_k}}{s + \omega_{p_k}} \tag{12}$$

where K is gain, ω_{z_k} are zeros and ω_{p_k} are poles of the filter [29]. This approximate transfer function has $2N + 1$ poles and zeros. The poles, zeros and gain are calculated below, respectively [29]:

$$\omega_{p_k} = \omega_b \left(\frac{\omega_h}{\omega_b} \right)^{\frac{k+N+\frac{1}{2}+\frac{\alpha}{2}}{2N+1}} \tag{13}$$

$$\omega_{z_k} = \omega_b \left(\frac{\omega_h}{\omega_b} \right)^{\frac{k+N+\frac{1}{2}-\frac{\alpha}{2}}{2N+1}} \tag{14}$$

$$K = \left(\frac{\omega_h}{\omega_b} \right)^{\frac{-\alpha}{2}} \prod_{k=-N}^{N} \frac{\omega_{p_k}}{\omega_{z_k}} \tag{15}$$

As mentioned above, in this study, the FOPTID, FOPID, and PTID controllers are applied to the robotic manipulator. These fractional order controllers are implemented by using Oustaloup Recursive Approximation [30]. In this study, the value of N is chosen as 5, and the frequency range is chosen as: $\{\omega_b, \omega_h\} = \{10^{-2}, 10^{+2}\}$ rad/s.

3.2. Fractional Order Controllers

Due to the simple design and construction of integer order PID controllers, they are still widely used in many industrial applications. The PID controller comprises three coefficients: proportional coefficient (K_p), integral coefficient (K_i) and derivative coefficient (K_d) to produce the control action. The transfer function of PID can be stated as:

$$C_{PID}(s) = K_p + K_i \frac{1}{s} + K_d s \tag{16}$$

On the other hand, due to the highly non-linear and uncertain dynamics of the robotic manipulator, the trajectory tracking control problem is quite difficult. Therefore, in general, conventional integer order PID controllers are not suitable for providing high-performance trajectory tracking control in precision operations. In order to design a robust control strategy which can improve stability and tracking performance, fractional order-based controller is considered by applying fractional calculus to the conventional PID control approaches.

A FOPID controller is depicted by five parameters. In comparison to the conventional PID controllers, FOPID controllers have two more parameters in which the orders of the integral part λ and derivative part μ are non-integer. These additional parameters bring more flexibility to the design of the controller and also may lead to obtaining an enhanced dynamic performance. The structure of the FOPID is shown in Figure 2 which $E(s)$ and $U(s)$ represent the error and the control signals, respectively.

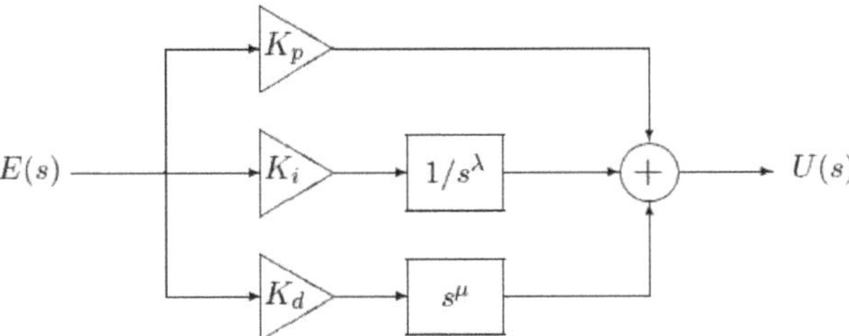

Figure 2. FOPID controller structure.

The transfer function of the FOPID controller is given below:

$$C_{FOPID}(s) = K_p + K_i \frac{1}{s^\lambda} + K_d s^\mu \tag{17}$$

One of the different control strategies based on Fractional Order Calculus is the PTID control method. The design of the PTID controller has been proposed quite recently in [25], and, in fact, it is a modified version of the TID controller. On the other hand, the only difference between the TID controller from conventional PID is that its proportional parameter is replaced with a tilted one having a transfer function $s^{-1/n}$ [12]. Thanks to these modifications, the PTID controller can achieve better disturbance rejection and reduce the effects of the system parameter changes for the closed-loop system as compared to the PID controller.

The structure of the PTID controller is presented in Figure 3. As shown in the figure, the proportional term, K_p, is added to the TID controller. Therefore, the transfer function of the PTID controller is:

$$C_{PTID}(s) = K_p + K_t \frac{1}{s^{1/n}} + K_i \frac{1}{s} + K_d s \tag{18}$$

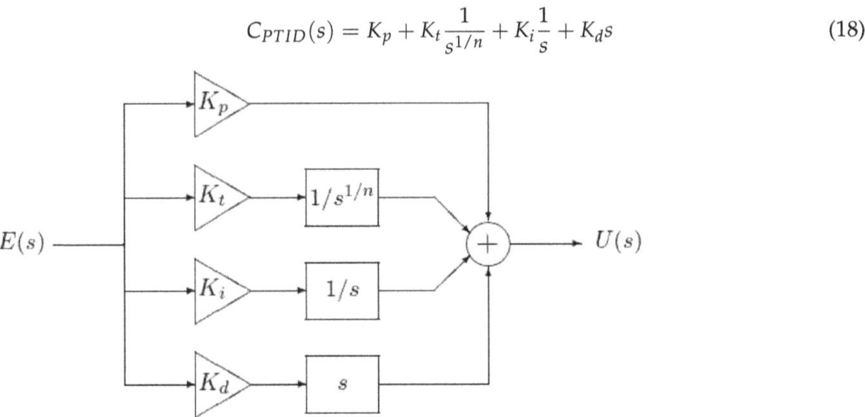

Figure 3. PTID controller structure.

As examined in the works [25,26,31,32] in the literature, FO controllers are more stable and useful in various control applications. Also, the TID controllers are able to reject disturbances, respond quickly and be consistent with uncertainties in linear and nonlinear control implementations. Moreover, they have several tuning parameters. Thus, superior performance for both of them can be obtained in control implementations. Considering their great qualities, a FOPTID controller is proposed as a hybrid controller of FOPID and PTID in [25]. The structure of the FOPTID controller is shown in Figure 4. FOPTID controller has the non-integer order of integral and derivative coefficients of the PTID controller. By defining λ and μ as the integral and derivative non-integer orders, respectively, the transfer function of the FOPTID controller is:

$$C_{FOPTID}(s) = K_p + K_t \frac{1}{s^{1/n}} + K_i \frac{1}{s^{\lambda}} + K_d s^{\mu} \tag{19}$$

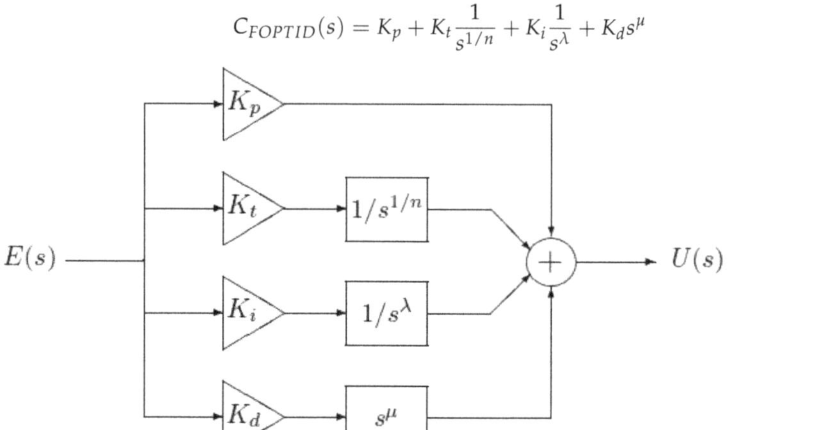

Figure 4. FOPTID controller structure.

In this study, the effect of the FOPTID controller on the trajectory-tracking control of the robotic manipulator will be evaluated. In this regard, an efficient FOPTID controller can be obtained by tuning with the hybrid optimization algorithm GWO–PSO, which will be presented in the next section.

4. Optimization Tasks

The parameters of the proposed controller have been optimized with a hybrid GWO–PSO algorithm by considering a specific objective criterion. The superior and contribution of the GWO–PSO-based FOPTID controller has been demonstrated by comparing the results with those offered by PID, FOPID and PTID control strategies tuned by the same algorithm.

4.1. Optimization Algorithm

4.1.1. Particle Swarm Optimization (PSO) Algorithm

Particle swarm optimization is a population-based stochastic optimization method that was first proposed in 1995 to obtain the best results on nonlinear numerical problems by modeling the movements of living swarms [33]. The PSO algorithm finds out an optimal solution among the randomly distributed particles in a swarm. Essentially, each particle within the swarm indicates a potential solution with its particular velocity and position. In this context, for each iteration of the PSO algorithm, the velocity and position of the particles are updated according to the following expressions, respectively:

$$v_i^{k+1} = \xi v_i^k + \varphi_1 rand_1 \left(pbest_i - p_i^k \right) + \varphi_2 rand_2 \left(gbest - p_i^k \right) \tag{20}$$

$$p_i^{k+1} = p_i^k + v_i^{k+1} \tag{21}$$

In these equations, v_i^k is the velocity of the ith particle for the k iteration, p_i^k is the position of the ith particle for the k iteration, ξ represents the inertial weight function, $\varphi_{1,2}$ represents the learning factors, and $rand_{1,2}$ represents the random number values assigned in the range of [0, 1]. In addition, $pbest_i$ is the coordinates that provide the best solution that particle i has achieved so far. $gbest$ is also the coordinates that provide the best solution obtained over all particles. Figure 5 shows the two-dimensional motion of one of the particles depending on the terms defined above.

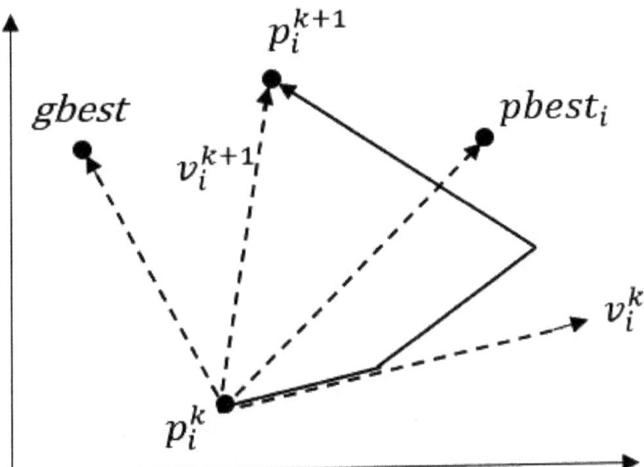

Figure 5. The updating process of velocity and position for each particle.

4.1.2. Gray Wolf Optimization (GWO) Algorithm

As a swarm-based optimization method, inspiration for Gray Wolf Optimization comes from the behavior and the hunting strategy of the grey wolves in nature. Based on the social hierarchy as depicted in Figure 6, gray wolves are classified into four groups as alpha (α), beta (β), delta (δ) and omega (ω). As seen from the figure, the social hierarchy goes down from top to bottom, and the leading group consists of alpha wolves. Beta wolves help alpha wolves in making decisions. As the third level, the delta wolves' mission is to submit to alpha and beta wolves but control the omega wolves. The least priority wolves are the omegas, which must follow the leading grey wolves [34].

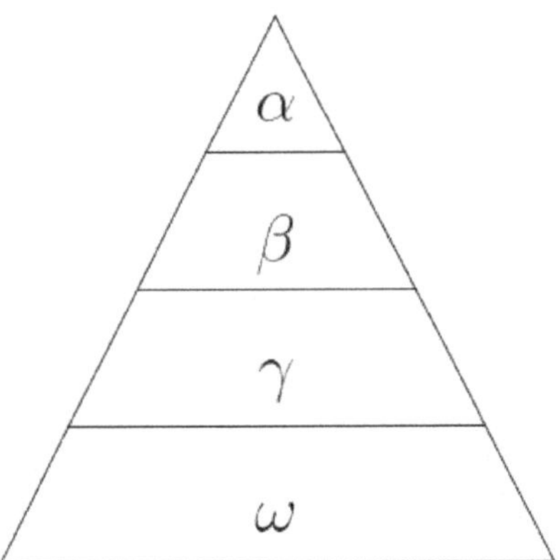

Figure 6. Hierarchy of grey wolves.

In the GWO algorithm, the encircling behaviour of the grey wolves is modeled with the below equations:

$$\vec{D} = \left| \vec{C} \times \vec{X}_p(k) - \vec{X}(k) \right| \tag{22}$$

$$\vec{X}(k+1) = \vec{X}_p(k) - \vec{A} \times \vec{D} \tag{23}$$

In these equations, k is the number of iterations and the $\vec{X}$ and $\vec{X}_p$ are the position vectors of the grey wolves and the prey, respectively. $\vec{A}$ and $\vec{C}$ are the coefficient vectors and calculated as given below:

$$\vec{A} = \vec{a} \times (2 \times \vec{r}_1 - 1) \tag{24}$$

$$\vec{C} = 2 \times \vec{r}_2 \tag{25}$$

where $\vec{a}$ is linearly decreased from 2 to 0 through iteration steps and $\vec{r}_1$ and $\vec{r}_2$ are random vectors within $[0, 1]$.

In the GWO algorithm, hunting and encircling prey are modelled by the following equations:

$$
\begin{aligned}
\vec{D}_\alpha &= \left| \vec{C}_1 \times \vec{X}_\alpha - \vec{X}(k) \right| \\
\vec{D}_\beta &= \left| \vec{C}_2 \times \vec{X}_\beta - \vec{X}(k) \right| \\
\vec{D}_\delta &= \left| \vec{C}_1 \times \vec{X}_\delta - \vec{X}(k) \right|
\end{aligned}
\tag{26}
$$

$$
\begin{aligned}
\vec{X}_1 &= \left| \vec{X}_\alpha - \vec{A}_1 \vec{D}_\alpha \right| \\
\vec{X}_2 &= \left| \vec{X}_\beta - \vec{A}_2 \vec{D}_\beta \right| \\
\vec{X}_3 &= \left| \vec{X}_\delta - \vec{A}_3 \vec{D}_\delta \right|
\end{aligned}
\tag{27}
$$

$$\vec{X}(k+1) = \frac{\vec{X}_1 + \vec{X}_2 + \vec{X}_3}{3} \tag{28}$$

where $\vec{D}_\alpha$, $\vec{D}_\beta$, and $\vec{D}_\delta$ vectors represent the distances between the ω wolves and α, β and δ wolves, respectively. $\vec{X}_1$, $\vec{X}_2$, and $\vec{X}_3$ vectors represent the relative positions based on α, β and δ wolves, respectively. The updating process of positions for each group of wolves is also depicted in Figure 7.

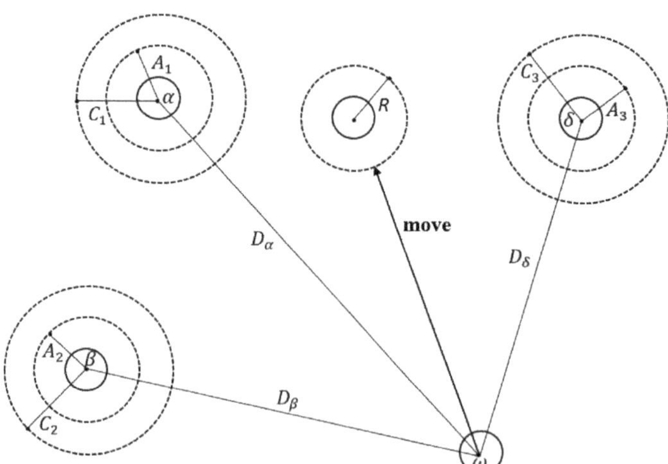

Figure 7. The position update process for grey wolves.

4.1.3. GWO–PSO Algorithm

In this work, GWO is hybridized with a PSO method to improve the progress of the GWO. The hybrid GWO–PSO has been seen as an effective optimization technique when searching for the best solution globally to an optimization problem [35]. The pseudo-code of the GWO–PSO algorithm is presented in Figure 8.

Initialize the positions of particles in the swarm

    and the positions of wolves in the population.

while the maximum iteration number is not reached

    Run GWO: Update of each wolf position.

    Obtain three best ones among all search agents.

    Run PSO by using the best values found by GWO.

    Return the positions modified by PSO back to the GWO.

end

Figure 8. Pseudo code of GWO–PSO.

The best positions of the grey wolves obtained at the end of the GWO–PSO algorithm represent the parameters of the controllers for each joint of the robot as follows:

$$\text{PID} : \left\{ K_p, K_i, K_d \right\}$$

$$\text{FOPID} : \left\{ K_p, K_i, K_d, \lambda, \mu \right\}$$

$$\text{PTID} : \left\{ K_p, K_i, K_d, K_t, n \right\}$$

$$\text{FOPTID} : \left\{ K_p, K_i, K_d, K_t, \lambda, \mu, n \right\}$$

4.2. Objective Function

In this study, the objective function used in optimization of the controller parameters is chosen as ITAE (Integral of Time Absolute Error) for each joint of the 3-DOF robotic manipulator. Thus, the objective function is given as the following:

$$J_{ITAE} = \sum_{i=1}^{3} \int_0^t t|e_i(t)|dt \tag{29}$$

Here, t is the time and $e_i(t)$ is the trajectory error for joint i.

4.3. Proposed Control System Framework

The schematic diagram of the proposed control system is presented in Figure 9. The optimal parameters of PID, FOPID, PTID and FOPTID controllers are found by using the GWO–PSO algorithm. The number of maximum iteration is set to 100 in the algorithm. Moreover, the optimal controller parameters are obtained after 10 runs of the algorithm.

Based on the detailed literature review, during the optimization, the lower and upper boundaries of the parameters are set to $\left\{ K_p, K_i, K_d, K_t \right\} \in [0, 350]$, $\{\mu, \lambda\} \in [0, 2]$ and $\{n\} \in [0, 300]$.

In addition, the simulation time is adjusted differently in each simulation process according to the type of reference trajectory signal, with a fixed interval time of 0.001 s.

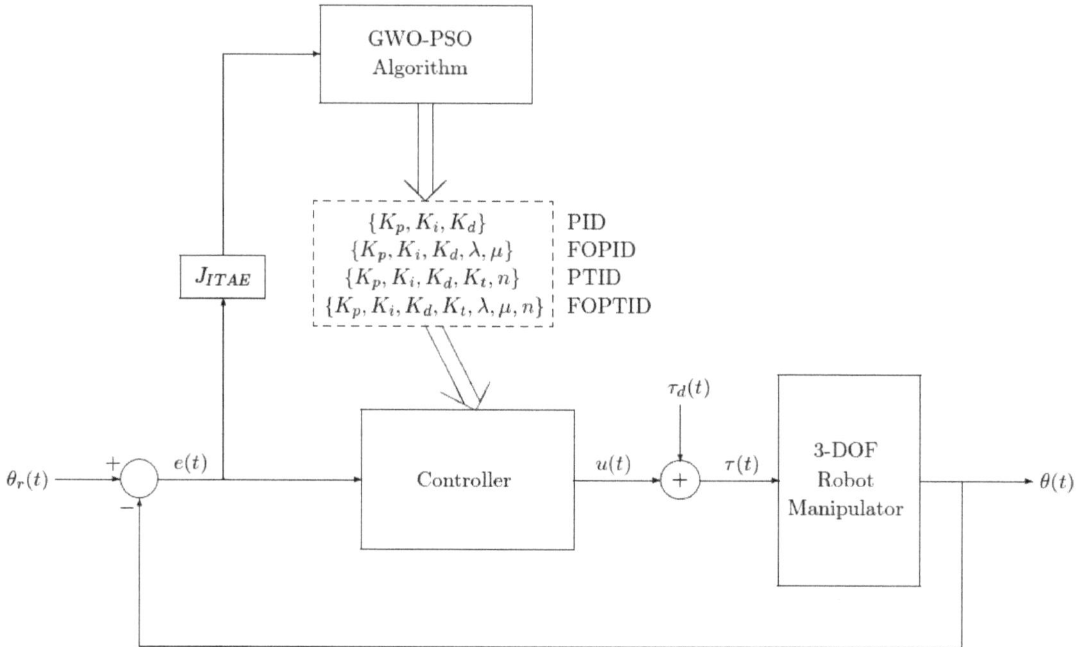

Figure 9. The schematic diagram of the proposed control system.

5. Simulation Results and Discussions

In this section, the robustness and effectiveness of the proposed control strategy have been comparatively verified on the first three-link of a 6-DOF serial robotic manipulator. The order of tasks for simulations is as follows: Firstly, tuning the parameters of the presented controllers by GWO–PSO and demonstrating the results from trajectory tracking; Secondly, testing the proposed control scheme by comparing with the PID, FOPID and PTID controllers and showing the results obtained from the different trajectory, internal and external disturbances. Design of the overall system model and the optimization with the GWO–PSO are simulated using MATLAB/SIMULINK environment, and also all simulations have been executed on a personal computer having an Intel CoreTM i5-7200U CPU @ 2.50 GHz processor and 8.0 GB RAM. Furthermore, a Ninteger toolbox [36] is used in the MATLAB environment as the approximator for simulating the fractional order terms of FOPID, PTID and FOPTID controllers.

5.1. Trajectory Tracking Analysis

In order to achieve the trajectory tracking in joint space, firstly, four control strategies (PID, FOPID, PTID and FOPTID) are optimized by GWO–PSO for each joint using the trajectory tracking evaluation (J_{ITAE}) with respect to the given path for the end-effector of the robot. During this tunning tasks, it is assumed that there is no friction. That means the values of F_{c_i} and F_{v_i} are taken as zero in Equation (3) of the mathematical model. The presence of friction will be taken into account in one of the robustness tests. After optimization, for comparing the tracking performance of the tuned controllers, the mean of absolute error (MAE) for each joint over the trajectory is computed as follows:

$$MAE_i = \frac{1}{N}\sum_{j=1}^{N}|e_i(j)| \tag{30}$$

where $e_i(j)$ is the trajectory error of jth sample of ith joint and N is the number of samples. At the end of conducting a total of 10 individual trials depending on the generated random numbers, eventually, the obtained optimal controller parameters and *MAE* values for the presented control strategies are given in Table 1. For analysis of the total tracking performance of the proposed approach, the comparison of the J_{ITAE} values based on four potential control approaches, namely PID, FOPID, PTID and FOPTID controllers, are illustrated in Figure 10. In addition, the reference path of the end-effector, the corresponding reference and system output trajectories in each joint are depicted in Figure 11 for a better view of tracking the reference trajectories of each joint based on the tuned controllers.

Table 1. Comparison of the optimized parameters of the controllers and MAE values for each joint.

Joint	Controller	K_t	K_p	K_i	K_d	μ	λ	n	MAE
1	PID	-	203.8760	0.0127	132.5981	-	-	-	2.0832
	FOPID	-	271.4936	0.0124	132.2961	1.0381	0.0756	-	1.9153
	PTID	236.3371	349.7559	0.0122	298.3974	-	-	299.9889	1.8705
	FOPTID	80.0347	349.7559	21.2705	273.0510	0.9257	0.3053	268.3995	1.9048
2	PID	-	325.0161	0.0130	79.4103	-	-	-	3.0791
	FOPID	-	333.5564	298.1256	148.4613	1.0962	0.0308	-	3.0235
	PTID	298.1256	20.5604	0.0131	93.6091	-	-	233.8233	3.0426
	FOPTID	90.5690	348.9735	221.4909	179.8575	1.0533	0.0104	220.1078	2.9837
3	PID	-	251.7546	295.1566	25.5284	-	-	-	1.9342
	FOPID	-	296.9951	80.5790	311.0399	0.5549	0.6426	-	1.1130
	PTID	340.8880	290.3104	0.0121	50.3965	-	-	132.3825	1.2832
	FOPTID	318.2374	29.3824	7.6325	145.1791	0.6516	1.0140	280.9249	1.1771

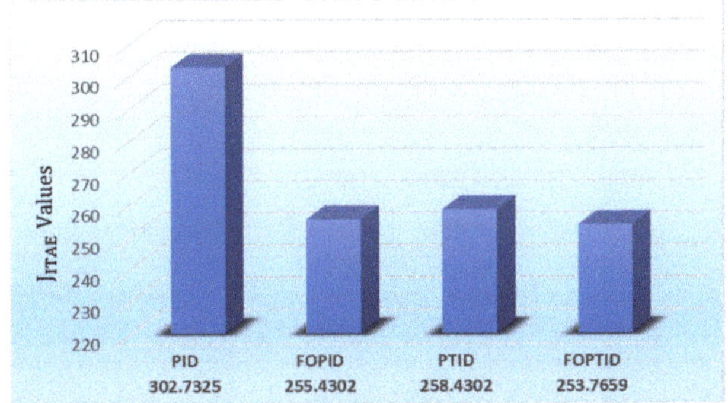

Figure 10. Comparison of J_{ITAE} values for PID, FOPID, PTID and FOPTID control strategies.

From Figure 10, it is revealed that the proposed FOPTID approach tuned by GWO–PSO has the smallest J_{ITAE} value. The improvement in J_{ITAE} is resulting from the introduction of fractional operators in the FOPTID controller, which adds extra design variables. Therefore, the proposed control approach is able to maintain relatively higher trajectory tracking accuracy when compared to the PID, FOPID and PTID approaches.

As shown in Figure 11, all of the actual joint positions can track the desired joint trajectories by using the tuned controllers. It is inferred that a remarkable tracking performance for all joints is achieved by the PTID and FOPTID controllers. Furthermore, from Figures 10 and 11 and Table 1, it can be seen that the proposed FOPTID method has better control performance in comparison with the existing controllers under a more complex joint trajectory tracking task.

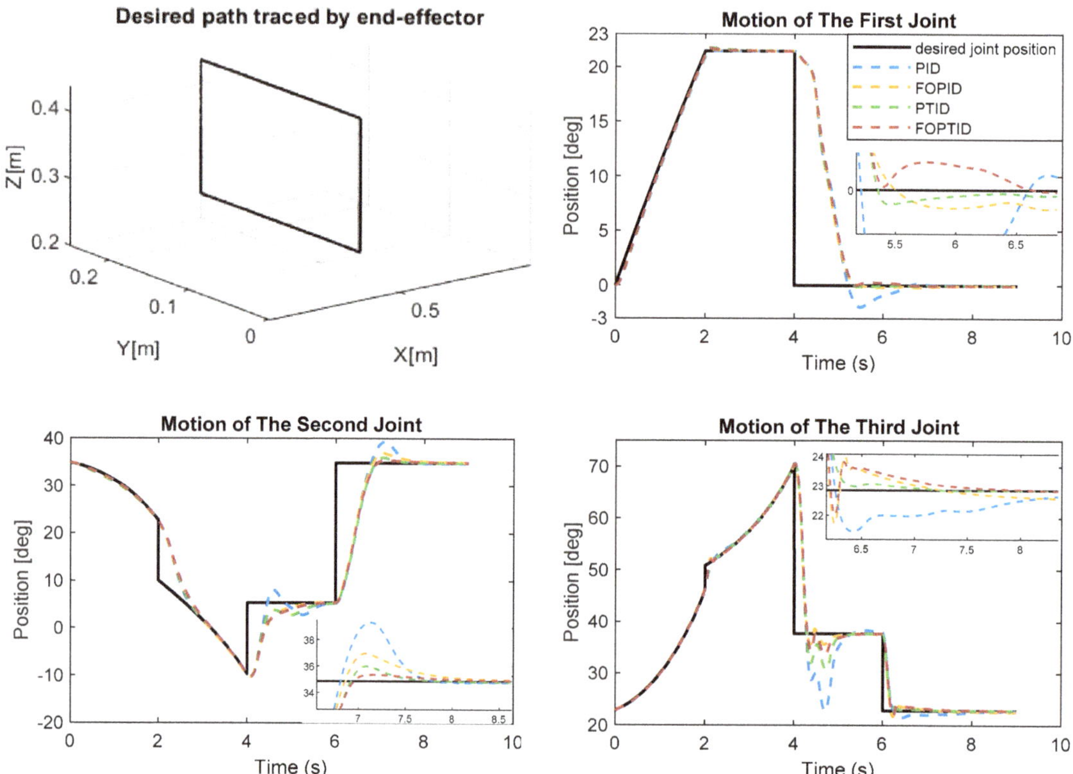

Figure 11. Trajectory tracking performance of PID, FOPID, PTID and FOPTID controllers for each joint.

5.2. Robustness Testing: Different Trajectory

In actual situations, the effectiveness of the tuned controllers is investigated under different joint trajectories or paths traced by the end-effector of the robot in task space. The desired path and actual path in task space and the corresponding joint trajectories are shown in Figure 12 for observing the improvements in tracking errors based on the tuned controllers. Moreover, in order to demonstrate a quantitative comparison among the results, the MAE values of the joint errors (e_j) and the root mean square (RMS) values of the control signals (τ_j) for each joint are calculated as shown in Figure 13.

As shown in Figure 12, the end-effector trajectory tracking based on the FOPTID controller exhibits almost the same results as the end-effector tracking based on the PTID controller. On the other hand, when using the proposed FOPTID controller tuned by GWO–PSO, the trajectory tracking precision is relatively higher as compared to the other tuned controllers. From joint space, it can be observed that all trajectory tracking controllers can make the robot track the joint reference trajectory. However, the TID-based trajectory tracking control approaches can accurately track the change in the joint angle and maintain stability.

As can be observed from Figure 13, the designed controllers yield almost the same RMS control action values for joint 1, while the TID-based controllers produce smaller MAE error values as compared to others. In addition, compared to the controllers (PID and PTID or FOPID and FOPTID), the TID-based controllers need lesser applied torque for tracking the desired joint trajectories than the PID-based controllers.

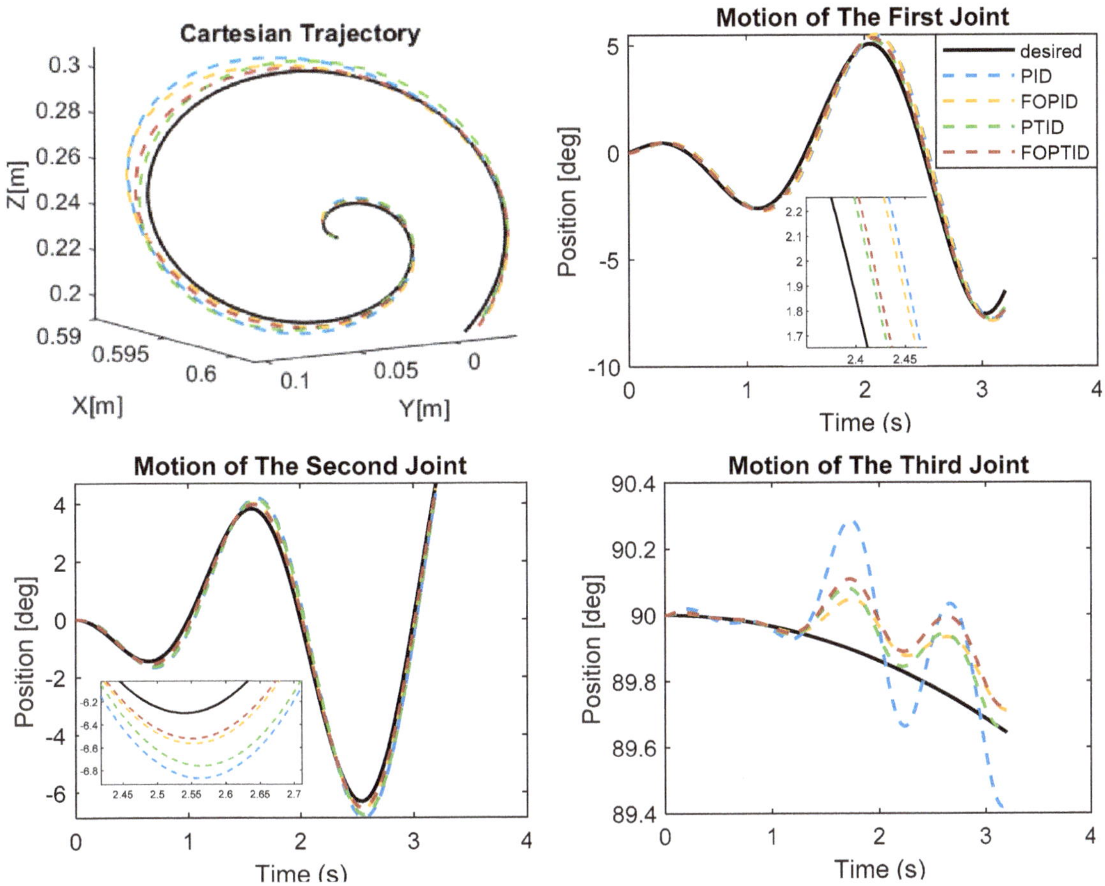

Figure 12. Trajectory tracking performance in task and joint space for PID, FOPID, PTID and FOPTID controller schemes.

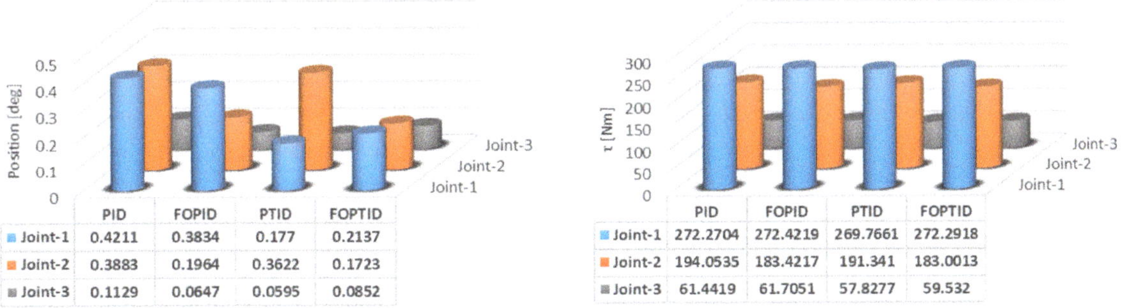

	PID	FOPID	PTID	FOPTID
Joint-1	0.4211	0.3834	0.177	0.2137
Joint-2	0.3883	0.1964	0.3622	0.1723
Joint-3	0.1129	0.0647	0.0595	0.0852

	PID	FOPID	PTID	FOPTID
Joint-1	272.2704	272.4219	269.7661	272.2918
Joint-2	194.0535	183.4217	191.341	183.0013
Joint-3	61.4419	61.7051	57.8277	59.532

Figure 13. Joint position MAE values and control signal RMS values.

5.3. Robustness Testing: Disturbance Rejection

To verify the disturbance rejection ability of the TID-based controllers, a sinusoidal torque signal is added to the control signal. This external disturbance, applied to each joint, is given as follows:

$$\tau_d(t) = \begin{cases} 0\ [Nm], & t < 2 \\ \begin{cases} 250\sin(t) + 250\ [Nm] \\ 350\sin(t) + 350\ [Nm], \\ 450\sin(t) + 450\ [Nm] \end{cases} & t \geq 2\ \text{for joint 1, 2 and 3, top-down.} \end{cases} \tag{31}$$

The desired path and actual path in joint space are shown in Figure 14 for each joint. Furthermore, the associated MAE values are illustrated in the same figure.

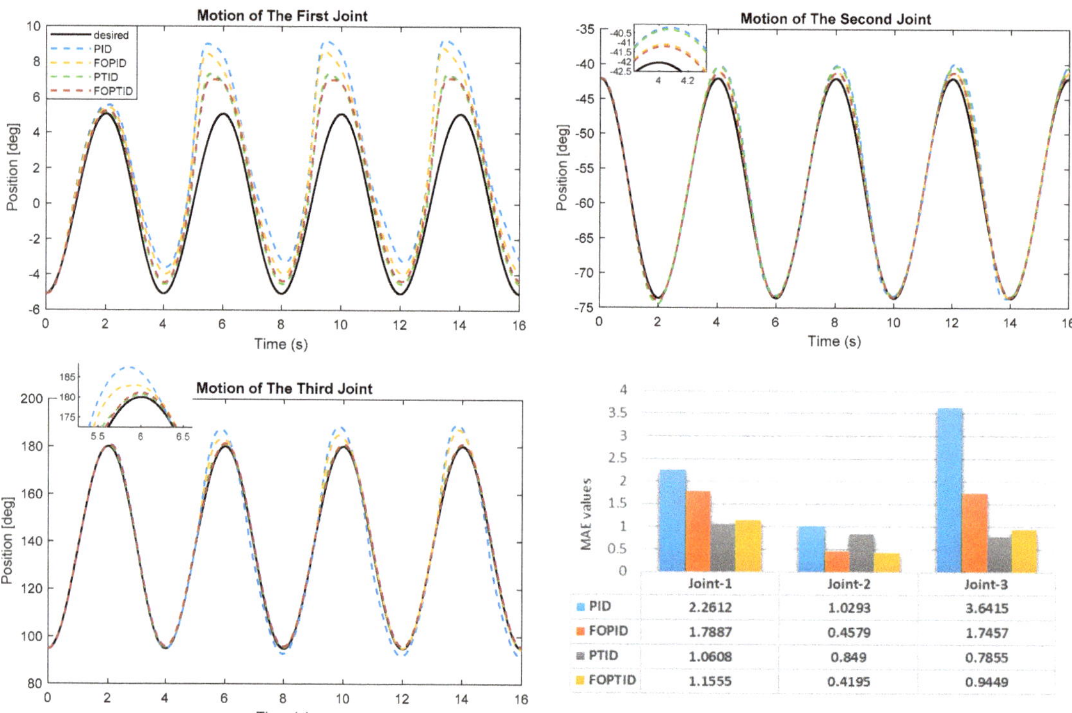

Figure 14. Comparison of disturbance rejection abilities and MAE values of joint angle tracking based on the designed controllers.

As can be seen from Figure 14, under sinusoid disturbance, better trajectory tracking is achieved with the TID-based controllers (PTID and FOPTID) as compared to the PID and FOPID controllers, which are quite obvious in trajectories and the amplitude of the heading angles. On the other hand, the difference between the trajectory tracking accomplished by means of the PTID and FOPTID controllers is almost small in each joint. Regarding MAE values demonstrated in Figure 14, the TID-based controllers produce a smaller tracking error in each joint than the other controllers. These simulation results reveal that although the applied perturbation directly affects the trajectory tracking error, the proposed robust FOPTID controller can exhibit better performance with a higher trajectory tracking accuracy against sinusoid disturbance than the other controllers and also maintain the error trajectories of each joint inside a compact set.

5.4. Robustness Testing: Friction Compensation

In order to demonstrate the robustness of the TID-based controllers and also compare them with other designed controllers in the presence of joint friction, the friction model, including Coulomb plus viscous friction, is adopted for each joint in the practical friction compensation of the 3-DOF robot manipulator. The friction parameters related to the Coulomb plus viscous friction model are given in Table 2 for each joint. The time profiles of joint positions and corresponding tracking errors based on PID, FOPID, PTID and FOPTID controllers are illustrated in Figure 15 for each joint under friction.

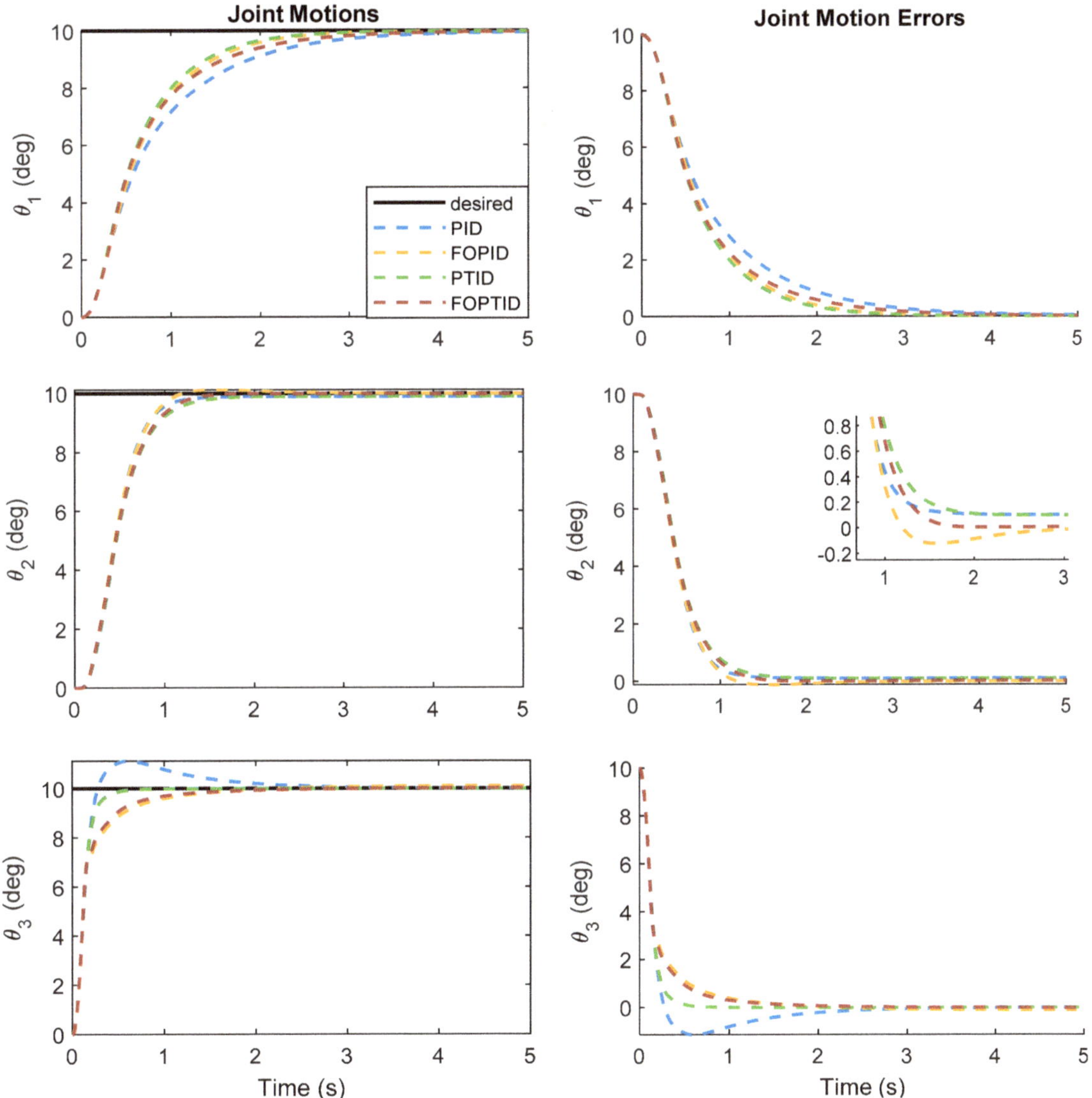

Figure 15. Joint tracking profiles and corresponding error profiles with respect to the presented controllers under friction.

Table 2. Coulomb plus viscous friction model parameters.

Friction Parameters	Joint-1	Joint-2	Joint-3	Unit
F_c	0.5	1.5	2.5	Nm
F_v	5.5	1.5	3.5	Nm/(rad/s)

In accordance with Figure 15, the convergence of the tracking error when using the PTID control scheme under friction for joints 1 and 3 is faster than the other three control schemes. Moreover, the TID-based controllers, which ensure the stability of the whole system, are robust against the defined frictions. The disturbance and joint tracking error performance of the FOPTID controller outperforms that of the FOPID controller for joint 2.

Numerical results related to the J_{ITAE} and MAE values with respect to the tracking errors of three joints are depicted in Figure 16 for the presented controllers under friction. From the figure, it can be observed that a remarkable performance is achieved by the TID-based controllers for a set point tracking task, in spite of added friction. Especially according to Figure 16, the PTID control strategy has good evaluation indicators with a smaller value of the J_{ITAE}. Moreover, compared with the PID, FOPID and FOPTID controller, the MAE values are reduced by 20.2% and 44.5% for joints 1 and 3, respectively. As for joint 2, the FOPID controller can decrease the MAE values by 12.1%. However, the MAE value of the FOPID controller is almost close to the value of the FOPTID controller. Therefore, the TID-based controllers can make the 3-DOF robotic arm achieve a good tracking effect in the presence of friction.

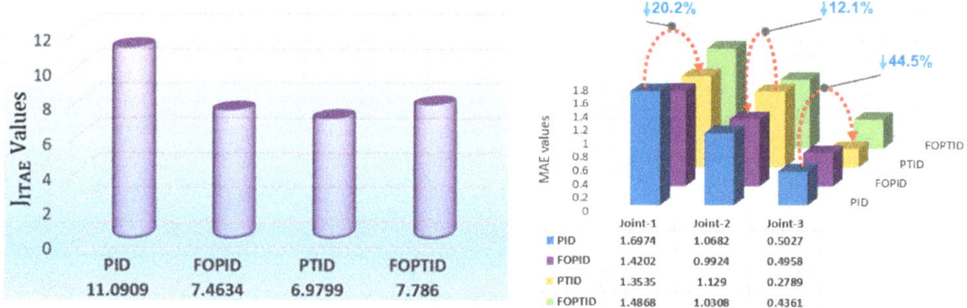

Figure 16. Comparison diagram of tracking error evaluation indicators based on J_{ITAE} and MAE values for the presented controllers.

For assessing the control effort of the tuned controllers in all robustness tests, the control efforts of each joint for the presented controllers are shown in Figure 17. In addition, RMS values of the control signals generated by the controllers are illustrated in Table 3.

Table 3. RMS values of the control signals for robustness tests.

Robustness Test	Joint	PID	FOPID	PTID	FOPTID
Different trajectory	Joint-1	272.2704	272.4219	269.7661	272.2918
	Joint-2	194.0535	183.4217	191.3410	183.0013
	Joint-3	61.4419	61.7051	57.8277	59.5320
Disturbance rejection	Joint-1	173.1580	154.1611	158.5644	157.0675
	Joint-2	328.8085	267.0632	245.4667	241.7147
	Joint-3	48.0855	39.8093	36.7112	35.8962
Friction compensation	Joint-1	110.4676	122.6951	124.7802	134.4499
	Joint-2	184.6817	141.4255	143.2402	141.6841
	Joint-3	68.1430	99.2728	74.6087	93.4725

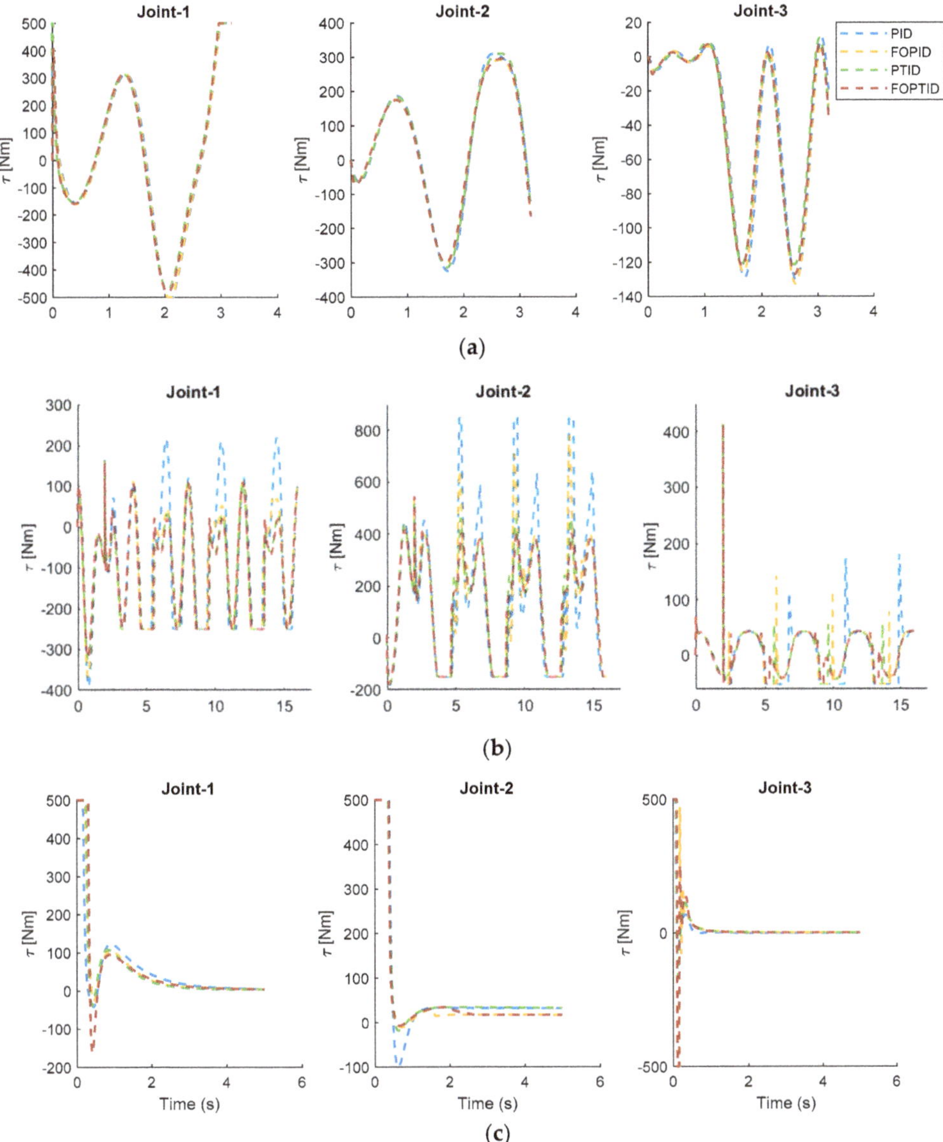

Figure 17. Simulation results of the tuned controllers for the torques of each joint based on the different trajectory (**a**), disturbance rejection (**b**) and friction compensation (**c**) tests.

As can be observed from Figure 17 and Table 3, the minimum RMS values are achieved by the FOPTID controller for all the two kinds of perturbations in the tests. On the other hand, for the friction compensation test, the FOPTID and PTID controllers attain much better tracking performance as compared to the others at the cost of a larger amplitude control signal. As a result, the TID-based proposed control approaches have exhibited considerably stable, robust, and superior tracking performance against different trajectories, disturbance and friction.

6. Conclusions

In this study, the 3-DOF robotic manipulator has been taken as the research object. The dynamic model of the robot manipulator has been presented and also the friction model has been added to the dynamics for robustness analysis. In order to enhance the trajectory tracking accuracy of the robot joint, TID-based control strategies such as PTID and FOPTID control techniques have been presented and also compared with PID based control strategies such as FOPID control in the case of different robustness tasks. Moreover, the TID and PID-based tracking controllers have been designed in joint space with the GWO–PSO algorithm to obtain the best controller parameters. Finally, different simulations have been performed to determine which controllers ensure directly that the actual joint trajectory can converge to the reference joint angles regardless of any disturbances and frictions.

The main outcomes of this study are stated as follows:

- TID-based controllers, as well as PID-based controllers, have been tuned by GWO–PSO with minimization of the objective function J_{ITAE} for the trajectory tracking control of the robot joints. Compared to the results from the tuned controllers, the proposed FOPTID control strategy achieved better performance than the other tuned controllers at the robot joints.
- For the purpose of observing the stability of the designed controllers, a different trajectory was applied to the robot joints. The simulation results showed that PTID and FOPTID control schemes can track the change in the joint angle more accurately and maintain stability as compared to PID and FOPID control schemes. As well, TID-based controllers required lesser applied torque for tracking the desired joint trajectories than the PID based controllers.
- As examined controller robustness in the presence of external disturbance applied to each joint, the proposed FOPTID controller was more capable of dealing with the disturbance in all joints during the reference trajectory tracking as compared to the PID, FOPID and PTID controllers. Accordingly, the effectiveness of the proposed controller was verified for disturbance rejection.
- As compared to the designed controllers in terms of reducing the effect of joint friction, a remarkable performance was achieved by both PTID and FOPTID for a set point tracking task. From the simulation results, it could be inferred that the TID-based control schemes have significantly reduced the means of absolute joint errors.

TID-based control strategies, which have received a considerable amount of interest and attracted the attention of many researchers because of their potential advantages and applications in many fields, will be considered as future research on the control of a real robotic manipulator.

Funding: This research received no external funding.

Data Availability Statement: No data was used for the research described in the article.

Conflicts of Interest: The author declares no conflict of interest.

References

1. Mirza, M.A.; Li, S.; Jin, L. Simultaneous learning and control of parallel Stewart platforms with unknown parameters. *Neurocomputing* **2017**, *266*, 114–122. [CrossRef]
2. Oustaloup, A. From fractality to non-integer derivation through recursivity, a property common to these two concepts: A fundamental idea from a new process control strategy. In Proceedings of the 12th IMACS World Congress, Paris, France, 18–22 July 1998.
3. Oustaloup, A.; Moreau, X.; Nouillant, M. The CRONE suspension. *Control Eng. Pract.* **1996**, *4*, 1101–1108. [CrossRef]
4. Oustaloup, A.; Sabatier, J.; Lanusse, P. From fractal robustness to CRONE control. *Fract. Calc. Appl. Anal.* **1999**, *2*, 1–30.
5. Podlubny, I. *Fractional-Order Systems and Fractional-Order Controllers*; Tech. Rep. UEF-03-94; Slovak Academy of Sciences Institute of Experimental Physics, Department of Control Engineering, University of Technology: Kosice, Slovakia, 1994.

6. Bingul, Z.; Karahan, O. Fractional PID controllers tuned by evolutionary algorithms for robot trajectory control. *Turk. J. Electr. Eng. Comput. Sci.* **2012**, *20*, 1123–1136.

7. Angel, L.; Viola, J. Fractional order PID for tracking control of a parallel robotic manipulator type delta. *ISA Trans.* **2018**, *79*, 172–188. [CrossRef]

8. Al-Mayyahi, A.; Aldair, A.A.; Chatwin, C. Control of a 3-RRR Planar Parallel Robot Using Fractional Order PID Controller. *Int. J. Autom. Comput.* **2020**, *17*, 822–836. [CrossRef]

9. Zhang, J.; Wang, Y.; Che, L.; Wang, N.; Bai, Y.; Wang, C. Workspace analysis and motion control strategy of robotic mine anchor drilling truck manipulator based on the WOA-FOPID algorithm. *Front. Earth Sci.* **2022**, *10*, 1253.

10. Sharma, R.; Gaur, P.; Mittal, A.P. Performance analysis of two-degree of freedom fractional order PID controllers for robotic manipulator with payload. *ISA Trans.* **2015**, *58*, 279–291. [CrossRef]

11. Dumlu, A.; Erenturk, K. Trajectory Tracking Control for a 3-DOF Parallel Manipulator Using Fractional-Order $PI^\lambda D^\mu$ Control. *IEEE Trans. Ind. Electron.* **2014**, *61*, 3417–3426. [CrossRef]

12. Lurie, B.J. Three-Parameter Tilt-Integral-Derivative (TID). U.S. Patent 5,371,670, 6 December 1994.

13. Sain, D.; Swain, S.K.; Mishra, S.K. TID and I-TD controller design for magnetic levitation system using genetic algorithm. *Perspect. Sci.* **2016**, *8*, 370–373. [CrossRef]

14. Sahu, R.B.; Panda, S.; Biswal, A.; Sekhar, G.T.C. Design and analysis of tilt integral derivative controller with filter for load frequency control of multi-area interconnected power systems. *ISA Trans.* **2016**, *61*, 251–264. [CrossRef] [PubMed]

15. Gnaneshwar, K.; Padhy, P.K. Robust Design of Tilted Integral Derivative Controller for Non-integer Order Processes with Time Delay. *IETE J. Res.* **2021**. [CrossRef]

16. Bhuyan, M.; Das, D.C.; Barik, A.K.; Sahoo, S.C. Performance Assessment of Novel Solar Thermal-Based Dual Hybrid Microgrid System Using CBOA Optimized Cascaded PI-TID Controller. *IETE J. Res.* **2022**. [CrossRef]

17. Xue, D.; Chen, Y. A comparative introduction of four fractional order controllers. In Proceedings of the 4th World Congress on Intelligent Control and Automation, Shanghai, China, 10–14 June 2002; Volume 4, pp. 3228–3235.

18. Morsali, J.; Zare, K.; Hagh, M.T. Comparative performance evaluation of fractional order controllers in LFC of two-area diverse-unit power system with considering GDB and GRC effects. *J. Electr. Syst. Inf. Technol.* **2018**, *5*, 708–722. [CrossRef]

19. Topno, P.N.; Chanana, S. Differential evolution algorithm-based tilt integral derivative control for LFC problem of an interconnected hydro-thermal power system. *J. Vib. Control* **2018**, *24*, 3952–3973. [CrossRef]

20. Sharma, M.; Prakash, S.; Saxena, S.; Dhundhara, S. Optimal fractional-order tilted-integral-derivative controller for frequency stabilization in hybrid power system using salp swarm algorithm. *Electr. Power Compon. Syst.* **2021**, *48*, 1912–1931. [CrossRef]

21. Sharma, M.; Prakash, S.; Saxena, S. Robust Load Frequency Control Using Fractional-order TID-PD Approach via Salp Swarm Algorithm. *IETE J. Res.* **2021**. [CrossRef]

22. Lu, C.; Tang, R.; Chen, Y.Q.; Li, C. Robust tilt-integral-derivative controller synthesis for first-order plus time delay and higher-order systems. *Int. J. Robust Nonlinear Control* **2023**, *33*, 1566–1592. [CrossRef]

23. Mohamed, E.A.; Ahmed, E.M.; Elmelegi, A.; Aly, M.; Elbaksawi, O.; Mohamed, A.A. An Optimized Hybrid Fractional Order Controller for Frequency Regulation in Multi-Area Power Systems. *IEEE Access* **2020**, *8*, 213899–213915. [CrossRef]

24. Ahmed, E.M.; Mohamed, E.A.; Elmelegi, A.; Aly, M.; Elbaksawi, O. Optimum Modified Fractional Order Controller for Future Electric Vehicles and Renewable Energy-Based Interconnected Power Systems. *IEEE Access* **2021**, *9*, 29993–30010. [CrossRef]

25. Choudhary, R.; Rai, J.N.; Arya, Y. Cascade FOPI-FOPTID controller with energy storage devices for AGC performance advancement of electric power systems. *Sustain. Energy Technol. Assess.* **2022**, *53 Pt C*, 102671. [CrossRef]

26. Yanmaz, K.; Mengi, O.O.; Sahin, E. Advanced STATCOM Control with the Optimized FOPTID-MPC Controller. *IETE J. Res.* **2022**. [CrossRef]

27. Bingül, Z.; Karahan, O. Dynamic identification of Staubli RX-60 robot using PSO and LS methods. *Expert Syst. Appl.* **2011**, *38*, 4136–4149. [CrossRef]

28. Ortigueira, M.D. *Fractional Calculus for Scientists and Engineers*; Springer: Berlin, Germany, 2011.

29. Oustaloup, A.; Levron, F.; Mathieu, B.; Nanot, F.M. Frequency-band complex noninteger differentiator: Characterization and synthesis. *IEEE Trans. Circuits Syst.-I Fundam. Theory Appl.* **2000**, *47*, 25–39. [CrossRef]

30. Hegedus, E.T.; Birs, I.R.; Ghita, M.; Muresan, C.I. Fractional-Order Control Strategy for Anesthesia–Hemodynamic Stabilization in Patients Undergoing Surgical Procedures. *Fractal Fract.* **2022**, *6*, 614. [CrossRef]

31. Behera, M.K.; Saikia, L.C. Anti-windup filtered second-order generalized integrator-based spontaneous control for single-phase grid-tied solar PV-H_2/Br_2 redox flow battery storage microgrid system. *J. Energy Storage* **2022**, *55B*, 105551. [CrossRef]

32. Ramoji, S.K.; Saikia, L.C. Optimal Coordinated Frequency and Voltage Control of CCGT-Thermal Plants with TIDF Controller. *IETE J. Res.* **2021**. [CrossRef]

33. Kennedy, J.; Eberhart, R.C. Particle swarm optimization. In Proceedings of the IEEE International Conference on Neural Networks, Perth, Australia, 27 November–1 December 1995; Volume IV, pp. 1942–1948.

34. Mirjalili, S.; Mirjalili, S.M.; Lewis, A. Grey Wolf Optimizer. *Adv. Eng. Softw.* **2014**, *69*, 46–61. [CrossRef]

35. Shaheen, M.A.M.; Hasanien, H.M.; Alkuhayli, A. A novel hybrid GWO-PSO optimization technique for optimal reactive power dispatch problem solution. *Ain Shams Eng. J.* **2021**, *12*, 621–630. [CrossRef]
36. Valério, D.; Da Costa, J.S. NINTEGER: A non-integer control toolbox for MATLAB. In Proceedings of the Fractional Differentiation and Its Applications, Bordeaux, France, 19–20 July 2004.

 *fractal and fractional*

Article

Fuzzy Fractional Order PID Tuned via PSO for a Pneumatic Actuator with Ball Beam (PABB) System

Mohamed Naji Muftah [1,2,*], Ahmad Athif Mohd Faudzi [1,3,*], Shafishuhaza Sahlan [1,3] and Shahrol Mohamaddan [4]

1   Faculty of Electrical Engineering, Universiti Teknologi Malaysia, Skudai 81310, Malaysia
2   Department of Control Engineering, College of Electronics Technology, Bani Walid P.O. Box 38645, Libya
3   Centre for Artificial Intelligence and Robotics (CAIRO), Universiti Teknologi Malaysia,
    Kuala Lumpur 51400, Malaysia
4   College of Systems Engineering and Science, Shibaura Institute of Technology, Saitama 337-8570, Japan
*   Correspondence: namohamed@graduate.utm.my (M.N.M.); athif@utm.my (A.A.M.F.)

Abstract: This study aims to improve the performance of a pneumatic positioning system by designing a control system based on Fuzzy Fractional Order Proportional Integral Derivative (Fuzzy FOPID) controllers. The pneumatic system's mathematical model was obtained using a system identification approach, and the Fuzzy FOPID controller was optimized using a PSO algorithm to achieve a balance between performance and robustness. The control system's performance was compared to that of a Fuzzy PID controller through real-time experimental results, which showed that the former provided better rapidity, stability, and precision. The proposed control system was applied to a pneumatically actuated ball and beam (PABB) system, where a Fuzzy FOPID controller was used for the inner loop and another Fuzzy FOPID controller was used for the outer loop. The results demonstrated that the intelligent pneumatic actuator, when coupled with a Fuzzy FOPID controller, can accurately and robustly control the positioning of the ball and beam system.

Keywords: IPA system; system identification technique; Fuzzy FOPID controller; PSO algorithm; PABB system

Citation: Muftah, M.N.; Faudzi, A.A.M.; Sahlan, S.; Mohamaddan, S. Fuzzy Fractional Order PID Tuned via PSO for a Pneumatic Actuator with Ball Beam (PABB) System. *Fractal Fract.* **2023**, *7*, 416. https://doi.org/10.3390/fractalfract7060416

Academic Editor: António Lopes

Received: 31 March 2023
Revised: 19 May 2023
Accepted: 20 May 2023
Published: 23 May 2023

1. Introduction

The pneumatic system is a commonly used actuator in industrial automation, offering benefits such as affordability, natural cooling, environmental safety, and simplicity [1,2]. These systems have a broad range of applications, from simple processes to complex ones, such as those found in production lines, aeronautics, and the automotive industry [3,4]. Their popularity is due to their durability, ease of maintenance, and safety [5]. However, pneumatic systems have a significant downside, namely, their non-linear behavior due to the compressibility of air, friction between the piston and cylinder, and discontinuous flow through control valves [6,7]. Additionally, modeling these systems dynamically is challenging because their air dynamics are often based on empirical assumptions [8]. Achieving precise positioning of pneumatic actuators is also challenging. To expand their range of uses, pneumatic systems must possess the ability to achieve rapid response times and precise positioning control.

System identification (SI) differs from the theoretical approach by relying on observational analysis, rather than fundamental laws of nature, to determine its concepts. SI is a method that can be used to model systems and estimate unknown parameters, as well as to linearize systems to mitigate the limitations of mathematical models [8]. Moreover, this approach proves particularly suitable for complex systems or processes, especially within real-world, practical settings. The goal of SI is to develop a mathematical model that can describe the behavior of a system based on measured input–output data. The process of SI typically involves collecting input–output data from the system and then using these data to estimate the model parameters. The estimated model can then be used to analyze the system's behavior, predict its response to new inputs, or design a controller to achieve desired performance. The models used in SI can range from simple linear models to more complex non-linear models, and the accuracy of the model depends on the quality and quantity of the input–output data and the complexity of the model. SI is widely used in various fields, including engineering, economics, and biology, to model and control complex systems. It is important to consider the constraints and limitations of the system when designing a controller for practical applications to avoid damaging the system or its components, as well as reducing the control system's performance.

Numerous researchers have been conducting extensive studies in this field, focusing on developing different control strategies for achieving precise pneumatic motion control. These strategies include proportional-integral-derivative (PID) control [1,9,10], sliding mode control (SMC) [11,12], adaptive control [13], fuzzy control [14], and predictive control [15,16].

For this research, the system identification (SI) approach was utilized to obtain the model of the pneumatic system. Additionally, a new control method called fuzzy fractional order proportional integral derivative (Fuzzy FOPID) controller was proposed. The study demonstrated that fractional proportional-integral-derivative (FOPID) controllers offer greater accuracy and flexibility in feedback system adjustment, which can be used to meet more rigorous specifications related to stability phase, gain margins, maximum sensitivity, and performance set point tracking and load disturbance rejection than what is achievable with the conventional PID controller [17–19]. Many researchers have adopted FOPID controllers in recent years because they provide additional features that enhance the durability and success of the system in various applications. Moreover, the investigation results revealed that FOPID is used as a controller in many systems such as motor control systems [20], robotics systems [21], and time-delay systems [22].

Various optimization algorithms have been proposed for tuning controller parameters, including Genetic Algorithm (GA) [23], Cuckoo Search Algorithm (CS) [24], Grey Wolf Optimization (GWO) [25], Gradient-Based Optimization (GBO) [26], and Particle Swarm Optimization (PSO) [27]. GA is a population-based search algorithm inspired by biological evolution. CS is a population-based search algorithm inspired by the behavior of cuckoo birds. GWO is a population-based search algorithm inspired by the social hierarchy of gray wolves. SCA is a population-based search algorithm that simulates the sine and cosine functions. GBO uses the gradient of the cost function to iteratively update the parameter values. PSO is a population-based search algorithm that simulates the movement and interaction of particles. These optimization algorithms have been applied to various control systems and their effectiveness in improving control performance has been demonstrated. However, the choice of algorithm depends on the specific characteristics of the system and the control objectives, and a combination of different algorithms may be necessary to achieve the desired performance. Among these algorithms, PSO was selected for controlling the converter in this study. PSO is inspired by the dynamics of animals moving in groups and builds a solution to the problem by simulating swarm communications [28]. When continuous variables are present, the PSO algorithm presents an effective solution for optimization problems [29].

This research introduces a unique approach to designing a smart pneumatic actuator system through the use of a fuzzy logic control structure. The primary goals and contributions of this study include:

- Developing a two-input-one-output fuzzy controller for the intelligent pneumatic actuator system and assessing its performance in the positioning system. This design incorporates FOPID, which is connected to the output terminal of the fuzzy controller to produce the proposed Fuzzy FOPID controller.
- Utilizing the Particle Swarm Optimization (PSO) technique to identify the optimal values for the suggested controller parameters. Seven parameters are adjusted to achieve the best dynamic behavior for the Fuzzy FOPID controller.

- Validating the superiority of the proposed design by comparing the results obtained from simulations and real-world environments with those of the Fuzzy FOPID.
- Developing a Pneumatic Actuated Ball and Beam System and implementing the Fuzzy FOPID controller on the system.
- Validating the performance of the position controller through both simulation and real-time experiments.

The article is organized into six principal sections. In the second section, the authors describe the process of modeling the IPA system and PABB system. The third section provides a detailed overview of the inner loop controller designs, including Fuzzy PID and Fuzzy FOPID, as well as outer loop designs such as FOPI-FOPD and Fuzzy FOPID. The fourth section of the article focuses on the PSO algorithm. In Section 5, the authors present and analyze the results obtained from the simulation and real-time experiments. Lastly, Section 6 offers a summary of the conclusions drawn from the study.

2. System Modelling

This research aims to create two plant designs: the Intelligent Pneumatic Actuator (IPA) and the ball and beam (BB) system. The IPA model was built using the system identification approach, while the BB model was constructed using mathematical models. These two designs will be utilized in the development of an Intelligent Pneumatic Actuated Ball and Beam System, referred to as IPABBS.

2.1. Intelligent Pneumatic Actuator (IPA) System
2.1.1. IPA Experimental Setup

The setup for the pneumatic cylinder used in this research is shown in Figure 1. The cylinder consists of a guiding rod, an optical encoder, a pressure sensor, and two on/off solenoid valves labeled *V1* and *V2*. The cylinder has a precision of 0.09 mm and can extend up to 200 mm in length. Its operating pressure is 0.6 MPa, with only one chamber controlling the cylinder while the second chamber maintains a constant pressure of 0.6 MPa. Table 1 provides a summary of the different movements that the cylinder can perform.

The movements of the cylinder stroke are based on the conditions of two solenoid valves (*V1* and *V2*). When both valves are off, the cylinder stops. When V1 is off and V2 is on, the cylinder retracts. Conversely, when *V1* is on and *V2* is off, the cylinder extends. Finally, when both valves are on, the cylinder does not move.

The DAQ system and SHC68-68-EPM cable are used to establish communication between the personal computer and the pneumatic actuator system, as depicted in Figure 2.

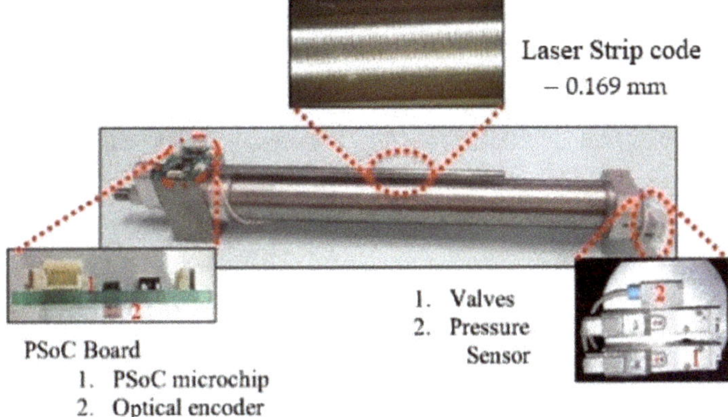

Figure 1. Pneumatic cylinder parts.

Table 1. The PABBS parameters.

Quantity	Value
Beam Length (l)	0.5 m
Pneumatic Actuator Stroke Length (h)	0–200 mm
Angle (α)	Depends on h
The Ball Mass (m)	0.04012 kg
The Ball Radius (R)	0.0107 m
Ball's Moment of Inertia (J)	1.8373 e^{-6}
Gravitational Acceleration (g)	9.8 ms^{-2}

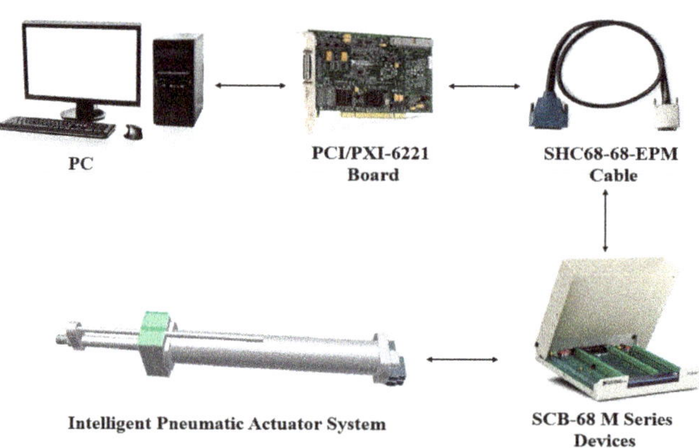

Figure 2. The experimental setup for the IPA system.

2.1.2. System Identification of IPA

To develop an accurate mathematical model of the pneumatic system, a system identification technique was employed in this study. Data were collected through experimentation, resulting in 1600 measurements of input and output data with a sample time (ts) of 10 ms. The collected data was divided into two sets of 800 samples each, with the first set being used for training and the second set for validation. Figure 3 illustrates the plot of input and output data obtained from the real-time experiment.

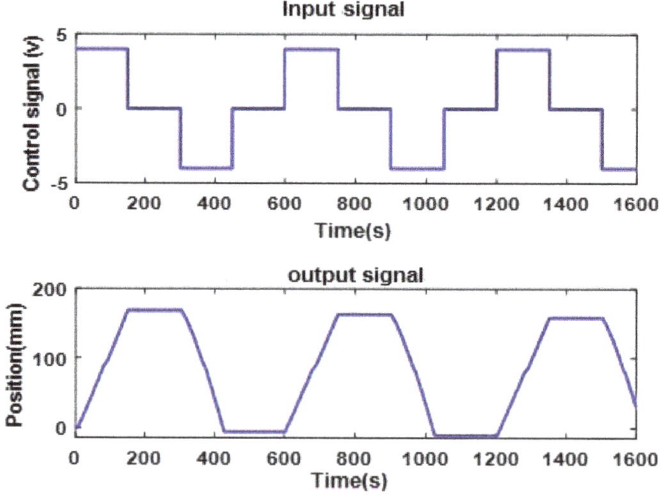

Figure 3. The measured input and output data of the system.

To represent the real system in this study, the ARX331, a third-order linear Autoregressive with Exogenous Input (ARX) model with the order na = 3, nb = 3, and nk = 1, was used. Equation (1) presents the discrete state space equation of the linear third-order ARX.

$$A = \begin{bmatrix} 1.555 & -0.3957 & -0.1593 \\ 1 & 0 & 0 \\ 0 & 1 & 0 \end{bmatrix} \quad B = \begin{bmatrix} 1 \\ 0 \\ 0 \end{bmatrix} \tag{1}$$

$$C = \begin{bmatrix} 0.008 & 0.002 & -0.0012 \end{bmatrix} \quad D = 0$$

Figure 4 displays a comparison between the measured values of the system (represented by a black line) and the output of the simulation model (represented by a blue line). The simulation model was generated using the System Identification Toolbox and has a best fit of 90.75%. The remaining loss of 9.25% could be attributed to factors such as dead zone, air leakage, and friction present in the pneumatic system. The model plant is deemed acceptable since all its poles and zeros are located within the unit circle, as illustrated in Figure 5. Thus, the model is stable and capable of delivering good performance.

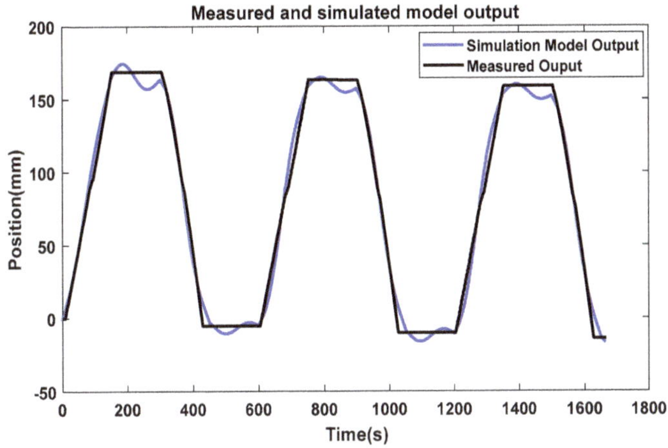

Figure 4. The measured and simulated model output.

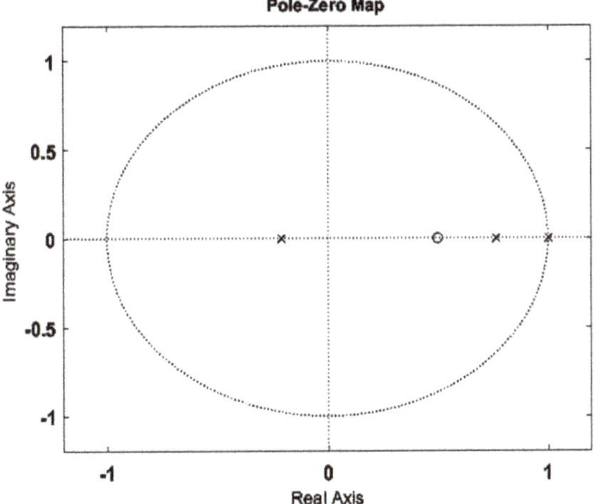

Figure 5. The zeros–poles plot for the model.

2.2. Pneumatic Actuated Ball and Beam (PABB) System

2.2.1. Mathematical Model of Pneumatic Actuated Ball and Beam (PABB) System

The objective of the research is to achieve precise ball placement by regulating the stroke length of the Intelligent Pneumatic Actuator (IPA). In instances where the ball is in an unstable state, adjusting the angle of the beam by moving the pneumatic actuator helps to stabilize the ball. The ball's position is determined by the voltage reading of the resistance sensor, while the beam's angle, which depends on the pneumatic actuator stroke, is determined by the encoder's position. However, controlling the velocity and acceleration of the ball is challenging due to the friction coefficient between the ball and the beam, and directly controlling the stroke of the PA is difficult due to its nonlinearity.

To develop an appropriate controller for the system, it is necessary to derive the system's dynamics equation. Figure 6 illustrates that torque is applied through the right pneumatic actuator at the pivot on the left end, causing the beam to rotate vertically along the *y*-axis. The ball moves horizontally along the *x*-axis as the beam moves up and down.

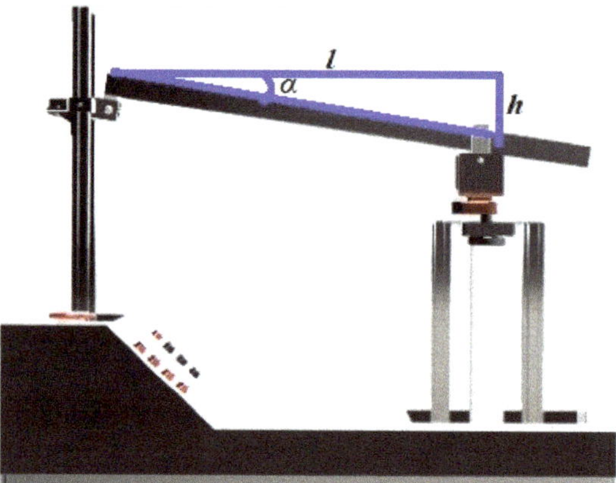

Figure 6. The PABB System.

To streamline and make the model more manageable, all frictional forces have been disregarded. Furthermore, it has been presumed that the ball and the beam remain in constant contact and that there is no slipping during the ball's rolling on the beam. The recommended system parameters can be found in Table 1.

The Lagrangian method has been widely used in model-based research on ball and beam systems with motors [30,31], and it is also used in this study to derive the equation of motion for the ball and beam system. By neglecting friction forces and assuming continuous contact between the ball and beam with no slippage, the resulting Lagrangian equation of motion for the ball can be expressed as follows:

$$\left(\frac{J_b}{r^2} + m\right)\ddot{x} + mg\sin\alpha - mx\left(\dot{\alpha}\right)^2 = 0 \tag{2}$$

Linearization of this equation about the beam angle, $\alpha \approx 0$, gives the following linear approximation of the system:

$$\left(\frac{J_b}{r^2} + m\right)\ddot{x} = -mg\alpha \tag{3}$$

The beam angle can be expressed as in Equation (4).

$$\alpha = \sin^{-1}\frac{h}{l} \tag{4}$$

Equation (4) is linearized using a simple approach where the values of h and l are already known. Figure 7 shows the graph of all possible values obtained by substituting the given values into Equation (4).

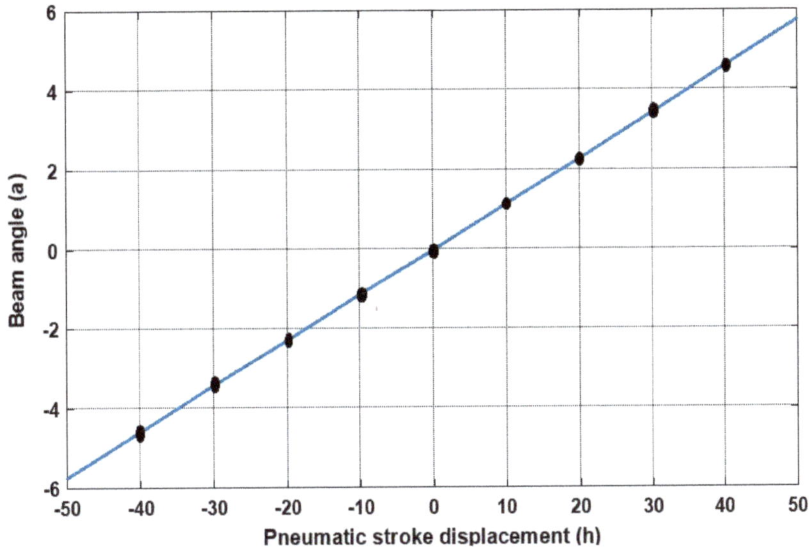

Figure 7. Beam angle vs. Pneumatic stroke displacement graph.

By applying the equation of a straight line to the data presented in Figure 7, one obtains the following equation.

$$\alpha = 0.115\,h \tag{5}$$

Then, by substituting Equation (5) into Equation (3), we obtain

$$\left(\frac{J_b}{r^2} + m\right)\ddot{x} = -0.115\,mgh \tag{6}$$

Taking the Laplace transform of Equation (6), we find

$$\left(\frac{J_b}{r^2} + m\right)X(s)s^2 = -0.115\,mg\,H(s) \tag{7}$$

Rearrange Equation (7), and the transfer function from the pneumatic actuator (H) to the ball position (X) will be obtained.

$$\frac{X(s)}{H(s)} = \frac{-0.115\,mg}{\left(\frac{J_b}{r^2} + m\right)s^2} \tag{8}$$

2.2.2. PABB Experimental Setup

Figure 8 shows the experimental setup designed for the IPA application. The setup consists of several components, including the PABBS structure, compressor, servo-pneumatic actuator, pressure regulator, stainless-steel ball, position sensor (to measure the location of the ball), DAQ card for communication, and PC with MATLAB software.

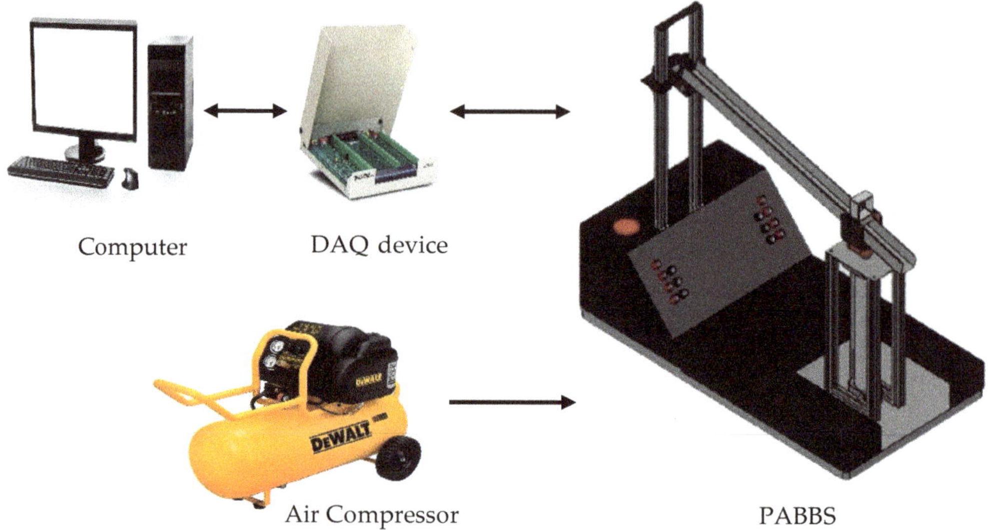

Figure 8. The experimental setup for the PABBS system.

3. Controller Design

This section addresses the controllers' design for both the IPA plant and the ball and beam plant. The control system design for the proposed system comprises two feedback loops: one for the IPA (inner loop) and another for ball position control (outer loop). The inner loop is responsible for controlling the pneumatic stroke length, h, which in turn adjusts the beam angle, α. The inner loop controller is designed with the objective of precisely controlling the IPA's position. On the other hand, the outer loop utilizes the inner feedback loop to regulate the ball's position.

3.1. Intelligent Pneumatic Actuator Controller Design—Inner Loop

This study utilized Fuzzy FOPID and Fuzzy PID controllers to control the intelligent pneumatic actuator, which relies on fuzzy logic control implemented using the fuzzy logic toolbox within the MATLAB/Simulink platform. The fundamental components of fuzzy logic, as presented in Figure 9, include the fuzzifier, rule base, inference engine mapping, and de-fuzzifier [32].

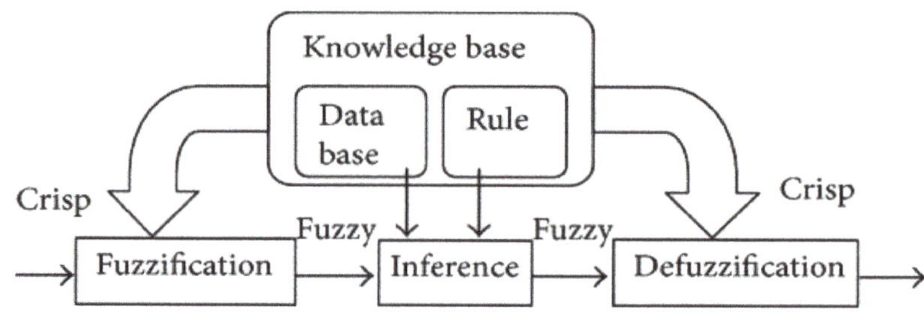

Figure 9. The basic structure of the fuzzy controller.

Figure 10 displays the use of five triangular membership functions, namely Large Negative (LN), Small Negative (SN), Zero (Z), Small Positive (SP), and Large Positive (LP), for input 1 and input 2. The range of MF for input 1 is between −10 and 10, and for input 2 is between −5 and 5, respectively. The output for the fuzzy design is singular, with a linear value, and each variable's value is V2 = −255, V2k = −100, off = 0, V1k = 200, and

V1 = 255. Table 2 shows that 25 rule bases are required to generate the controller's fuzzy output. These rules are derived from a detailed analysis of the dynamic behavior of the pneumatic actuator under investigation because the controller's performance is dependent on them. Furthermore, this design uses the Sugeno-type inference system for fuzzification and the Centroid tool for defuzzification. Figure 11 displays the surface viewer of the Fuzzy controller.

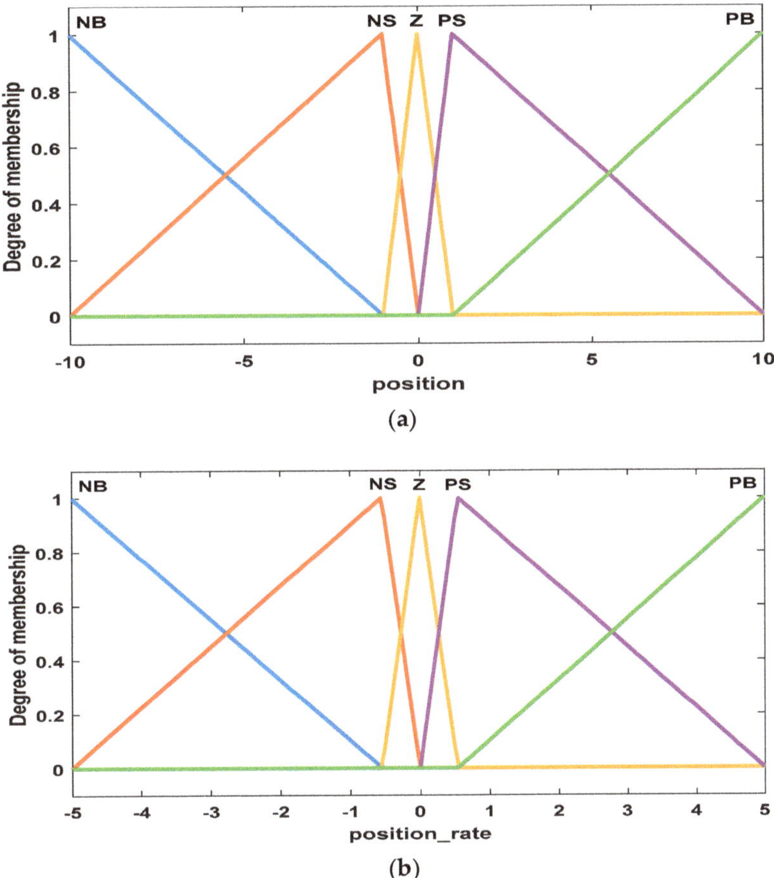

Figure 10. Membership function of (**a**) input1, (**b**) input2.

Table 2. Linguistic rules of the fuzzy controller design.

Error rate,	Error, e(t)				
Δe(t)	NB	NS	Z	PS	PB
NB	NB	NB	NB	PB	PB
NB	NB	NS	NS	PS	PB
NB	NB	NS	Z	PS	PB
NB	NB	NS	PS	PS	PB
NB	NB	NB	PB	PB	PB

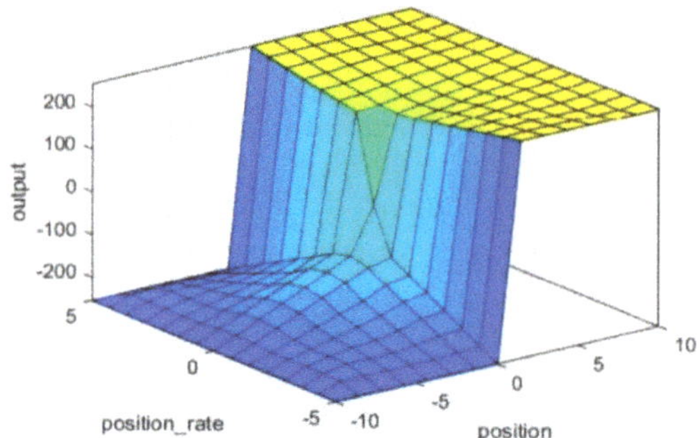

Figure 11. The Fuzzy control surfaces.

3.1.1. The Fuzzy PID (Fuzzy PID) Controller

The PID controller has remained a popular choice in recent times because of its ease of development and installation, as well as its ability to perform well in the presence of system uncertainties [33]. The transfer function for the PID controller can be expressed as shown in Equation (9):

$$C(s) = \frac{U(s)}{E(s)} = K_p + \frac{K_i}{S} + K_d S \tag{9}$$

The system takes three input signals: the error signal, the error derivative signal, and the controller control signal. The error signal is generated by comparing the desired and actual positions of the pneumatic stroke. The error derivative signal is produced by differentiating the error signal, and both signals are inputted into the fuzzy logic controller block and PID controller. The control signal generated by the Fuzzy PID controller is then fed into the pneumatic actuator block to adjust the position of the pneumatic stroke.

The Fuzzy PID controller includes five parameters: K_p, K_i, K_d, K_1, and K_2. The PID parameters (K_p, K_i, and K_d) determine the proportional, integral, and derivative gains of the controller. K_1 and K_2 are the error gain and error rate gain, respectively, which are used to scale the input signals to the fuzzy logic controller block. These parameters can be adjusted to customize the performance of the controller according to the specific requirements of the system. Figure 12 illustrates the structure of a Fuzzy PID controller.

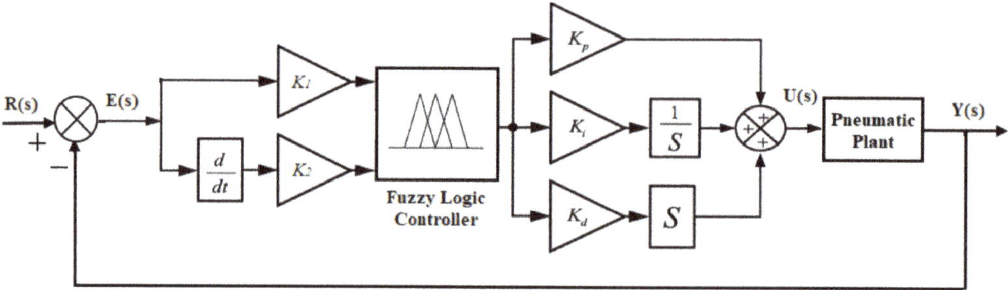

Figure 12. The structure of a Fuzzy PID controller.

3.1.2. The Fuzzy Fractional Order PID (Fuzzy FOPID) Controller

The Fuzzy Fractional Order PID (Fuzzy FOPID) controller is a type of controller used for regulating the position tracking of systems. It combines fuzzy logic with fractional order PID (FOPID) control to improve the system's performance. FOPID control is a type of PID control that uses fractional order calculus to define the proportional, integral, and derivative terms [17]. The fractional order calculus provides more flexibility in designing the controller to meet specific system requirements. The transfer function for the FOPID controller can be expressed as shown in Equation (10) [18]:

$$C(s) = \frac{U(s)}{E(s)} = K_p + \frac{K_i}{S^\lambda} + K_d\,S^\mu \tag{10}$$

The Fuzzy FOPID controller is characterized by seven unknown parameters, namely K_p, K_i, K_d, λ, μ, K_1, and K_2. K_p represents the proportional gain, K_i represents the integral gain, K_d represents the derivative gain, λ represents the fractional-order integral, and μ represents the fractional-order derivative. K_1 and K_2 are the error gain and error rate gain, respectively, which scale the input signals to the fuzzy logic controller block. The architecture of a Fuzzy FOPID controller is depicted in Figure 13.

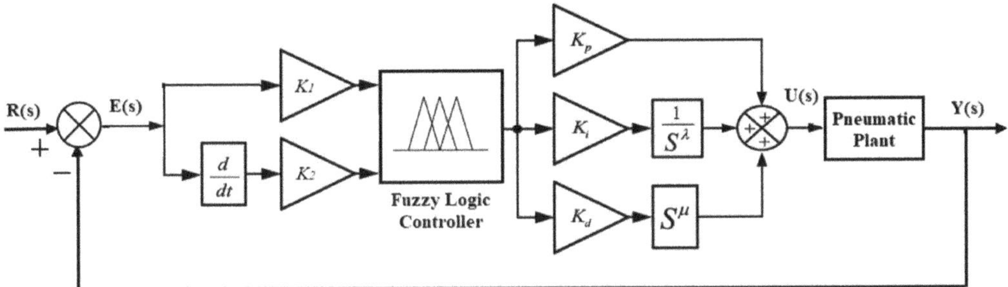

Figure 13. The structure of the Fuzzy FOPID controller.

The effectiveness of the Fuzzy FOPID controller is verified through simulations and real-time experiments. These tests demonstrate that the controller can effectively regulate the position tracking of the system and improve its performance compared to traditional PID controllers. The Fuzzy FOPID controller can be customized to meet the specific requirements of the system by adjusting the scales of fuzzy logic and FOPID.

3.2. PABBS Controller Design—Outer Loop

In this study, two types of controllers were used to regulate the PABB system: Fractional-Order PI–Fractional-Order PD (FOPI-FOPD) controller and Cascade Fuzzy FOPID (CF-FOPID) controller.

3.2.1. Fractional-Order PI–Fractional-Order PD (FOPI-FOPD) Controller Design

The structure proposed in this paper for the FOPI-FOPD controller is depicted in Figure 14. The controller consists of two fractional-order controllers, namely FOPI and FOPD, which are cascaded in series. The control signal for the system is given by Equation (11):

$$C(s) = \frac{U(s)}{E(s)} = \left(K_p + \frac{K_i}{S^\lambda}\right)\cdot\left(K_{p1} + K_d\,S^\mu\right) \tag{11}$$

In order to improve the controller's performance by reducing errors and enhancing transient responsiveness, six parameters need to be adjusted: K_p, K_i, λ, $Kp1$, K_d, and μ. K_p, K_i, and λ represent the proportional gain, integral gain, and fractional-order integral for the FOPI controller, while K_{p1}, K_d, and μ are the proportional gain, derivative gain, and fractional-order derivative, respectively, for the FOPD controller. Figure 14 illustrates the structure of a FOPI-FOPD controller. Adjusting these parameters allows for the optimization of the controller's response to changes in the system's behaviour.

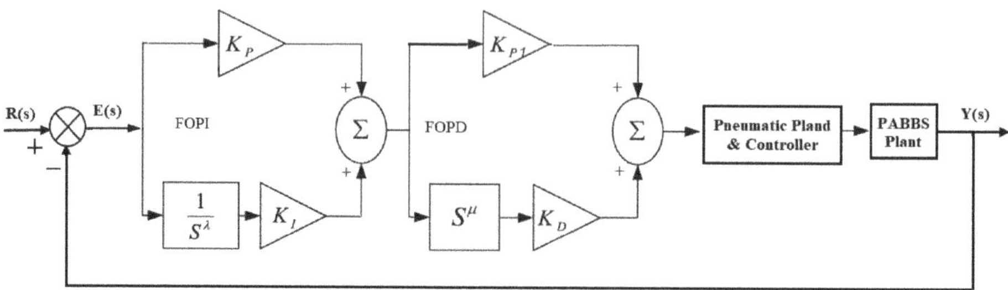

Figure 14. The structure of the FOPI-FOPD controller.

3.2.2. Cascade Fuzzy Fractional-Order PID (CFFOPID) Controller Design

Cascade Fuzzy Fractional-Order PID controller design is a control strategy that combines fuzzy logic with fractional-order calculus to achieve more accurate and flexible control of complex systems. The controller is designed using a cascade structure that allows for the division of the control problem into simpler sub-problems. The basic structure of a CFFOPID controller consists of two stages: the fuzzy logic controller (FLC) stage and the fractional-order PID (FOPID) controller stage. Figure 15 illustrates the structure of a CFFOPID controller.

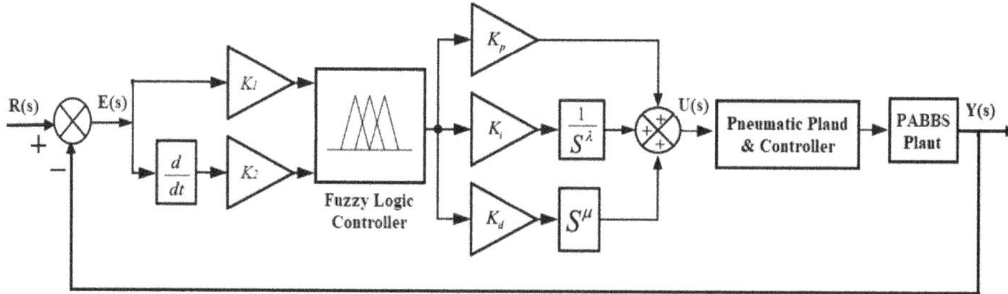

Figure 15. The structure of the CFFOPID controller.

The CFFOPID controller design involves adjusting several parameters, including the gains for the FLC stage, the gains for the FOPID stage, and the fractional-order parameters for the integral and derivative terms.

4. Particle Swarm Optimization Algorithm

In 1995, James and Russell proposed the Particle Swarm Optimization (PSO) algorithm, which is inspired by the collective behavior of birds and has a random probability distribution [34]. PSO is a powerful optimization technique that is particularly effective for solving nonlinear optimization problems. In 1998, an improved version of PSO was introduced by adding an inertia weight coefficient to enhance its performance [35]. PSO is a rule-based algorithm that incorporates both the individual and collective behavior of birds [29,36].

In this approach, every particle in the swarm searches for the best position by continuously updating its location based on its knowledge of the best position it has found so far, as well as the global best position within the swarm. This method is formulated using Equations (12) and (13) for the optimization process [37].

$$v_i(k+1) = W\,v_i(k) + C_1 R_1(g_{best} - x_i(k)) + C_2 R_2(p_{best} - x_i(k)) \tag{12}$$

$$x_i(k+1) - x_i(k) + v_i(k+1) \tag{13}$$

$$i = 1, 2, \ldots, n$$

where v_i is the *i*th particle velocity, x_i is the *i*th particle position, k is the iteration number, C_1 and C_2 are the cognitive and social coefficients, w is the inertia weight factor, R_1 and R_2 are random variables of from 0 to 1, p_{best}, i is the individual best position of particle i, g_{best} is the best global position of all the particles in the swarm, and n is the number of birds (particles).

If the condition in (14) is met, then the position is updated through (15):

$$f(x_{ik}) < f(p_{best}) \tag{14}$$

$$x_{ik} = p_{best} \tag{15}$$

where f performs the minimization objective fitness function. The flowchart of the PSO algorithm is illustrated in Figure 16.

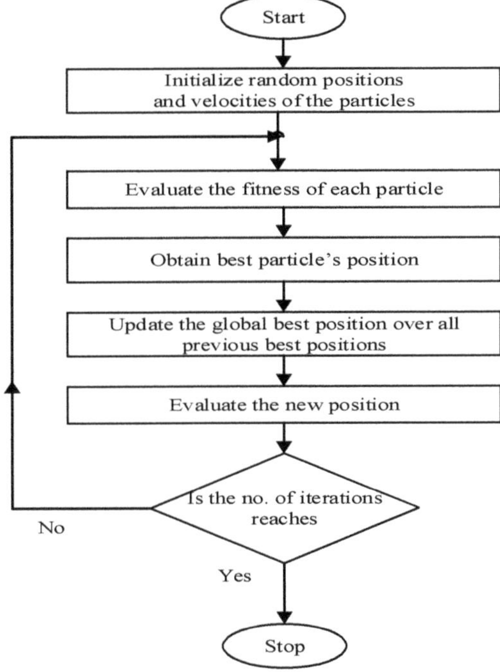

Figure 16. Flow chart of the PSO algorithm.

Typically, optimization methods involve evaluating system performance based on various fitness criteria such as Integral Absolute Errors (IAE), Integral Square Errors (ISE), and Integral Time Square Errors (ITSE). These fitness criteria take into account parameters such as overshoot, rising time, settled time, steady-state error, and the overall tightness of the control system [38]. In this study, the ITSE fitness function as shown in Equation (16) is used to evaluate the performance of the system's output response:

$$ITSE = \int_0^{\infty} t\, e^2(t).dt \tag{16}$$

Table 3 summarizes the parameters of the PSO algorithm used in this study.

Table 3. The parameters of the PSO algorithm.

Parameter	No. Iteration	No. Particles	Social Coefficient	Cognitive Coefficient	Inertia Weight
Value	30	10	1.42	1.42	0.9

The optimization process for the Fuzzy FOPID-PSO system is illustrated in Figure 17, which includes a block diagram of the process. To find the optimal values of the seven controller parameters, a MATLAB program is used. The program applies a minimization algorithm to search the domain of the particles' position and velocity, and the optimal values of the FOPID controller are obtained in 30 iterations.

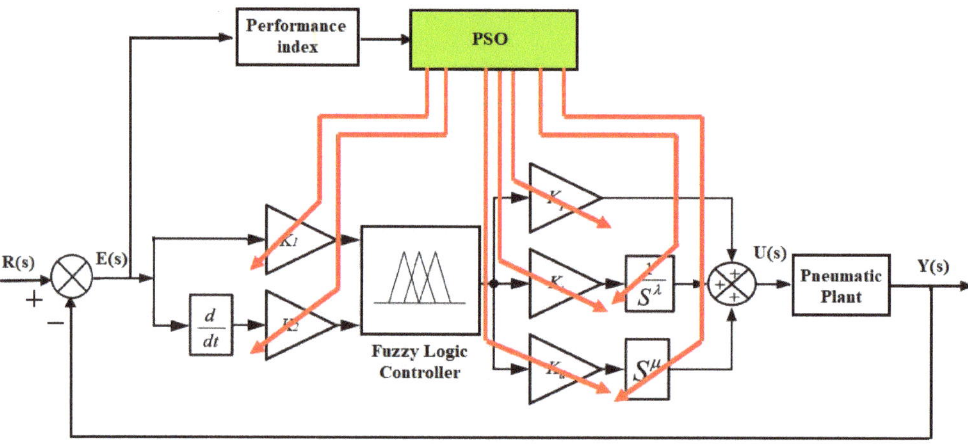

Figure 17. The block diagram of the optimized Fuzzy FOPID controller.

5. Results and Discussion

This part presents the simulation and real-time experimental results for the position control of IPA and PABB systems. The simulation was carried out using a mathematical model of the system, while the real-time experiments were conducted using physical prototypes of the systems. The results of the simulation and experimental tests were compared to evaluate the performance of the position control system. The discussion section provides an analysis of the results and discusses the potential applications of the position control system for IPA and PABB systems.

5.1. Position Control for IPA System

MATLAB-Simulink was utilized as the platform for this research, and Figure 18 displays the Simulink block diagram used for simulation. The controller block in this diagram consists of either a Fuzzy FOPID controller or a Fuzzy PID controller, and the IPA model is represented by Equation (1). Figure 19, on the other hand, illustrates the Simulink block diagram utilized for the real-time experiment setup. The block diagram design consists of five parts, namely the input (position-setpoint), controller, DAQ configuration (I/O), performance index, and output. The input signal used in this experiment is the same as in the simulation, where identical parameters of Fuzzy FOPID and Fuzzy PID controllers were implemented in the real-time experiment for validation purposes.

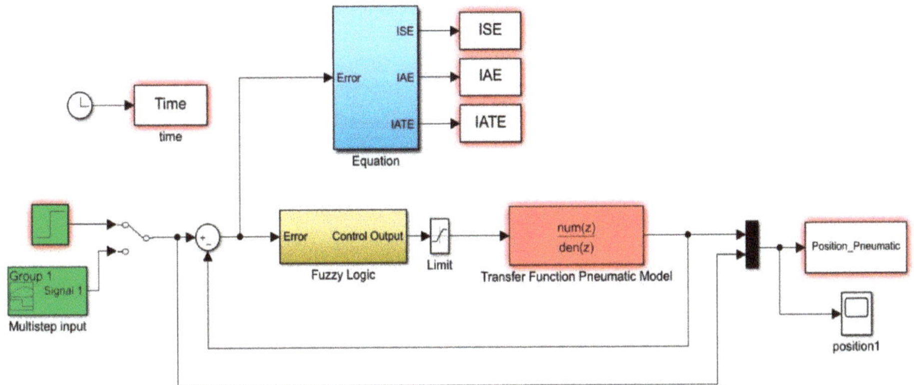

Figure 18. Simulink diagram for IPA simulation.

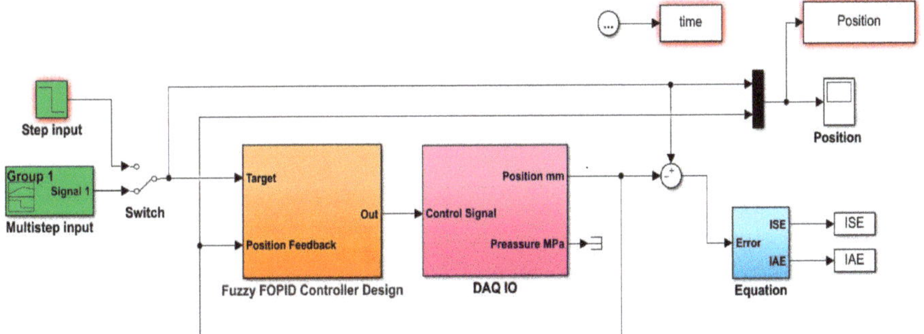

Figure 19. Simulink diagram for IPA real-time experiment.

The ITAE values obtained through successive generations of PSO using the Fuzzy FOPID controller and Fuzzy PID controller are illustrated in Figures 20 and 21, respectively. Table 4 provides a summary of the optimal parameter values found via PSO for both controllers (Fuzzy FOPID and Fuzzy PID).

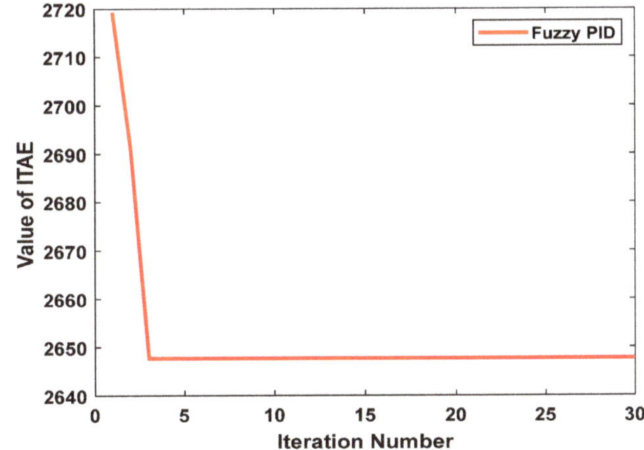

Figure 20. The value of ITAE in successive generations of the PSO Fuzzy PID.

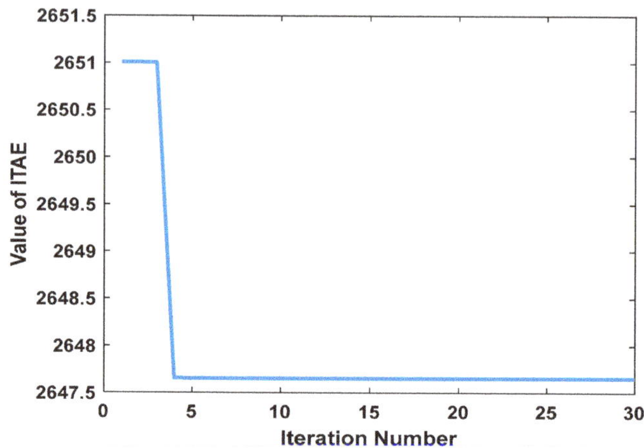

Figure 21. The value of ITAE in successive generations of the PSO Fuzzy FOPID.

Table 4. The optimal values of the controllers.

Criteria	K_1	K_2	K_p	K_i	λ	K_d	μ
Fuzzy FOPID	0.0001	0.0495	25	1	0.1	10	0.1
Fuzzy PID	0.0002	0.05	25	0.1	-	1	-

5.1.1. Simulation Performances of the IPA Positioning System

In this study, the step and multistep trajectories were utilized as reference input signals to evaluate the IPA system's positioning control. Simulation tests were conducted for position step responses with a duration of 15 s and multistep responses with a duration of 25 s, and all the control parameters mentioned in Table 4 were utilized to develop the control system. Two different positioning distances were examined using the step input test, namely mid-stroke (100 mm) and near fully extended (150 mm), where only one valve (Valve 1) of the IPA system was utilized to extend the cylinder stroke. In contrast to the step input test, the multistep input test involved both extension and retraction of the IPA cylinder stroke. Therefore, both Valve 1 and Valve 2, were employed to control the extension and retraction of the IPA cylinder stroke. In Figure 22, the simulation responses of the Fuzzy FOPID controller for step inputs at 100 mm (mid-stroke) and 150 mm (nearly fully extended) are illustrated. On the other hand, Figure 23 displays the multistep response for the Fuzzy FOPID controller. Similarly, Figure 24 showcases the simulation responses of the Fuzzy PID controller for step inputs at 100 mm (mid-stroke) and 150 mm (nearly fully extended), while Figure 25 demonstrates the multistep response for the Fuzzy PID controller. A summary of all the data presented in these figures has been compiled in Table 5.

The data presented in Figures 22–25 along with the information provided in Table 5, indicate that as the distance increased, the rise time (Tr) and settling time (Ts) of the IPA positioning system also increased steadily. However, the employment of Fuzzy PID resulted in a slightly longer response time compared to Fuzzy FOPID, causing a slower system. To meet the requirements of wider applications, the IPA system should be capable of achieving both fast speed response and accurate positioning control. The inclusion of fractional order parameters in the control signal to the IPA valves using Fuzzy FOPID resulted in a faster and more aggressive speed response compared to Fuzzy PID. Additionally, Fuzzy PID control resulted in a slight overshoot of 0.00022% and 0.00031% for distances of 100 mm and 150 mm, respectively. Furthermore, this study evaluated three performance indices (IAE, ISE, and ITAE) and found that all the controllers produced similar outcomes in terms of the step response. The values of the performance indices for both Fuzzy FOPID and Fuzzy PID controllers were relatively comparable to each other. Specifically, for IAE, the total error ranged from approximately 39 to 41.1; for ISE, it was around $(26–26.6) \times 10^2$; for ITAE, the Fuzzy FOPID controller had a value of 10.54, while the Fuzzy PID controller had a value of 25.42. On the other hand, Fuzzy FOPID control did not exhibit any overshoot for all distances. All control strategies demonstrated zero steady-state error (e_{ss}) for all distances, indicating that the control system can accurately track the IPA positioning system. This further confirms the effectiveness of the control strategies in achieving accurate positioning control.

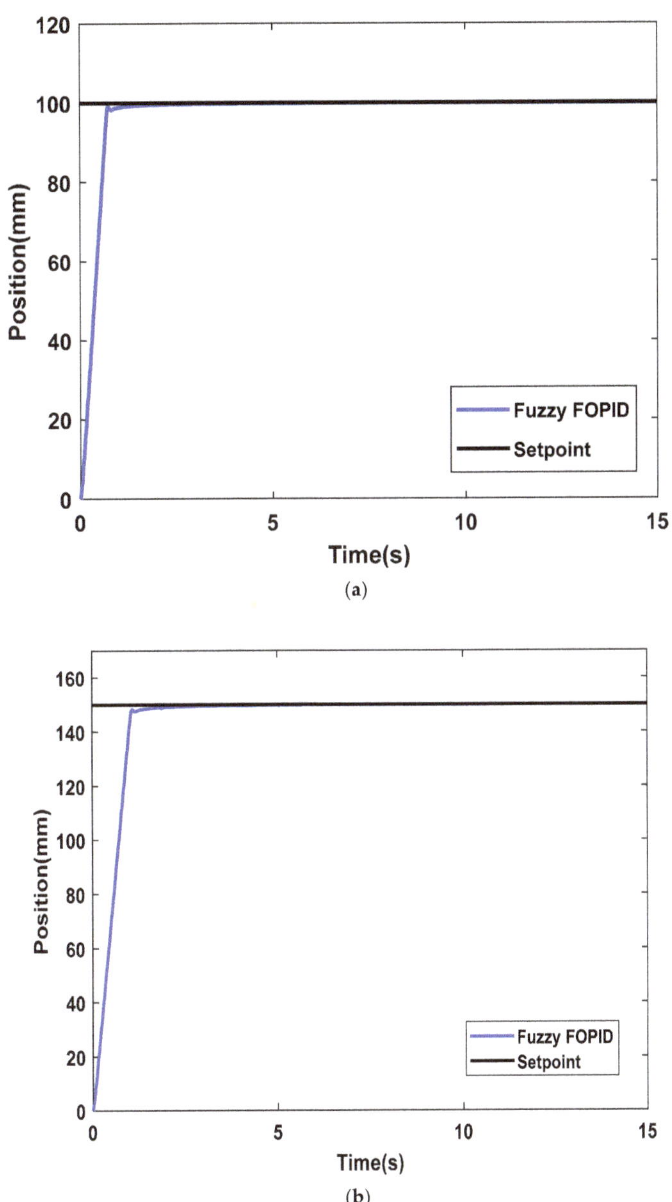

Figure 22. Step response for Fuzzy FOPID simulation at (**a**) 100 mm, (**b**) 150 mm.

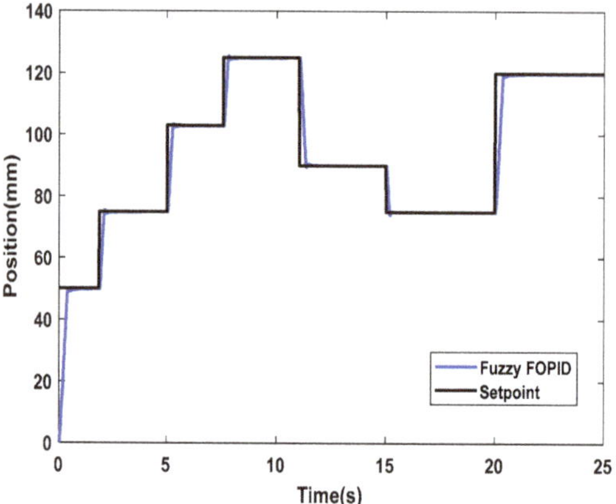

Figure 23. Multistep response for Fuzzy FOPID simulation.

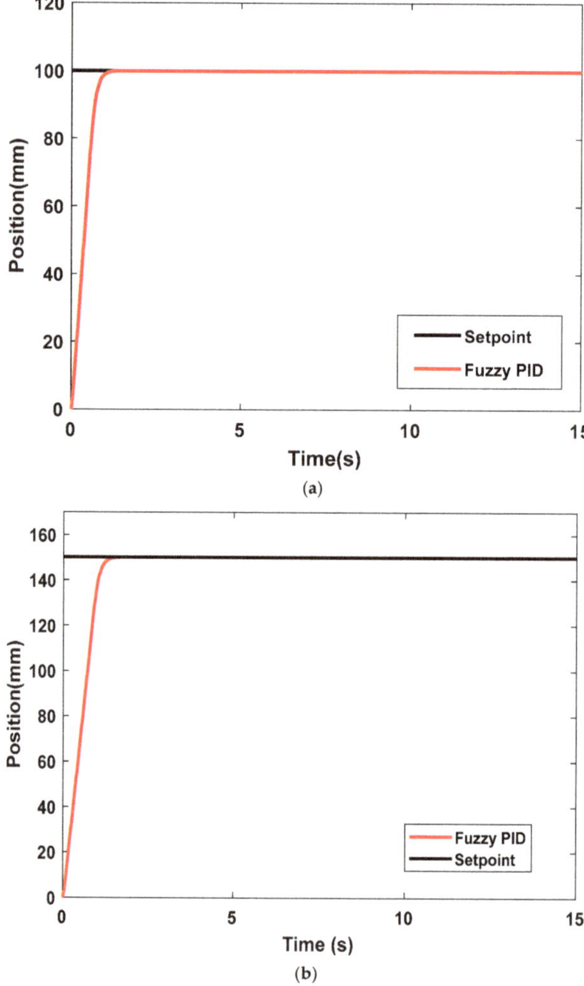

Figure 24. Step response for Fuzzy PID simulation at (**a**) 100 mm, (**b**) 150 mm.

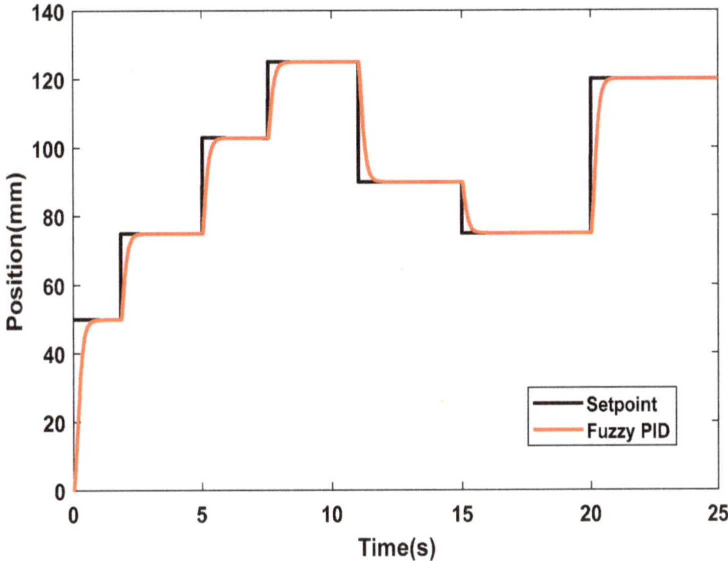

Figure 25. Multistep response for Fuzzy PID simulation.

Table 5. Summary of the step response performances using Fuzzy FOPID and Fuzzy PID.

Distance (mm)	Transient Performance	Fuzzy FOPID	Fuzzy PID
	Tr (s)	0.5616	0.6953
	Ts (s)	0.7188	1.0349
100	OS (%)	0	0.00022
	e_{ss}	0	0
	Tr (s)	0.8400	0.9510
	Ts (s)	1.0602	1.2002
150	OS (%)	0	0.00031
	e_{ss}	0	0

5.1.2. Experimental Validation Performances of the IPA Positioning System

The research assessed the efficacy of the suggested control approach by conducting several experiments, such as positioning control at different distances and examining the system's robustness to load changes. The performance of the proposed control method was evaluated in each experiment and juxtaposed with established techniques used for comparable pneumatic plant systems to identify enhancements, especially in the IPA positioning system's transient response. A sampling time of 10 ms was used for experimentation, and MATLAB/Simulink was used to develop the proposed control strategy. The simulation test used the same controller parameters to validate the results.

Two position distances (100 mm and 150 mm) and two directions of the cylinder position (horizontal and vertical) were utilized for comparison, and the step signal was applied as the input signal. Each test was carried out for 20 s, and the controller parameters used in the simulation test were the same. The performance of the Fuzzy FOPID system's transient response, including the rise time (Tr), settling time (Ts), overshoot (OS), and steady-state error (e_{ss}), in controlling the IPA positioning system at all distances were then compared with that of the Fuzzy PID. Figure 26 offers a comparative analysis of the step tests conducted with the cylinder positioned horizontally, and the distance of the position was adjusted between fully retracted (0 mm) and nearly fully retracted (100 mm). Similarly, Figure 27 provides a comparative view of the step tests, with the cylinder position being horizontal and the position distance adjusted between fully retracted (0 mm) and nearly fully extended (150 mm). The findings from both figures are summarized in Table 6.

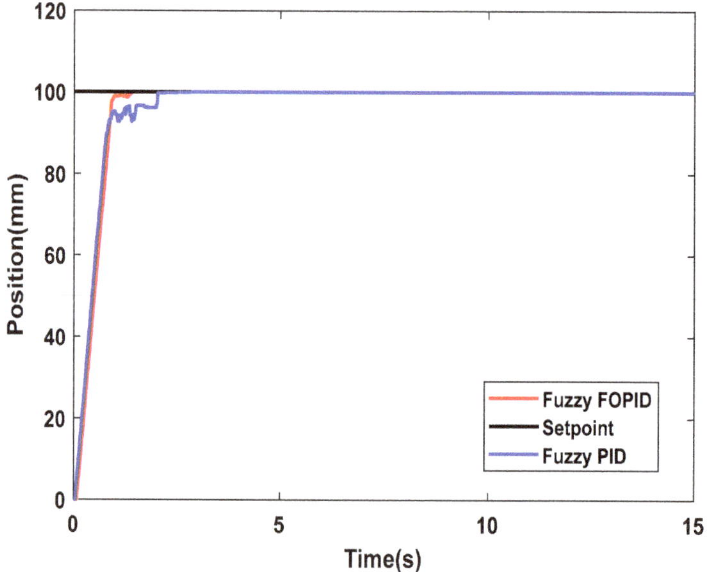

Figure 26. Experimental step response at 100 mm for the horizontal position.

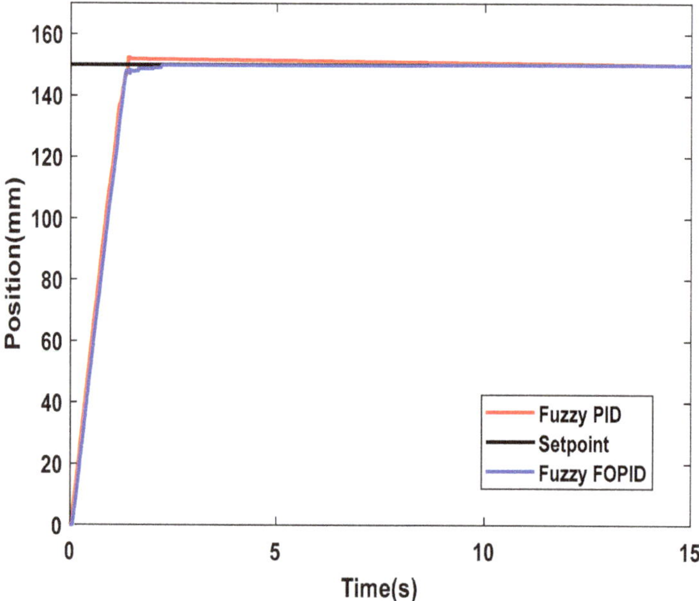

Figure 27. Experimental step response at 150 mm for the horizontal position.

Table 6. Summary of the experimental response using different control strategies for the horizontal position.

Distance (mm)	Transient Performance	Fuzzy FOPID	Fuzzy PID
100	Tr (s)	0.6577	0.7418
	Ts (s)	0.9544	1.0337
	OS (%)	0	0.00022
	e_{ss}	0	0
150	Tr (s)	1.0468	1.0301
	Ts (s)	1.3497	1.3741
	OS (%)	0	0.00031
	e_{ss}	0	0

Figure 28 compares the step tests for the vertical cylinder position, with the position distance ranging from fully retracted (0 mm) to almost fully retracted (100 mm), whereas Figure 29 compares the step tests for the vertical cylinder position, with the position distance ranging from fully retracted (0 mm) to almost fully extended (150 mm). The outcomes from Figures 9 and 10 are consolidated in Table 7 for reference.

The experimental results presented in Figures 26–29, as well as Tables 6 and 7, illustrate a noticeable increase in both rise time (Tr) and settling time (Ts) of all control strategies as the distance to be covered by the cylinder stroke increases. In other words, the farther the distance, the longer it takes for the control strategies to achieve their steady-state value. The comparison of the control strategies reveals that Fuzzy FOPID outperforms the others in achieving precise control of the IPA cylinder stroke at both 100 mm and 150 mm positioning distances. Furthermore, the Fuzzy FOPID controller outperformed conventional controllers by demonstrating no overshoot and steady-state inaccuracy. In addition, the IAE, ISE, and ITAE of the Fuzzy FOPID are less than those of the Fuzzy PID. The Fuzzy FOPID controller improved the system's transient response by 12.78%, 24.25%, and 100% in rise time, settling time, and overshoot, respectively, compared to the results obtained from the Fuzzy PID controller. As a result, the outcomes obtained from utilizing the Fuzzy FOPID controller are deemed satisfactory. The controller exhibited a significant enhancement in the transient response performance by offering a faster response without overshooting for all position distances.

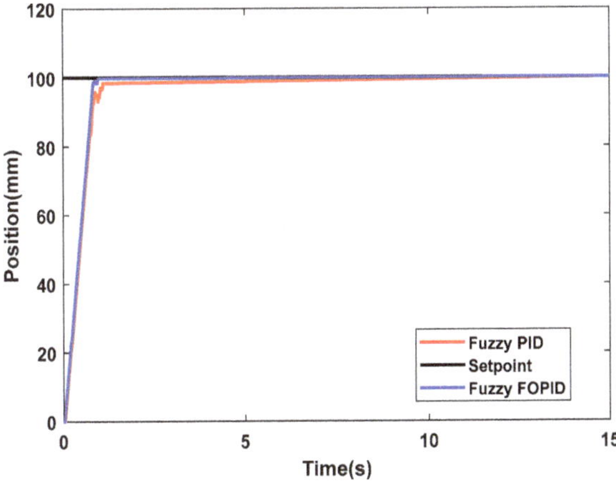

Figure 28. Experimental step response at 100 mm for the vertical position.

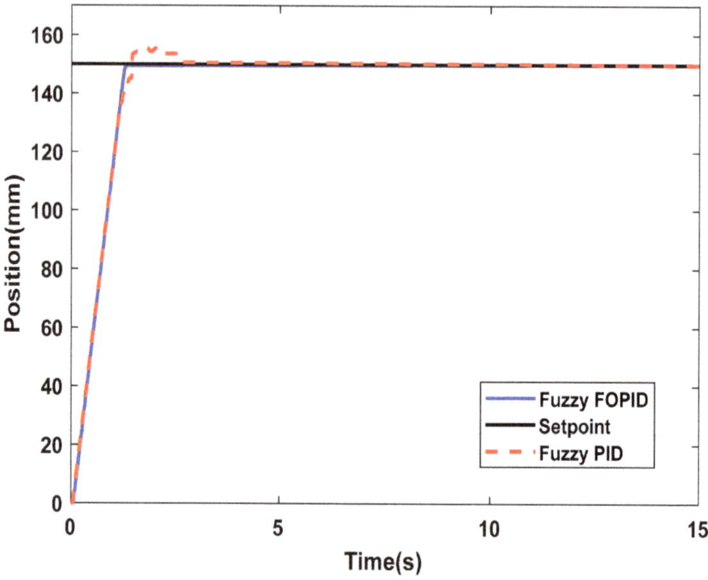

Figure 29. Experimental step response at 150 mm for the vertical position.

Table 7. Summary of the experimental response using different control strategies for the vertical position.

Distance (mm)	Transient Performance	Fuzzy FOPID	Fuzzy PID
100	Tr (s)	0.6372	0.7306
	Ts (s)	0.8381	1.1076
	OS (%)	0	0.00022
	e_{ss}	0	0
150	Tr (s)	0.9874	1.0196
	Ts (s)	1.2726	1.5662
	OS (%)	0	0.00031
	e_{ss}	0	0

Moreover, the IPA system's positioning performance for horizontal and vertical positions using Fuzzy PID and Fuzzy FOPID strategies with a multistep trajectory as the input signal is depicted in Figures 30 and 31, respectively.

When developing a controller, robustness is a crucial factor that must be considered. A controller is considered robust if it can compensate for any changes in the system due to external loads. This study examined the effect of varying loads on controller robustness. The IPA positioning system was subjected to a step response with mid-stroke positions (100 mm and 150 mm), and different external loads (1 kg, 3 kg, 6 kg, and 8 kg) were attached to the end of the cylinder stroke for each test, which lasted 20 s. The controller parameters used in this test were the same as those used in the unloading condition. The experimental performance of the vertical IPA positioning system's response to varying loads, controlled by the Fuzzy FOPID, is illustrated in Figures 32 and 33. Moreover, the performance details are presented in Table 8.

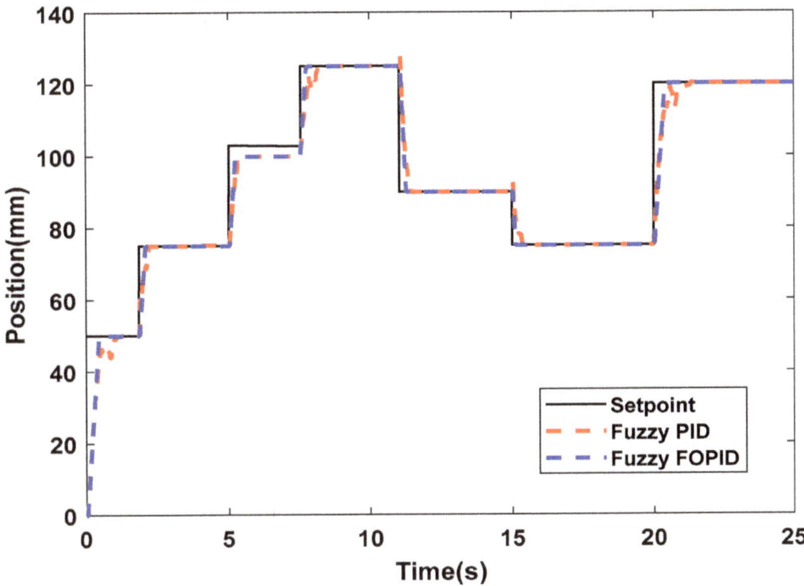

Figure 30. Experimental multistep response for the horizontal position.

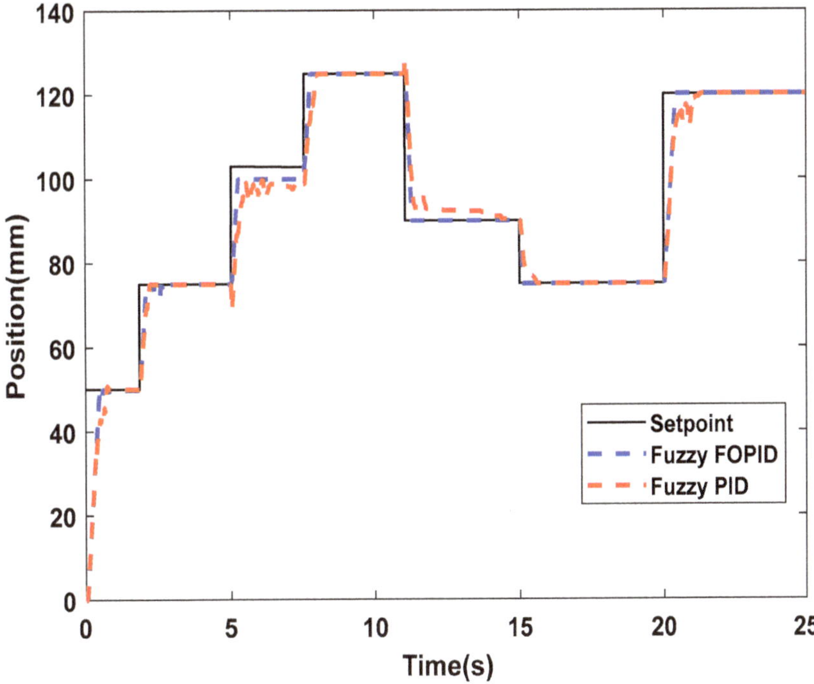

Figure 31. Experimental multistep response for the vertical position.

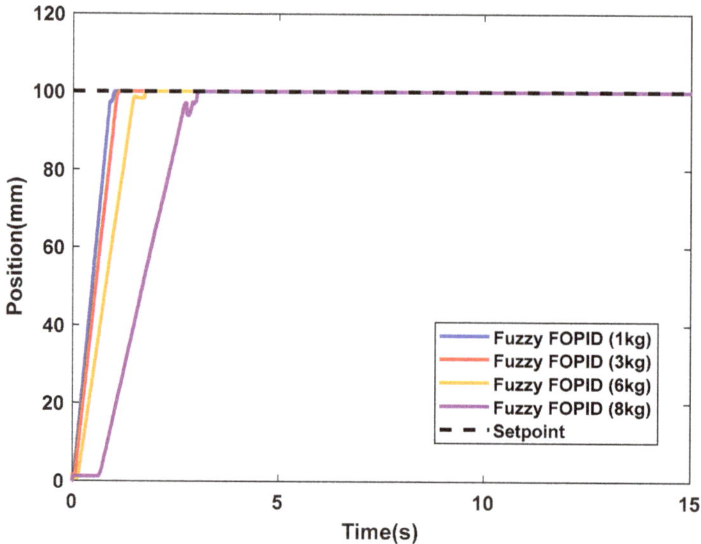

Figure 32. Experimental step response based on load variation at 100 mm.

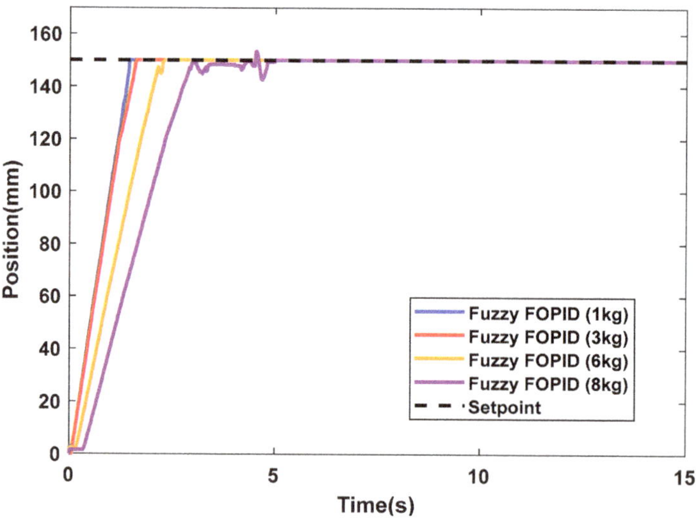

Figure 33. Experimental step response based on load variation at 150 mm.

Table 8. Summary of the experimental response using different loads.

Loads (kg)	Fixed Position at 100 mm		Fixed Position at 150 mm	
	Rise Time Tr (s)	Settling Time Ts (s)	Rise Time Tr (s)	Settling Time Ts (s)
1	0.7199	0.9763	1.1513	1.4590
3	0.8290	1.0781	1.2435	1.6047
6	1.1002	1.4944	1.6076	2.2613
8	1.6912	3.0097	2.0625	4.7827

Furthermore, Figure 34 demonstrates the positioning performances of the IPA system using the Fuzzy FOPID strategy, considering the multistep trajectory as an input signal to the IPA system, with load variations.

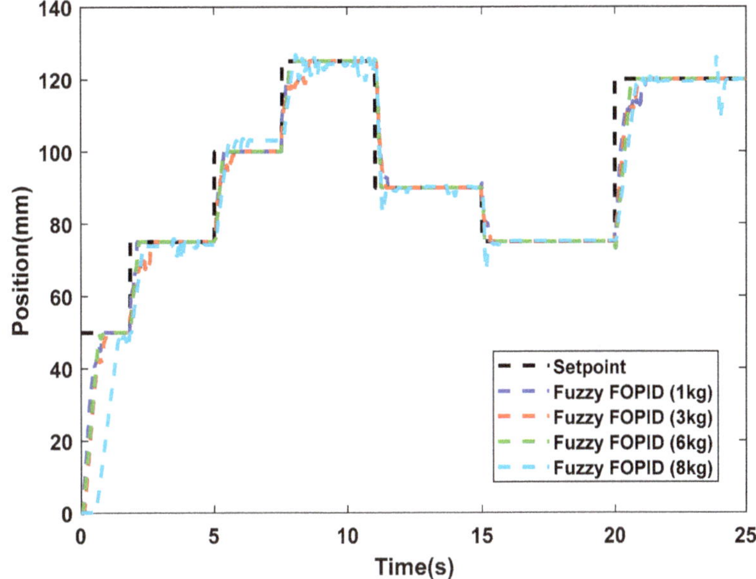

Figure 34. Experimental multistep response based on load variation.

Table 8 and Figure 34 present the results of the experimental evaluation of the performance of the pneumatic actuator controlled via Fuzzy FOPID under different loads and fixed positions. The data presented in the table show the rise time and settling time for the IPA system when the load is varied from 1 kg to 8 kg at fixed positions of 100 mm and 150 mm. The data show that the performance of the system decreases as the load increases, particularly for the fixed position at 150 mm. At 6 kg, the rise time and settling time for both fixed positions are still within acceptable limits. However, at 8 kg, the rise time and settling time for the fixed position at 150 mm are significantly higher, indicating a deterioration in performance. This suggests that the maximum load capacity and performance of the pneumatic actuator should be considered when designing and implementing a control system for the actuator.

5.2. PABB System Application Results

This research employed MATLAB-Simulink as the platform, and the Simulink block diagrams used for the PABB system simulation and the real-time experiment setup are shown in Figures 35 and 36, respectively. The controller block in Figure 35 comprises a Fuzzy FOPID or a FOPI-FOPD controller, while the PABB system model is represented by Equation (1). Figure 36 shows the block diagram design, which includes five parts: input (position-setpoint), controller, DAQ configuration (I/O), performance index, and output. The input signal used in the experiment is the same as in the simulation, and identical parameters of Fuzzy FOPID and FOPI-FOPD controllers were used for validation purposes.

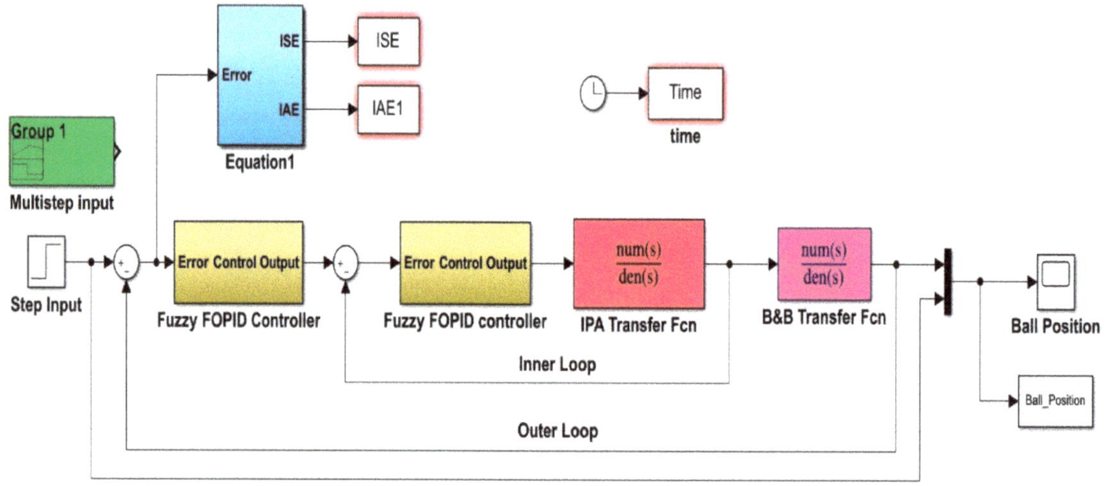

Figure 35. Simulink diagram for PABB system simulation.

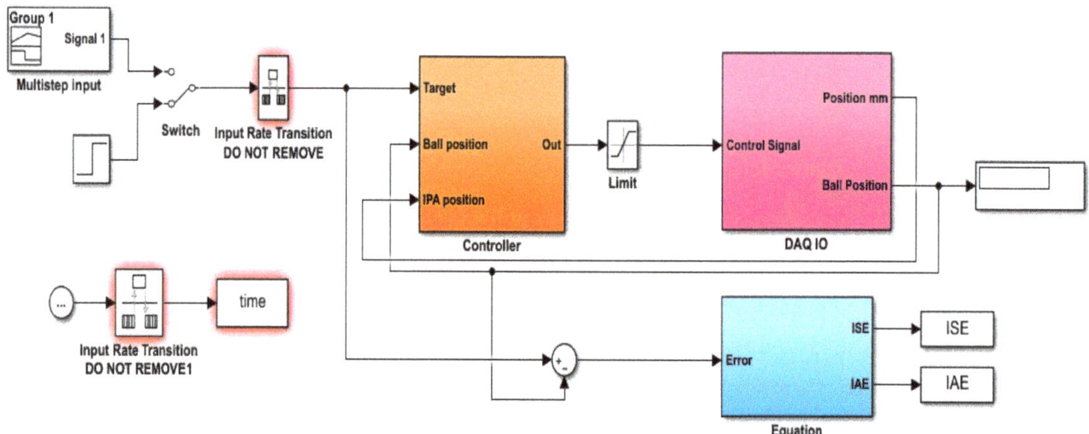

Figure 36. Simulink diagram for PABB system real-time experiment.

5.2.1. Simulation Performances of the PABB System

The PABB system's positioning control was evaluated using step and multistep trajectories as reference input signals in this study. Simulation tests were conducted for both position step responses (15 s) and multistep responses (25 s), with all the control parameters mentioned in Table 9 utilized to develop the control system.

Table 9. The optimal values of Fuzzy FOPID and FOPI-FOPD controllers.

Criteria	K_1	K_2	K_p	K_{pi}	K_{pd}	K_i	λ	K_d	μ
Fuzzy FOPID	0.0495	0.0001	0.1	-	-	0.15	1	1	0.1
FOPI-FOPD	-	-	-	1	1	0.01	0.1	1	1

Figure 37 shows the simulation results, which indicate that both the FOPI-FOPD and Fuzzy FOPID controllers are capable of tracking the target input. The Fuzzy FOPID controller achieves the setpoint faster than the FOPI-FOPD controller. Both controllers exhibit consistent and rapid responses while regulating the movement of the ball, with no overshoot and only minor steady-state error. In addition, the FOPI-FOPD controller had an ISE value of 8.233×10^4, an IAE value of 403.4, and an IATE value of 423. The Fuzzy FOPID controller had a slightly lower ISE value of 6.036×10^4, an IAE value of 252.6, and an IATE value of 875, indicating better performance compared to the FOPI-FOPD controller. The Fuzzy FOPID controller outperforms the FOPI-FOPD controller, as evidenced by the better results displayed in Table 10.

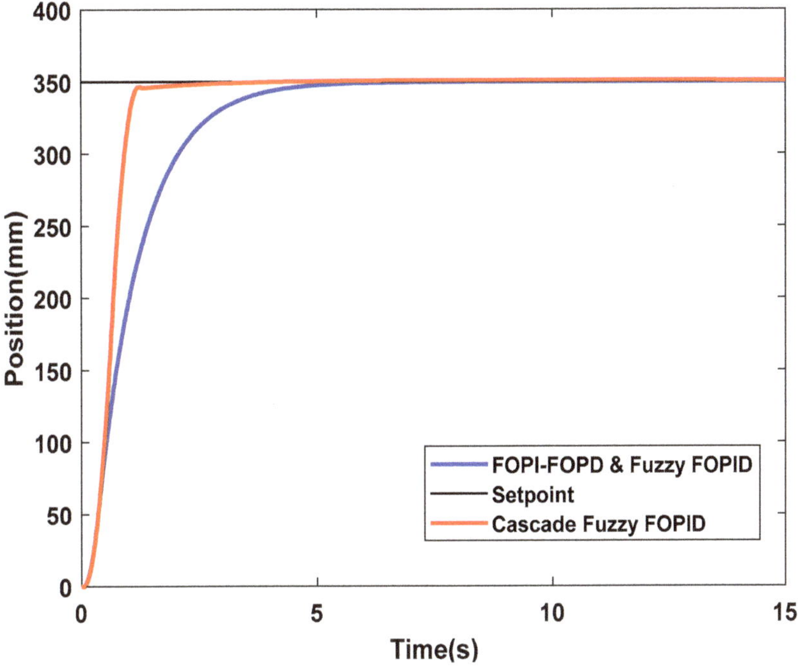

Figure 37. Simulated step responses for Fuzzy FOPID and FOPI-FOPD.

Table 10. Summary of the simulated response of the PABB system.

Controllers	Performance Index			
	Rise Time Tr (s)	Settling Time Ts (s)	Overshoot OS (%)	Steady-State Error e_{ss} (%)
FOPI-FOPD	2.091	3.910	0.0055	0
Cascade Fuzzy FOPID	0.6682	1.1359	0.2768	0

Following that, a multistep input was employed to assess the stability and performance of the controllers. Both the FOPI-FOPD and Fuzzy FOPID controllers performed almost identically, smoothly controlling the ball. The results of the multistep response are shown in Figure 38.

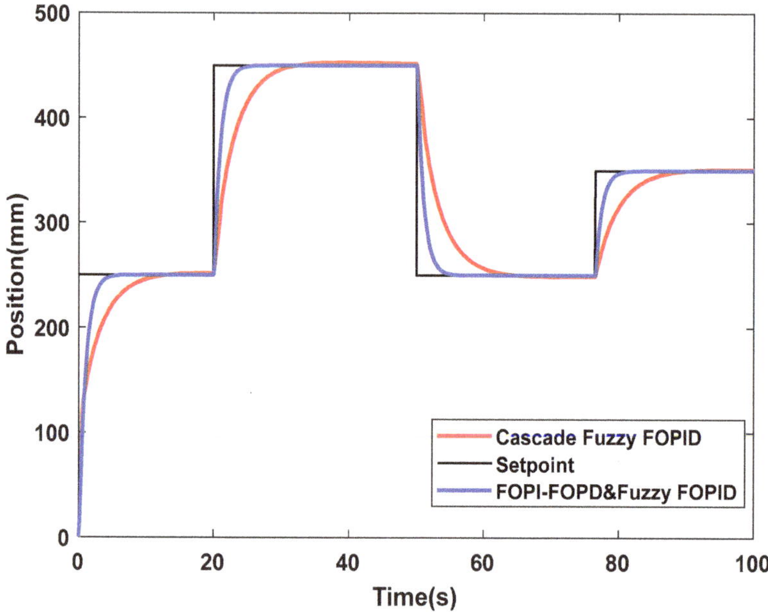

Figure 38. Simulated multistep responses for Fuzzy FOPID and FOPI-FOPD.

5.2.2. Experimental Validation Performances of the PABB System

To prevent the ball from moving too quickly, which could complicate regulation, the pneumatic actuator stroke, h, is restricted to a range of +50 mm (upward) and −50 mm (downward). At the outset of the experiment, h is set to 100 mm. This configuration ensures that the pneumatic system can execute the necessary up-and-down movements.

The controllers are designed using a simulation environment and a mathematical model of the PABB system. The experimental verification of the FOPI-FOPD controller and the Cascade Fuzzy FOPID controller is demonstrated in Figure 39. Table 11 provides a summary of the outcomes from the step response analysis, revealing that the ball's position control experiences increased oscillations as a result of the pneumatic movement's impact. However, due to several factors such as changes in nonlinear characteristics caused by air compressibility, valve dead zone issues, high friction forces, and noise generated by position sensors, the experimental results exhibit slight variations from the simulation outcomes. Despite these challenges, the PABB system yields a favorable outcome in which the primary objective of the system is accomplished. The proposed system can effectively regulate the position and balance of the ball.

Table 11. Summary of the experimental response of the PABB system.

Controllers	Performance Index			
	Rise Time Tr (s)	Settling Time Ts (s)	Overshoot OS (%)	Steady-State Error e_{ss} (%)
FOPI-FOPD Fuzzy FOPID	0.3710	10.7986	35.5380	1.3520
Cascade Fuzzy FOPID	0.8235	4.9381	9.8500	0

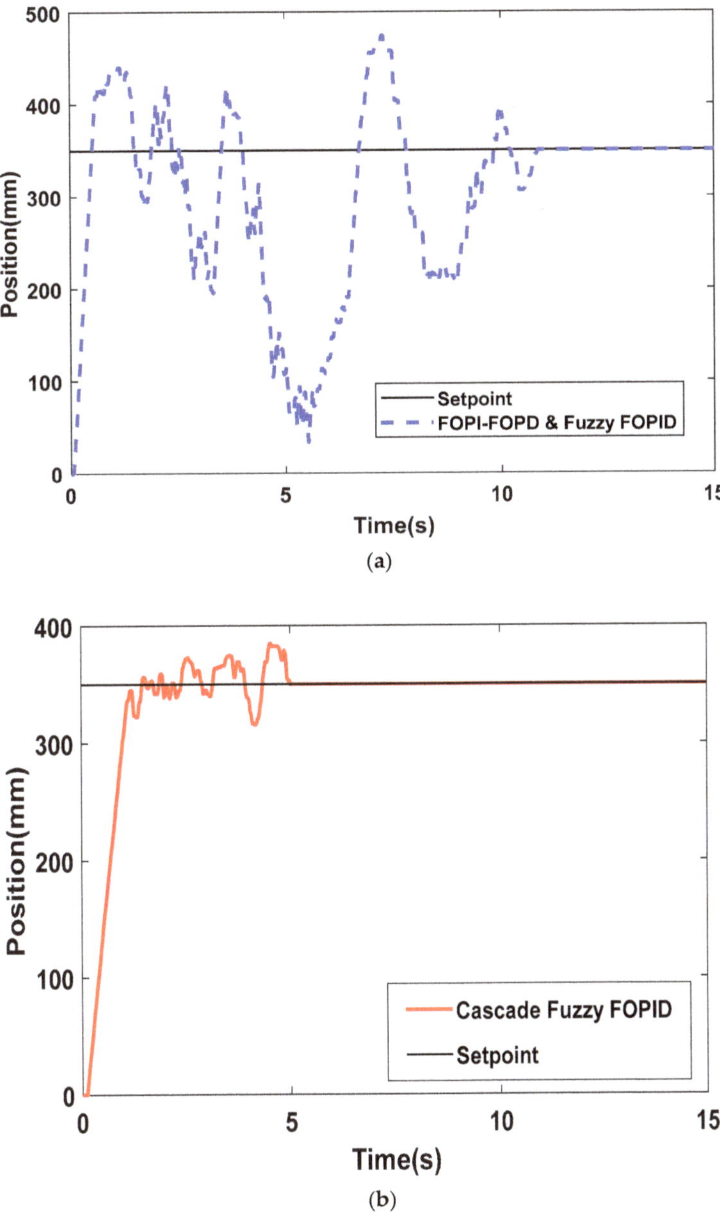

Figure 39. Experimental step responses for (**a**) Fuzzy FOPID and (**b**) FOPI-FOPD.

6. Conclusions

The primary objective of this study was to develop and evaluate a Fuzzy FOPID controller for an intelligent pneumatic actuator (IPA) system to achieve accurate positional control. The study utilized the ARX model to simulate the pneumatic system and the PSO technique to determine the optimal values for the seven controller parameters required to achieve the best dynamic behavior for the Fuzzy FOPID controller. The results indicated that the Fuzzy FOPID controller outperformed the Fuzzy PID controller in terms of stability, robustness, fast response, and zero steady-state error. The study then applied the model and controller of the pneumatic actuator to the pneumatically actuated ball and beam (PABB) system using two control loops for inner and outer positioning. The Cascade Fuzzy FOPID controller was found to provide a quick and smooth response in controlling the ball's motion. The study validated the performance of the position controller through both simulation and real-time experiments, which demonstrated the effectiveness of the proposed Fuzzy FOPID controller in achieving precise and stable positional control in pneumatic systems.

Author Contributions: Conceptualization, M.N.M., A.A.M.F., S.S. and S.M.; Methodology, M.N.M., A.A.M.F., S.S. and S.M.; Software, M.N.M.; Validation, M.N.M.; Formal analysis, M.N.M.; Investigation, M.N.M., A.A.M.F. and S.S.; Resources, M.N.M., A.A.M.F., S.S. and S.M.; Data curation, M.N.M., A.A.M.F., S.S. and S.M.; Writing—original draft, M.N.M.; Writing—review & editing, A.A.M.F., S.S. and S.M.; Supervision, A.A.M.F., S.S. and S.M.; Project administration, M.N.M. and A.A.M.F.; Funding acquisition, A.A.M.F., S.S. and S.M. All authors have read and agreed to the published version of the manuscript.

Funding: This research received no external funding.

Data Availability Statement: Not applicable.

Acknowledgments: The research has been carried out under program Research Excellence Consortium (JPT (BPKI) 1000/016/018/25 (57)) with the title Consortium of Robotics Technology for Search and Rescue Operations (CORTESRO) provided by Ministry of Higher Education Malaysia (MOHE). The authors also acknowledge Universiti Teknologi Malaysia (UTM) under vote no (4L930) for the facilities and support to complete this research.

Conflicts of Interest: The authors declare no conflict of interest.

References

1. Van Varseveld, R.B.; Bone, G.M. Accurate position control of a pneumatic actuator using on/off solenoid valves. *IEEE/ASME Trans. Mechatron.* **1997**, *2*, 195–204. [CrossRef]
2. Taghizadeh, M.; Najafi, F.; Ghaffari, A. Multimodel PD-control of a pneumatic actuator under variable loads. *Int. J. Adv. Manuf. Technol.* **2010**, *48*, 655–662. [CrossRef]
3. Rahmat, M.F.; Salim, S.N.S.; Faudzi, A.A.M.; Ismail, Z.H.; Samsudin, S.I.; Sunar, N.H.; Jusoff, K. Non-linear modeling and cascade control of an industrial pneumatic actuator system. *Aust. J. Basic Appl. Sci.* **2011**, *5*, 465–477.
4. Qian, P.; Pu, C.; Liu, L.; Lv, P.; Ruiz Páez, L.M. A novel pneumatic actuator based on high-frequency longitudinal vibration friction reduction. *Sens. Actuators A Phys.* **2022**, *344*, 113731. [CrossRef]
5. Saravanakumar, D.; Mohan, B.; Muthuramalingam, T. A review on recent research trends in servo pneumatic positioning systems. *Precis. Eng.* **2017**, *49*, 481–492. [CrossRef]
6. Faudzi, A.M.; Osman, K.; Rahmat, M.F.; Mustafa, N.D.; Azman, M.A.; Suzumori, K. Nonlinear mathematical model of an Intelligent Pneumatic Actuator (IPA) systems: Position and force controls. In Proceedings of the 2012 IEEE/ASME International Conference on Advanced Intelligent Mechatronics (AIM), Kaohsiung, Taiwan, 11–14 July 2012; pp. 1105–1110. [CrossRef]
7. Chen, H.M.; Shyu, Y.P.; Chen, C.H. Design and Realization of a Sliding Mode Control Scheme for a Pneumatic Cylinder X-Y Axles Position Servo System. *IET* **2010**, 416–421. [CrossRef]
8. Muftah, M.N.; Xuan, W.L.; Athif, A.; Faudzi, M. ARX, ARMAX, Box-Jenkins, Output-Error, and Hammerstein Models for Modeling Intelligent Pneumatic Actuator (IPA) System. *J. Integr. Adv. Eng.* **2021**, *1*, 81–88. [CrossRef]
9. Gentile, A.; Giannoccaro, N.I.; Reina, G. Experimental tests on position control of a pneumatic actuator using on/off solenoid valves. *Proc. IEEE Int. Conf. Ind. Technol.* **2002**, *1*, 555–559. [CrossRef]
10. Osman, K.; Faudzi, A.A.M.; Rahmat, M.F.; Mustafa, N.D.; Abidin, A.F.Z.; Suzumori, K. Proportional-integrative controller design of Pneumatic system using particle swarm optimization. In Proceedings of the 2013 IEEE Student Conference on Research and Development, Putrajaya, Malaysia, 16–17 December 2013; pp. 421–426. [CrossRef]
11. Tsai, Y.C.; Huang, A.C. Multiple-surface sliding controller design for pneumatic servo systems. *Mechatronics* **2008**, *18*, 506–512. [CrossRef]
12. Azahar, M.I.P.; Irawan, A.; Ismail, R.M.T.R. Self-tuning hybrid fuzzy sliding surface control for pneumatic servo system positioning. *Control Eng. Pract.* **2021**, *113*, 104838. [CrossRef]
13. Schindele, D.; Aschemann, H. Adaptive friction compensation based on the LuGre model for a pneumatic rodless cylinder. In Proceedings of the 2009 35th Annual Conference of IEEE Industrial Electronics, Porto, Portugal, 3–5 November 2009; pp. 1432–1437. [CrossRef]
14. Azman, M.A.; Osman, K.; Natarajan, E. Integrating Servo-Pneumatic Actuator with Ball Beam System based on intelligent position control. *J. Teknol.* **2014**, *3*, 73–79. [CrossRef]

5. Mu, S.; Goto, S.; Shibata, S.; Yamamoto, T. Intelligent position control for pneumatic servo system based on predictive fuzzy control. *Comput. Electr. Eng.* **2019**, *75*, 112–122. [CrossRef]

6. Osman, K.; Faudzi, A.; Athif, M.; Rahmat, M.F.; Hikmat, O.F.; Suzumori, K. Predictive Functional Control with Observer (PFC-O) Design and Loading Effects Performance for a Pneumatic System. *Arab. J. Sci. Eng.* **2015**, *40*, 633–643. [CrossRef]

7. Muftah, M.N.; Athif, A.; Faudzi, M.; Sahlan, S.; Shouran, M. Modeling and Fuzzy FOPID Controller Tuned by PSO for Pneumatic Positioning System. *Energies* **2022**, *15*, 3757. [CrossRef]

8. Podlubny, I. Fractional-order systems and PIΛDμ controllers. *IEEE Trans. Automat. Contr.* **1999**, *44*, 208–214. [CrossRef]

9. Sikander, A.; Thakur, P.; Bansal, R.C.; Rajasekar, S. A novel technique to design cuckoo search based FOPID controller for AVR in power systems. *Comput. Electr. Eng.* **2018**, *70*, 261–274. [CrossRef]

20. Rajasekhar, A.; Kumar Jatoth, R.; Abraham, A. Design of intelligent PID/PIΛDμ speed controller for chopper fed DC motor drive using opposition based artificial bee colony algorithm. *Eng. Appl. Artif. Intell.* **2014**, *29*, 13–32. [CrossRef]

21. Dumlu, A.; Erenturk, K. Trajectory tracking control for a 3-DOF parallel manipulator using fractional-order PIΛDμ control. *IEEE Trans. Ind. Electron.* **2014**, *61*, 3417–3426. [CrossRef]

22. Luo, Y.; Chen, Y. Stabilizing and robust fractional order PI controller synthesis for first order plus time delay systems. *Automatica* **2012**, *48*, 2159–2167. [CrossRef]

23. Altintas, G.; Aydin, Y. Optimization of Fractional and Integer Order PID Parameters using Big Bang Big Crunch and Genetic Algorithms for a MAGLEV System. *IFAC-PapersOnLine* **2017**, *50*, 4881–4886. [CrossRef]

24. Karahan, O. Design of optimal fractional order fuzzy PID controller based on cuckoo search algorithm for core power control in molten salt reactors. *Prog. Nucl. Energy* **2021**, *139*, 103868. [CrossRef]

25. Fu, W.; Wang, K.; Li, C.; Tan, J. Multi-step short-term wind speed forecasting approach based on multi-scale dominant ingredient chaotic analysis, improved hybrid GWO-SCA optimization and ELM. *Energy Convers. Manag.* **2019**, *187*, 356–377. [CrossRef]

26. Altbawi, S.M.A.; Mokhtar, A.S.B.; Jumani, T.A.; Khan, I.; Hamadneh, N.N.; Khan, A. Optimal design of Fractional order PID controller based Automatic voltage regulator system using gradient-based optimization algorithm. *J. King Saud Univ.—Eng. Sci.* **2021**, *in press.* [CrossRef]

27. Shouran, M.; Alsseid, A. Particle Swarm Optimization Algorithm-Tuned Fuzzy Cascade Fractional Order PI-Fractional Order PD for Frequency Regulation of Dual-Area Power System. *Processes* **2022**, *10*, 477. [CrossRef]

28. Rajesh, R. Optimal tuning of FOPID controller based on PSO algorithm with reference model for a single conical tank system. *SN Appl. Sci.* **2019**, *1*, 1–14. [CrossRef]

29. Bingul, Z.; Karahan, O. Comparison of PID and FOPID controllers tuned by PSO and ABC algorithms for unstable and integrating systems with time delay. *Optim. Control Appl. Methods* **2018**, *39*, 1431–1450. [CrossRef]

30. Osinski, C.; Silveira, A.L.R.; Stiegelmaier, C.; Bergamini, M.G.; Leandro, G.V. Control of ball and beam system using fuzzy PID controller. In Proceedings of the 2018 13th IEEE International Conference on Industry Applications (INDUSCON), Sao Paulo, Brazil, 12–14 November 2018; pp. 875–880. [CrossRef]

31. Šitum, Ž.; Trslić, P. Ball and beam balancing mechanism actuated with pneumatic artificial muscles. *J. Mech. Robot.* **2018**, *10*, 055001. [CrossRef]

32. Hannan, M.A.; Ghani, Z.A.; Hoque, M.M.; Ker, P.J.; Hussain, A.; Mohamed, A. Fuzzy logic inverter controller in photovoltaic applications: Issues and recommendations. *IEEE Access* **2019**, *7*, 24934–24955. [CrossRef]

33. El Ouanjli, N.; Motahhir, S.; Derouich, A.; El Ghzizal, A.; Chebabhi, A.; Taoussi, M. Improved DTC strategy of doubly fed induction motor using fuzzy logic controller. *Energy Rep.* **2019**, *5*, 271–279. [CrossRef]

34. Kennedy, J.; Eberhart, R. Particle Swarm Optimisation. In Proceedings of the ICNN'95—International Conference on Neural Networks, Perth, WA, Australia, 27 November–1 December 1995; pp. 1942–1948. [CrossRef]

35. Shi, Y.; Eberhart, R. A Modified Particle Swarm Optimizer. In Proceedings of the 1998 IEEE International Conference on Evolutionary Computation Proceedings, IEEE World Congress on Computational Intelligence (Cat. No.98TH8360), Anchorage, AK, USA, 4–9 May 1998; pp. 69–73. [CrossRef]

36. Solihin, M.I.; Tack, L.F.; Kean, M.L. Tuning of PID Controller Using Particle Swarm Optimization (PSO). *Int. J. Adv. Sci. Eng. Inf. Technol.* **2011**, *1*, 458. [CrossRef]

37. Mishra, A.K.; Das, S.R.; Ray, P.K.; Mallick, R.K.; Mohanty, A.; Mishra, D.K. PSO-GWO Optimized Fractional Order PID Based Hybrid Shunt Active Power Filter for Power Quality Improvements. *IEEE Access* **2020**, *8*, 74497–74512. [CrossRef]

38. Shouran, M.; Anayi, F.; Packianather, M.; Habil, M. Different Fuzzy Control Configurations Tuned by the Bees Algorithm for LFC of Two-Area Power System. *Energies* **2022**, *15*, 657. [CrossRef]

fractal and fractional

Article

Hybrid LSTM-Based Fractional-Order Neural Network for Jeju Island's Wind Farm Power Forecasting

Bhukya Ramadevi [1], Venkata Ramana Kasi [1] and Kishore Bingi [2,*]

[1] School of Electrical Engineering, Vellore Institute of Technology, Vellore 632014, India; bhukya.ramadevi2020@vitstudent.ac.in (B.R.); venkataramana.kasi@vit.ac.in (V.R.K.)

[2] Department of Electrical and Electronics Engineering, Universiti Teknologi PETRONAS, Seri Iskandar 32610, Malaysia

* Correspondence: bingi.kishore@utp.edu.my

Abstract: Efficient integration of wind energy requires accurate wind power forecasting. This prediction is critical in optimising grid operation, energy trading, and effectively harnessing renewable resources. However, the wind's complex and variable nature poses considerable challenges to achieving accurate forecasts. In this context, the accuracy of wind parameter forecasts, including wind speed and direction, is essential to enhancing the precision of wind power predictions. The presence of missing data in these parameters further complicates the forecasting process. These missing values could result from sensor malfunctions, communication issues, or other technical constraints. Addressing this issue is essential to ensuring the reliability of wind power predictions and the stability of the power grid. This paper proposes a long short-term memory (LSTM) model to forecast missing wind speed and direction data to tackle these issues. A fractional-order neural network (FONN) with a fractional arctan activation function is also developed to enhance generated wind power prediction. The predictive efficacy of the FONN model is demonstrated through two comprehensive case studies. In the first case, wind direction and forecast wind speed data are used, while in the second case, wind speed and forecast wind direction data are used for predicting power. The proposed hybrid neural network model improves wind power forecasting accuracy and addresses data gaps. The model's performance is measured using mean errors and R^2 values.

Keywords: wind power; speed; direction; fractional arctan function; LSTM; fractional-order neural network

Citation: Bhukya, R.; Kasi, V.R.; Bingi, K. Hybrid LSTM-Based Fractional-Order Neural Network for Jeju Island's Wind Farm Power Forecasting. *Fractal Fract.* **2024**, *8*, 149. https://doi.org/10.3390/fractalfract8030149

Academic Editor: Gani Stamov

Received: 1 February 2024
Revised: 17 February 2024
Accepted: 20 February 2024
Published: 5 March 2024

1. Introduction

Optimising the integration of wind energy into the power grid and ensuring grid stability relies heavily on accurate wind power prediction. Over the years, researchers have explored various techniques, including neural networks, machine learning, and deep learning methods, to enhance the precision and reliability of wind power predictions. Machine learning creates a generalised model from previous input data and output results, then predicts outcomes in the future using multiple learning methods. In machine learning approaches, artificial neural networks (ANNs) [1] and support vector machines (SVMs) [2] are commonly used. ANNs can predict non-linear data and analyse the correlation between impact data and wind power. Training ANNs requires a lot of data and time, while high-dimensional data limits computational speed, leading to local optimum solutions. SVM avoids these issues and generalises them effectively. In [3], the integration of least squares and SVM (LS-SVM) was used to estimate the wind power load, enhancing computation efficiency and predicting accuracy. Incorporating LS-SVM principles, Zhang et al. introduced modifications to the model that effectively minimised prediction errors [4]. In [5], the researchers developed a fuzzy neural network for wind power forecasting coupled with online risk assessment and, in addition, investigated the effectiveness and potential improvements in enhancing wind energy forecasting models. Jie Shi et al. combined the Hilbert–Huang transform with artificial intelligence (AI) for power forecasting and further explored the effectiveness of this integrated model in improving prediction accuracy and enhancing renewable energy integration [6]. In [7], the authors developed the application of radial basis function neural networks in wind power forecasting, incorporating probabilistic methods to enhance forecasting accuracy and uncertainty assessment. The researchers employed the empirical mode decomposition (EMD) model with a neural network to forecast wind power and speed [8]. Further, they investigated the effectiveness of EMD-based models in improving short-term wind forecasting accuracy. The authors in [9] created an emotional neural network technique for predicting weather patterns and wind power generation. Additionally, they emphasised that this method can be applied to real-world scenarios. The authors proposed Gaussian processes integrated with numerical weather prediction (NWP) and complex-valued ANN for day-ahead wind power forecasting and examined their effectiveness in optimising wind energy generation and improving prediction accuracy [10,11].

Jyotirmayee et al. presented a variational mode decomposition technique in combination with a multi-kernel regularised pseudo-inverse ANN for wind power forecasting [12]. The authors developed 3D convolutional neural networks (CNNs) for extracting numerical weather prediction data in wind power forecasting, investigated similar methods to enhance prediction accuracy, and considered the potential benefits of utilising 3D CNNs in this context [13]. In [14], the structured neural network model for predicting short-term wind power emphasises the developed model's effectiveness and potential in achieving accurate short-term predictions. The researchers in [15,16] implemented an ANN for predicting wind power's discrete wavelet-transform-based wind speed and highlighted the network's effectiveness and potential in improving renewable energy integration. Additionally, the researchers emphasised the need for further investigation into model enhancements to address uncertainties and improve forecasting precision. In addition, the authors of [17] explored the current machine learning techniques for power forecasting, identifying emerging trends, and highlighted the key challenges faced in this domain. AI has shown promise in enhancing wind power generation forecasting through hybrid approaches, but significant challenges still need to be addressed for practical implementation and improved accuracy [18]. Despite these obstacles, the prospects for AI-based forecasting in the renewable energy sector remain encouraging. The authors of [19] employed an ANN model to forecast the wind power generation of the Pawan Danawi wind farm in Sri Lanka and highlighted that the model could also be applied to the environmental and climatic conditions to identify the wind power potential of the area. Machine learning methods are more effective than statistical approaches in predicting non-linear wind power data due to their adaptability and self-learning capabilities. However, these models have limitations in expressing complex data. This is because of the advancement of big data technology. Deep learning algorithms can overcome these challenges by extracting higher-level abstract features from the original samples. This enables the discovery of complex rules in high-dimensional data.

Deep learning models have advanced significantly in recent years, and deep neural network (DNN) algorithms have been introduced [20–22]. Recurrent neural networks (RNNs) and LSTM networks are robust architectures for sequence data, demonstrating advantages in non-linear feature learning [23]. Thus, RNN and LSTM are the most often used deep learning models in wind power prediction research. The authors of [24] conducted wind power generation prediction using multivariate LSTM time series. The researchers of [21] implemented deep feature extraction and LSTM techniques for data-driven wind speed forecasting and explored the effectiveness of these techniques in improving wind speed predictions. In [25], the authors used data cleaning and feature extraction techniques for power prediction. In [26], the authors used machine learning algorithms such as light gradient boosting machines (GBMs) and LSTM networks for short-term wind forecasting of weather stations in India and also aimed to enhance wind energy prediction accuracy, contributing to efficient renewable energy integration and management. The authors of [27] implemented an ensemble approach combining algorithms, namely, deep learning and gradient descent, for wind power forecasting and explored the model's effectiveness in improving forecasting accuracy and reliability. The deep-learning-based methods in [28] were developed to generate accurate and reliable prediction intervals for wind power forecasting, addressing the multi-objective nature of the problem. The researchers in [29] employed a Seq2Seq wind power output prediction method developed using deep learning and a clustering algorithm to forecast wind power with NWP data and real-time historical wind data. Adam Kisvari et al. [30] applied a deep learning approach using a data-driven and a gated recurrent unit (GRU) to forecast wind power. Further, the researchers in [31] implemented a temporal convolution network (TCN)-based approach for day-ahead wind power forecasting and compared the implemented method with the LSTM and GRU models. In [32], the authors developed LSTM-based RNNs for wind power forecasting, focusing on variable selection techniques. The authors in [33] implemented the GRU neural network method for wind power prediction, utilising evolutionary network architecture search for optimisation. The researchers in [34,35] constructed multi-modal spatio-temporal neural networks and optimised deep autoregressive RNNs for multi-horizontal wind power forecasting.

The use of attention-based models has become more popular for predicting long-term series. In [36,37], the authors demonstrated self-attention's effectiveness in capturing complex patterns and dynamics, particularly in capturing long-distance dependencies within time-series data. Juan Ren et al. [38] developed the CNN-LSTM-LightGBM framework with an attention mechanism for short-term wind power forecasting, which aimed to enhance forecasting accuracy by efficiently capturing temporal dependencies and extracting relevant features from wind power data. In [39], the authors proposed wind power forecasting methods using variational mode decomposition, and LSTM attention networks showed the encoder–decoder structure's superiority over a dual attention–LSTM neural network in enhancing prediction performance. Lei Wang et al. constructed an advanced transformer model for ultra-short-term wind power prediction [40]. Nevertheless, challenges such as space–time complexity and input and output sequence limitations remain. Furthermore, Ref. [41] developed a novel method for ultra-short-term wind power prediction, addressing previous limitations through feature extractions. The approach shows promising results, improving prediction accuracy and addressing space–time complexity issues.

A fractional-order activation function is a specific activation function used in artificial neural networks. It allows the use of non-integer exponents to calculate the output of a neuron, which can improve the performance of specific neural networks. These activation functions possess unique properties that make them suitable for specific modelling tasks and data. They are beneficial for capturing long-range dependencies and non-linear relationships in data, which cannot be effectively handled by conventional activation functions such as arctan. The fractal nature of these activation functions is attributed to their self-similarity property, which enables them to capture complex patterns in data at multiple scales. Hence, they benefit time-series forecasting applications where the data may exhibit fractal-like behaviour [42]. Using fractional-order activation functions can enhance the performance of FONN model-based forecasting by allowing for more accurate and efficient modelling of complex non-linear relationships in the data. Additionally, these functions can help to reduce overfitting, a common problem in traditional neural networks. The use of fractional-order activation functions may increase the computational load during training. However, this can be offset by the improved performance and accuracy of the model, leading to faster and more efficient forecasting [43]. Therefore, it is justified to use fractional-order activation functions because of their ability to capture complex non-linear relationships in data, which can improve the efficiency and accuracy of FONN model-based forecasting.

Motivated by the above literature, this paper presents a new hybrid model that uses LSTM to forecast missing input data and FONN to predict generated wind power. The performance of the proposed approach is evaluated based on the wind data collected from Jeju Island's wind farm in three different island sites. The key contributions of this research are outlined below:

- The LSTM model is designed to predict missing input parameters, including wind speed and direction. Its performance is evaluated through root mean squared error (RMSE) assessment.
- The FONN model predicts wind power using the LSTM's forecast data and evaluates performance with a coefficient of determination (R^2) and mean squared error (MSE).
- The models developed were evaluated in two case studies involving missing data scenarios for specific parameters.

The subsequent sections of the manuscript are organised to explore the research comprehensively. Section 2 describes the dataset from Jeju Island in three different sites and presents the data visualisation and correlation analysis in various scenarios. Section 3 describes the proposed hybrid LSTM-based fractional-order neural network model for wind power forecasting. Section 4 shows the results and discussion of the proposed models' performance evaluation to handle the missing parameter data. Section 5 concludes the proposed work.

2. Dataset Description

In South Korea, Jeju Island has a prosperous wind energy landscape with advanced wind farms strategically placed throughout scenic terrain. These wind farms take advantage of the island's plentiful wind resources, significantly contributing to its renewable energy portfolio. Sites A, B, and C are among the top wind farms on Jeju Island, each with unique specifications and characteristics. Table 1 provides an overview of the data collection period, collection time interval, and detailed wind turbine specifications for Sites A, B, and C [44].

Table 1. Jeju Island's wind farm data and specifications.

Data Aspect	Site A	Site B	Site C
Data Collection Period	11 January 2014–25 January 2014	11 January 2014–20 January 2014	11 January 2014–25 January 2014
Collection Time Interval	10 min	10 min	10 min
Wind Turbine Specifications			
Model	U88	U50	U50
Output	2000 kW	750 kW	750 kW
Wind Speed	Up to 12 m/s	Up to 12.5 m/s	Up to 12.5 m/s
Rotor Speed Range	6–17.5 rpm	9–28 rpm	9–28 rpm
Voltage and Frequency	690 V/60 Hz	690 V/60 Hz	690 V/60 Hz
Rotor Diameter	88 m	50 m	50 m
Hub Height	80 m	50 m	50 m
Power Control	Pitch Regulation	Pitch Regulation	Pitch Regulation

The wind turbine specifications presented in the table highlight each site's customised design and engineering considerations. These specifications, which include the model, output, wind speed capacity, rotor dynamics, voltage, and power control, showcase Jeju Island's commitment to harnessing wind energy efficiently and sustainably. The island's wind farms are characterised by their meticulous data collection and cutting-edge turbine specifications, which testify to their dedication to renewable energy and their aspiration to create a cleaner and greener future.

As shown in Figure 1, the data from sites A, B, and C include wind power, direction, and speed, indicating the chaotic behaviour. There were 1080, 432, and 720 samples from sites A, B, and C, respectively. These samples' pair plots are shown in Figures 2 and 3. Figure 1 shows the pairwise relationships in a dataset, while Figure 3 shows correlation coefficients between wind direction, speed, and power at the three sites. As shown in the figure, the diagonal elements are one, indicating a perfect correlation with each variable. The off-diagonal elements in the figure show the correlation between the two parameters.

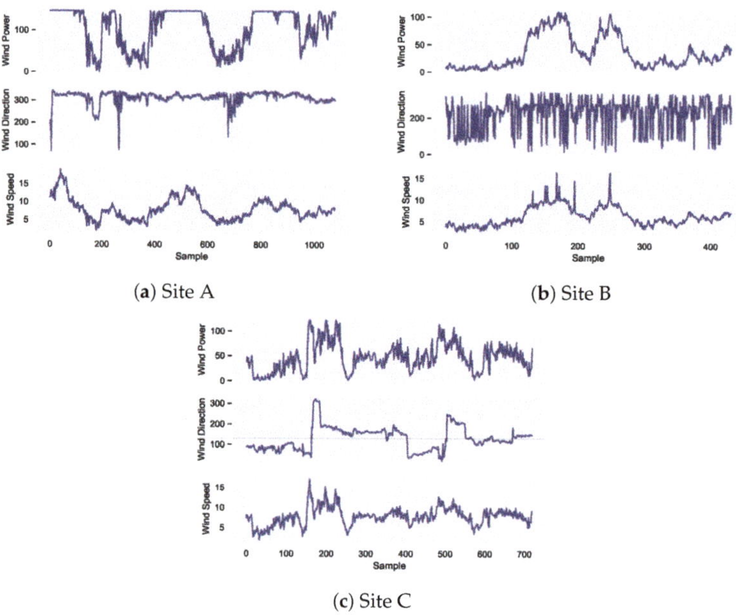

(**a**) Site A   (**b**) Site B

(**c**) Site C

Figure 1. The data from three wind farm sites on Jeju Island.

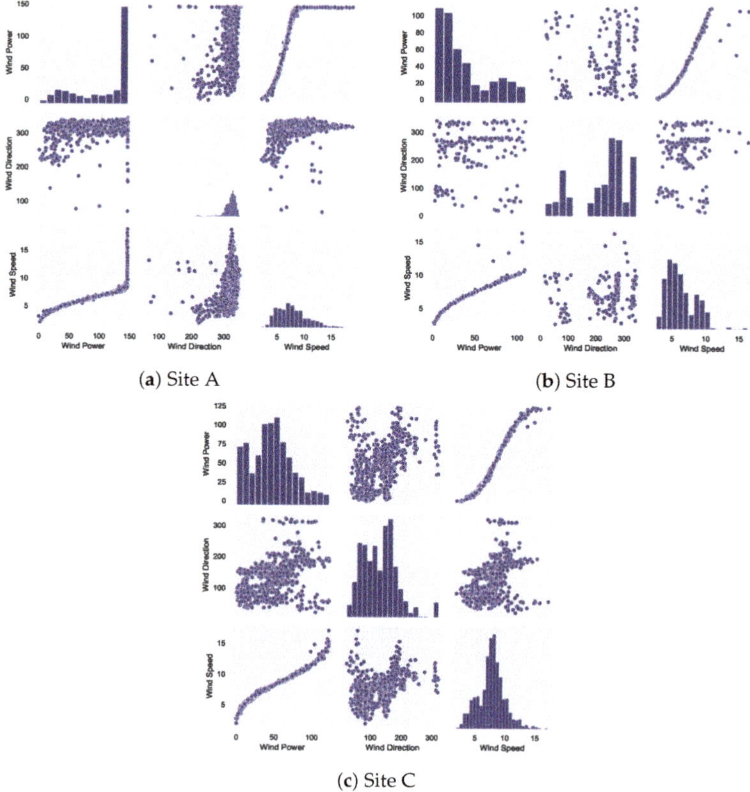

(**a**) Site A   (**b**) Site B

(**c**) Site C

Figure 2. Pairwise relationships in the data from three wind farm sites on Jeju Island.

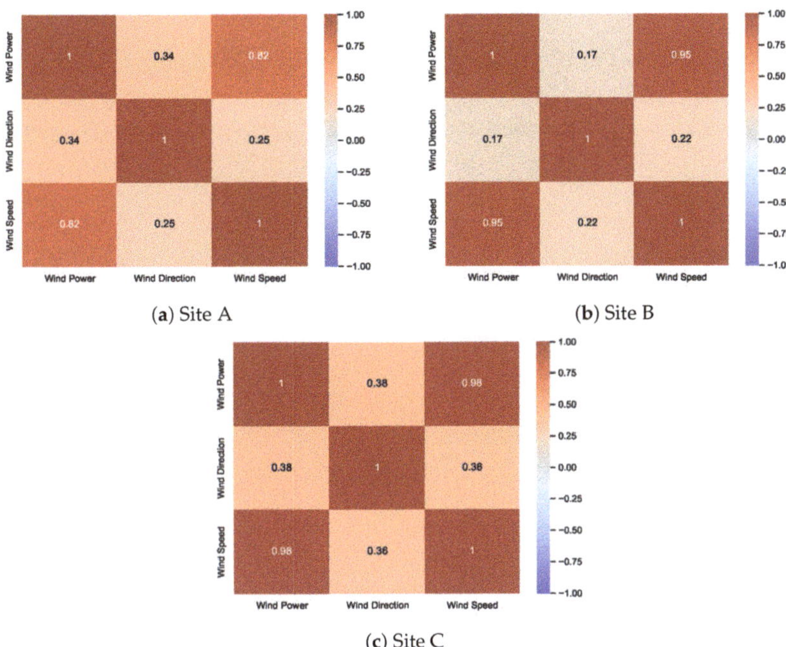

(a) Site A   (b) Site B

(c) Site C

Figure 3. Correlation analysis on the data from three wind farm sites on Jeju Island.

Utilising the Pearson correlation coefficient, a numerical measure ranging from -1 to 1, the correlation analysis shown in Figure 3 can quantify the strength and direction of relationships. This method is invaluable in providing essential insights into how changes in one parameter might correspond to changes in another. Upon examination of the wind farms at each site, it was discovered that a positive linear relationship exists between wind direction, speed, and power. The correlation coefficient between wind direction and speed at Site A is 0.25, at Site B it is 0.22, and at Site C it is 0.36, indicating a weak correlation. Similarly, the correlation coefficient between wind direction and power at Site A is 0.34, at Site B it is 0.17, and at Site C it is 0.38, indicating a weak correlation. Lastly, the correlation coefficient between wind speed and wind power at Site A is 0.82, at Site B it is 0.95, and at Site C it is 0.98, indicating a robust correlation. It is important to note that all correlation coefficients fall within the range of -1 to 1, demonstrating that the relationships are positive and linear. Additionally, it should be noted that the correlations observed at Site C are stronger than those at Site A, and the correlations at Site A are the most robust of the three sites.

2.1. Correlation Analysis of Wind Speed Parameter with Missing Data

The first case examines the correlation between wind direction, speed, and power across Sites A, B, and C, where the wind speed parameter has missing data. The correlation matrix for Site A, shown in Figure 4, presents a comprehensive view of these relationships. Notable correlations emerge, with wind power and direction exhibiting a moderate positive correlation of 0.31, indicating a tendency for increased wind direction to coincide with heightened wind power. Additionally, wind power and speed show a stronger positive correlation of 0.63, suggesting that higher wind power values correspond to elevated wind speed values. Conversely, wind direction and speed demonstrate a more modest positive correlation of 0.22. The correlation matrix for Site B mirrors these trends, showing modest positive correlations of 0.13, 0.21, and 0.23 between wind power and direction, wind power and speed, and wind direction and speed, respectively. At Site C, wind power and wind direction exhibit a moderately positive correlation of 0.31. In contrast, the correlation between wind power and speed is characterised by a relatively moderate positive correlation of 0.43. Similarly, wind direction and speed display a moderately positive correlation of 0.34. Together, these correlation matrices provide a comprehensive understanding of the relationships between wind power, direction, and speed, shedding light on the nature and strength of these associations and emphasising the need for thorough analysis, particularly in instances of missing data.

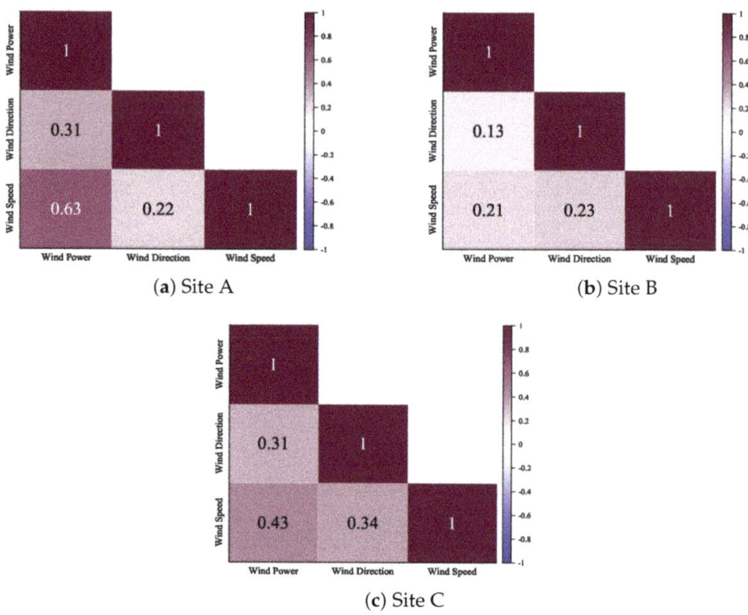

Figure 4. Correlation analysis of wind speed parameter with missing data.

2.2. Correlation Analysis of Wind Direction Parameter With Missing Data

The second case examines the correlation between wind direction, speed, and power across Sites A, B, and C, where the wind direction parameter has missing data. As shown in Figure 5, the correlation matrices for these sites reveal insightful connections. At Site A, wind power displays a moderate positive correlation with wind direction of 0.22 and a strong positive correlation with wind speed of 0.82. Additionally, a moderately positive correlation of 0.3 between wind direction and wind speed becomes evident. Site B's matrix unveils distinctive relationships. Wind power and wind direction exhibit a moderate negative correlation of −1.19, while wind power and wind speed share a robust positive correlation of 0.92. In contrast, wind direction and speed show a weak negative correlation of −1.084. Similarly, Site C's matrix reveals meaningful correlations. Wind power positively correlates with wind direction, with a correlation value of 0.19, and strongly correlates with wind speed, with a correlation value of 0.93. Further, wind direction and speed demonstrate a weak positive correlation of 0.081. In both scenarios, the correlation coefficients provide essential insights into the strength and nature of these relationships. Such understanding holds substantial implications for informed decision making and predictive analyses, particularly within renewable energy and meteorology.

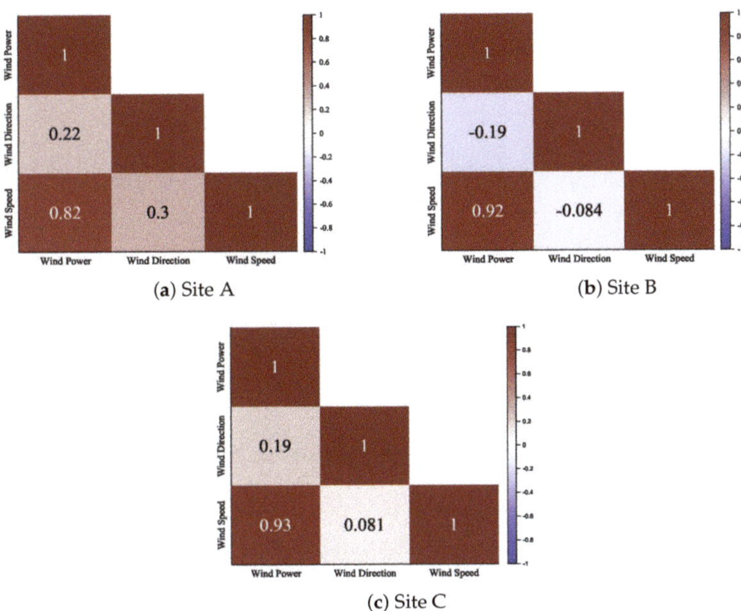

Figure 5. Correlation analysis of wind direction parameter with missing data.

3. Proposed Methodology

The proposed methodology outlines a step-by-step framework for predicting missing wind parameters and power generation within the Jeju Island wind farm context across different sites, as shown in Figure 6. This approach incorporates well-defined techniques to offer unique insights into predictive modelling and performance assessment. The initial step involves data collection and pre-processing. Specifically, the wind power, speed, and direction datasets from the Jeju Island wind farm at Sites A, B, and C are collected. This dataset provides the implementation for subsequent analysis and predictive modelling. Next, the correlation analysis is conducted with missing input data. This step considers two scenarios: missing wind speed data and missing wind direction data. The correlation analysis examines the complex relationships between wind speed, direction, and power under both scenarios. This analysis provides insights into how these parameters influence wind power generation. The detailed outcomes of this analysis are depicted in Figures 4 and 5.

In the third step of the process, an LSTM model is used to forecast the missing wind speed and direction data. The process begins with data normalisation within the range of −1 to +1, followed by the compilation of time-series data. The dataset is then partitioned into distinct subsets for training and testing. This step involves building and training the LSTM model, progressively improving its predictive capabilities by learning from the data. Continuous evaluation ensures that the model reaches the desired level of accuracy within the allocated number of training iterations. If this level is not achieved, adjustments are made through loss calculations and model updates. Once the desired level of proficiency is attained, the model is used to forecast missing wind speed and direction data. These predicted data are then compared against actual data for comprehensive analysis. The model's accuracy is evaluated through performance metrics such as the RMSE and visualisation of original and forecast waveforms.

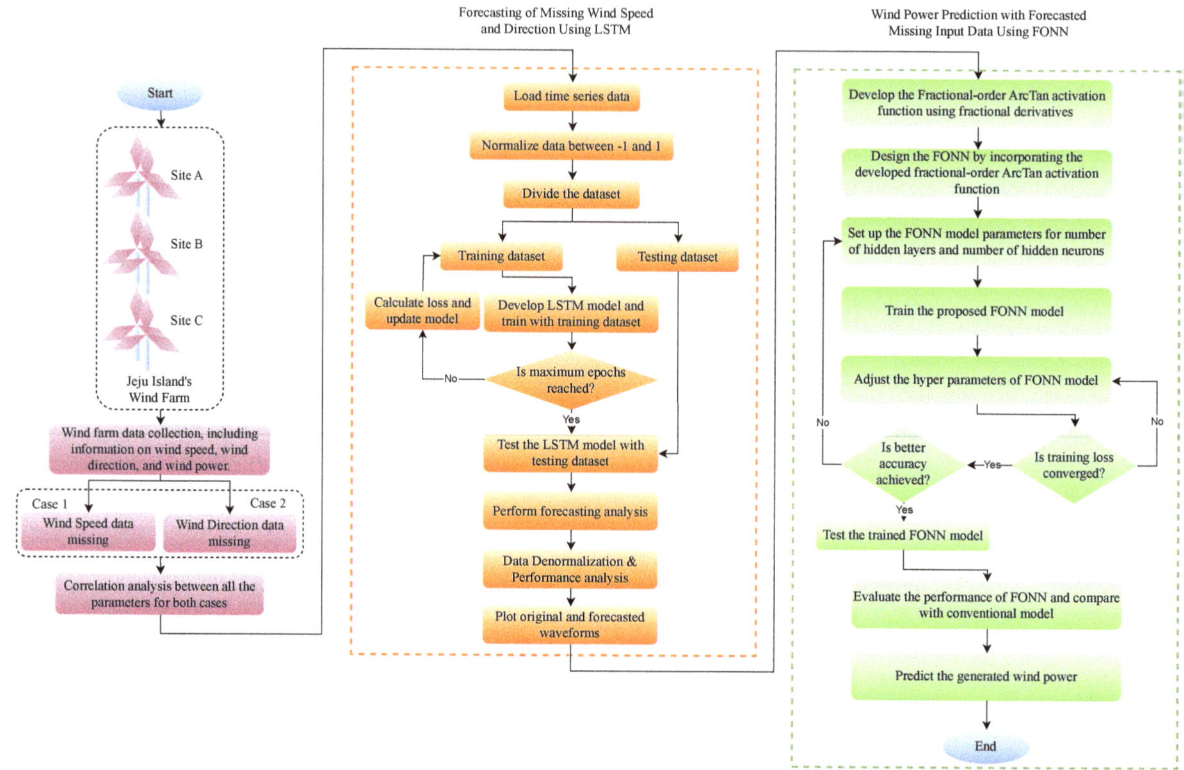

Figure 6. Flowchart of the proposed methodology.

The final stages of the methodology focus on predicting wind power using the FONN model. The process begins with developing a fractional-order arctan activation function using fractional derivatives. This newly developed function enhances the predictive capabilities of the FONN model. Subsequently, the FONN model is designed by integrating the developed fractional-order arctan function, rendering it adept at accurate wind power prediction. A precise parametrization of the model follows, encompassing the determination of the number of hidden layers and neurons. The FONN model is then iteratively trained to enhance its predictive performance. If necessary, model parameters are adjusted and retrained, ensuring continuous optimisation. A main evaluation criterion is whether the model achieves improved accuracy. Finally, the trained model is tested using a test dataset, and its performance is evaluated in terms of the coefficient of determination (R^2) and MSE, with comparisons made against a conventional neural network model. This comprehensive methodology offers a structured approach to predicting wind power and leveraging predictive modelling techniques for renewable energy applications. As shown in Figure 6 and as explained earlier, the first part of the methodology is the LSTM model's development for forecasting missing input data of wind speed and direction. The next part presents the FONN model that is used to make predictions of the generated wind power.

3.1. LSTM Model

The LSTM model's architecture for forecasting 20% of missing input data of wind speed and direction is demonstrated in Figure 7. The input to the LSTM consists of three time-steps x_{t-1}, x_t, and x_{t+1}. The architecture includes memory blocks comprising each memory cell, input, forget, and output dates, which will be explained below.

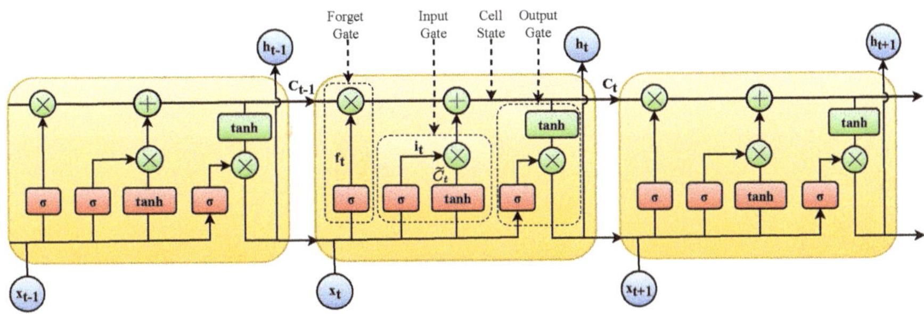

Figure 7. Architecture of LSTM model for forecasting missing input time-series data.

In the forget gate, f_t is computed using the sigmoid function; $\sigma(\cdot)$ determines the past information to forget:

$$f_t = \sigma(w_f \cdot [h_{t-1}, x_t] + b_f),\tag{1}$$

In the above equation, h_{t-1}, w_f, and b_f represent the previous cell's output, and the gate's weight and bias, respectively. The sigmoid function $\sigma(\cdot)$ output varies between 0 and 1, representing complete forgetting at 0 and full retention at 1.

As for the input gate i_t, it is also calculated based on the information to be stored in the cell state:

$$i_t = \sigma(w_i.[h_{t-1}, x_t] + b_i).\tag{2}$$

This gate's "tanh" layer adds weight to the cell state. The update equation for the memory cell, represented as $\tilde{C}_t$, is

$$\tilde{C}_t = tanh(w_C.[h_{t-1}, x_t] + b_C),\tag{3}$$

Here , w_C and b_C denote the memory cell's weight and bias.

Using the output gate O_t to determine the output information from the current cell, it can be calculated as

$$O_t = \sigma(w_O.[h_{t-1}, x_t] + b_O).\tag{4}$$

The current cell's output (h_t) will be calculated as

$$h_t = O_t \times tanh(C_t).\tag{5}$$

where C_t is the cell state and O_t is the output gate.

Two critical factors that influence the performance of an LSTM model are input delays and the number of hidden units. An LSTM network with 10 hidden units was trained to achieve adequate performance using a trial-and-error approach. Incomplete learning, limited generalisation, and underfitting can occur if the LSTM model is not sufficiently trained. Incomplete learning can cause suboptimal performance because the model fails to capture all underlying patterns. Limited generalisation means that the model needs to extend its predictions beyond the training data, leading to poor performance on missing data. On the other hand, underfitting causes poor performance on both training and test datasets. However, overtraining is unlikely due to the need for more learning from the data. Therefore, adequate training is crucial for accurate predictions. To achieve this, the Adam solver was introduced with variable learning and dropout rates of 0.005 and 0.2, respectively, over 1000 epochs.

3.2. FONN Model

The FONN model's architecture, designed to predict generated wind power using forecast missing wind direction and speed data with an LSTM model, is illustrated in Figure 8 [42]. In this configuration, there are 2 input nodes, 30 hidden nodes, and 1 output node, and their ratios are 2:30:1. The number of nodes in a hidden layer plays a significant role in determining the predictive capabilities of a neural network. Too few nodes may result in underfitting, while too many can lead to overfitting. In this case, a trial-and-error approach was used to determine that a hidden layer with 30 nodes would achieve satisfactory results. Within the architecture, bias values at the hidden and output layers were represented as "b", with values [30, 1]. For the output layer, the activation function selected was "Purelin". In contrast, the hidden layers' activation function "F" employed can vary between developed and standard tangential functions. This variation was assessed to determine the model's performance. For the training algorithm, the Levenberg–Marquardt algorithm was selected. This neural network training algorithm helps to fine-tune the model's parameters effectively. The model's performance evaluation was conducted using the performance measures outlined in the subsequent section, ensuring a comprehensive assessment of its predictive capabilities.

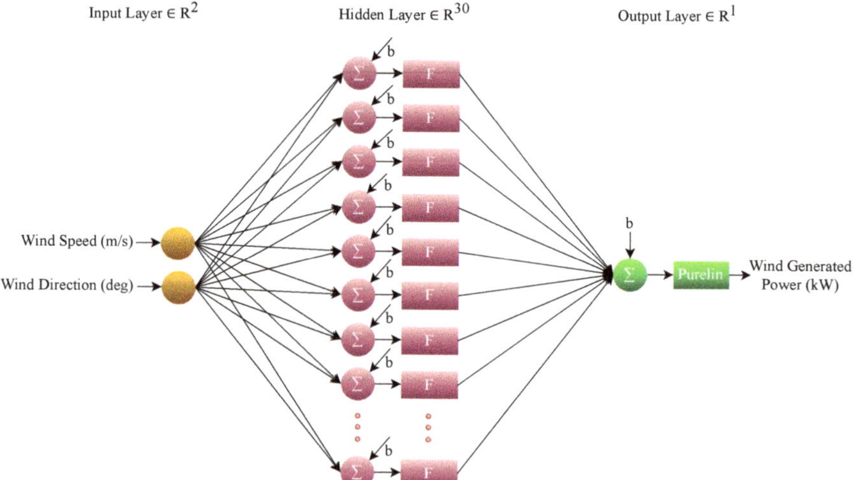

Figure 8. FONN model's architecture for generated wind power prediction.

3.3. Fractional-Order Tangential Activation Functions

The tansig activation function, also known as the hyperbolic tangent sigmoid, is commonly used in hidden layers for classification tasks [45]. This function maps input values from the range of $(-\infty, +\infty)$ to $(-1, 1)$. Its mathematical expression is given below [42]:

$$f(x) = \frac{2}{1 + e^{-2x}} - 1. \tag{6}$$

The tansig function is known to have a higher derivative compared to the sigmoid function. Additionally, its output mean is 0 when the average input values approach 0. These properties make the tansig function a valuable tool for training neural networks, as it can significantly improve convergence rates and expedite the training process. However, similar to the sigmoid function, tansig is also prone to the vanishing gradient problem [46]. Incorporating fractional-order derivatives into the tansig function to introduce a non-linear component can address the gradient problem. The fractional ordering of tansig can be derived by expressing Equation (6) using the MacLaurin series expansion as follows:

$$f(x) = \sum_{n=0}^{\infty} \frac{4^n (4^n - 1) B_{2n}}{(2n)!} x^{2n-1}. \tag{7}$$

The fractional ordering of the tansig activation function can be computed for an order $\alpha \in (0, 0.9)$ as follows [42]:

$$D^{\alpha} f(x) = g(x) = D^{\alpha} \sum_{n=0}^{\infty} \frac{4^n (4^n - 1) B_{2n}}{(2n)!} x^{2n-1},$$

$$g(x) = \sum_{n=0}^{\infty} \frac{4^n (4^n - 1) B_{2n} (2n - 1)!}{(2n)! \Gamma(2n - \alpha)} x^{2n-1-\alpha}. \tag{8}$$

Figure 9a shows the response of the fractional-order derivative of the tansig activation function for various values of α, as compared to the behaviour of the regular tansig function. The conventional tansig function has an S-shaped curve, similar to the sigmoid function and its variations. On the other hand, the fractional-order derivative of tansig has an S-shaped curve for lower values of α. However, for higher values of α, the function becomes non-linear due to the fractional ordering, which can help solve the vanishing gradient problem.

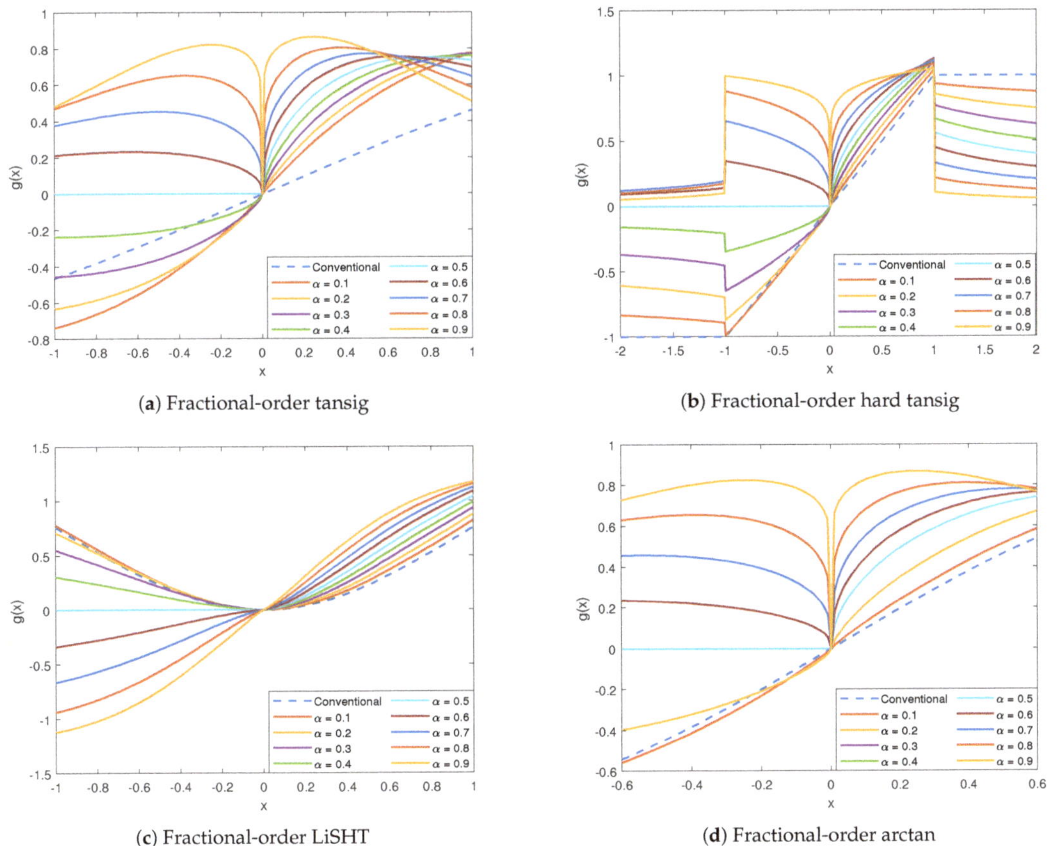

(a) Fractional-order tansig

(b) Fractional-order hard tansig

(c) Fractional-order LiSHT

(d) Fractional-order arctan

Figure 9. Response of FONN activation functions at hidden layer.

The hard tansig function is a commonly used version of the tansig activation function in deep learning applications. Unlike the tansig function, the hard tansig function is more efficient and computationally cheaper. It has a range of $[-1, 1]$ and is defined as follows [45,47]:

$$f(x) = \begin{cases} -1 & \text{if } x < -1, \\ x & \text{if } -1 \le x \le 1, \\ 1 & \text{if } x > 1. \end{cases} \tag{9}$$

The fractional-order derivative of the hard tansig function can be computed for an order $\alpha \in (0, 0.9)$ as follows [42]:

$$D^{\alpha} f(x) = g(x) = D^{\alpha} \begin{cases} -1 & \text{if } x < -1 \\ x & \text{if } -1 \le x \le 1, \\ 1 & \text{if } x > 1 \end{cases}$$

$$g(x) = \begin{cases} \frac{-1}{\Gamma(1-\alpha)} x^{-\alpha} & \text{if } x < -1 \\ \frac{1}{\Gamma(2-\alpha)} x^{1-\alpha} & \text{if } -1 \le x \le 1. \\ \frac{1}{\Gamma(1-\alpha)} x^{-\alpha} & \text{if } x > 1 \end{cases} \tag{10}$$

The comparison shown in Figure 9b highlights the response of the fractional-order derivative of the hard tansig activation function for different values of α orders compared to the conventional derivative. The analysis reveals that the functions take the α value derivative in specific intervals while exhibiting zero gradients in others. This aspect indicates that the vanishing gradient problem is less likely to occur in the fractional-order derivative of the hard tansig function as long as most of these units operate within the periods when the gradient is 1. Moreover, the analysis suggests that the fractional ordering has introduced non-linearity into the function, which will help resolve the vanishing gradient problem.

The LiSHT (linearly scaled hyperbolic tangent) is a popular activation function used in deep learning to address the "dead ReLU" issue. When the ReLU function is given negative input, it can become inactive, resulting in a zero gradient that prevents weight updates during backpropagation. As a result, to solve this problem, the LiSHT function multiplies the input with the element-wise hyperbolic tangent output. Additionally, since the hyperbolic tangent function has a range of $[-1, 1]$, negative gradients are not eliminated like with ReLU functions, which helps maintain the optimal learning for training deep neural networks. The LiSHT function can be computed by multiplying the tansig function with its input, as shown in [47].

$$f(x) = x \cdot \delta(x), \tag{11}$$

The following expression defines the tansig function $\delta(x)$, which can be found in Equation (6):

$$\delta(x) = \frac{2}{1 + e^{-2x}} - 1. \tag{12}$$

The LiSHT function in Equation (11) can be expressed using a MacLaurin series expansion as follows:

$$f(x) = \sum_{n=0}^{\infty} \frac{4^n (4^n - 1) B_{2n}}{(2n)!} x^{2n}. \tag{13}$$

The equation above enables computation of the fractional ordering of the LiSHT activation function for an order $\alpha \in (0, 0.9)$ as follows [42]:

$$D^{\alpha} f(x) = g(x) = D^{\alpha} \sum_{n=0}^{\infty} \frac{4^n (4^n - 1) B_{2n}}{(2n)!} x^{2n},$$

$$g(x) = \sum_{n=0}^{\infty} \frac{4^n (4^n - 1) B_{2n} \Gamma(2n + 1)}{(2n)! \Gamma(2n - \alpha + 1)} x^{2n - \alpha}. \tag{14}$$

The response of the fractional-order derivative of the LiSHT activation function for various α orders, compared with the conventional one, is shown in Figure 9c. The response indicates that the conventional LiSHT produces a positive output response. For lower α

values, the fractional ordering of LiSHT achieves similar behaviour. However, the response shows fractional ordering introduces more significant non-linearity than other activation functions for higher α values.

Additionally, the conventional arctan activation function is employed at the hidden layer in neural networks. However, its non-monotonic nature can pose optimisation challenges. Mathematically, the arctan function is expressed as [48]

$$f(x) = \tan^{-1}(x).\tag{15}$$

This equation can be expanded using the MacLaurin series as follows:

$$f(x) = \sum_{n=0}^{\infty} \frac{(-1)^n}{2n+1} x^{2n+1}.\tag{16}$$

The arctan function is enhanced by introducing fractional-order derivatives to tackle these challenges, improving its smoothness and optimisation potential within the FONN model. The fractional-order derivative of the arctan activation function for an order $\alpha \in (0, 0.9)$ can be computed as [49]

$$D^{\alpha} f(x) = g(x) = D^{\alpha} \sum_{n=0}^{\infty} \frac{(-1)^n}{2n+1} x^{2n+1},$$
$$g(x) = \sum_{n=0}^{\infty} \frac{(-1)^n \Gamma(2n+3)}{(2n+1)\Gamma(2n+2-\alpha)} x^{2n+1-\alpha}.\tag{17}$$

This enhancement results in smoother derivatives for the fractional-order arctan function, facilitating more effective gradient-based optimisation, which makes it better at capturing complex dynamics and long-range dependencies in wind power data, and its response at different α values is shown in Figure 9d. Compared to conventional functions, fractal activation functions like fractional-order arctan provide more flexibility in modelling non-linear systems. They are better at adapting to intricate patterns in renewable energy data.

Furthermore, the Purelin activation function is employed at the networks' output layer. It is a linear function that directly relates output to input, giving a response of kx for an input of x. The response of Purelin is shown in Figure 10. For $k = 1$, it functions as an identity. This function, with a hyperparameter k, is described as [50]

$$f(x) = kx.\tag{18}$$

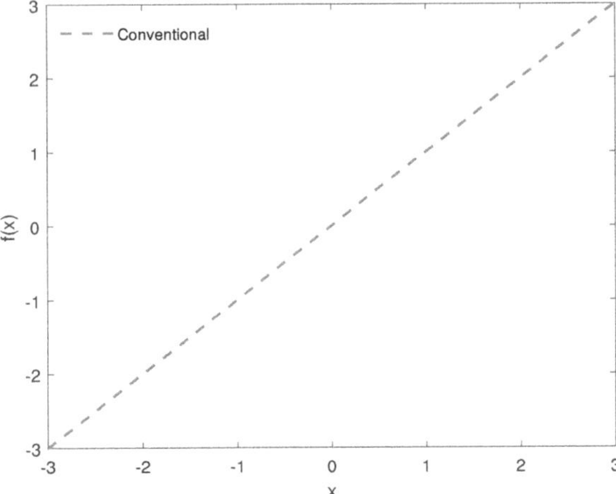

Figure 10. Response of Purelin activation function at output layer.

3.4. Performance Metrics

The MSE and RMSE are widely recognised performance metrics that assess the difference between predicted and actual values. Extensive studies have demonstrated their effectiveness as error measures in numerical prediction tasks [51,52]. The MSE is computed between actual (Y_i) and predicted $(\widehat{Y}_i)$ as follows:

$$\text{MSE} = \frac{1}{N} \sum_{i=1}^{N} (Y_i - \widehat{Y}_i)^2. \tag{19}$$

The RMSE provides an interpretable measure of the average forecasting error, which is computed as follows:

$$\text{RMSE} = \sqrt{\text{MSE}}. \tag{20}$$

Additionally, the coefficient of determination, denoted as R^2, is frequently used to show the predictive capability of forecasting methods in fitting actual data (Y_i), calculated as [52]

$$R^2 = 1 - \frac{\sum_{i=1}^{N} (Y_i - \widehat{Y}_i)^2}{\sum_{i=1}^{N} (Y_i - \overline{Y}_i)^2}. \tag{21}$$

where $\overline{Y}$ signifies the average of the predicted values. R^2 yields values ranging from 0 (indicating a poor match) to 1 (representing a perfect fit).

In all the above equations, 'N' represents the sample size and $\widehat{Y}_i$ denotes predicted values. These metrics provide valuable insights into numerical forecasting approaches' accuracy and predictive performance.

4. Results and Discussion

This section evaluates the LSTM model's accuracy in forecasting missing wind speed and direction data and the FONN model's performance in predicting wind power using the forecast data of missing wind speed and direction data in the Jeju Island wind farm for all sites under different cases.

4.1. Performance of LSTM Model

In Jeju Island's wind farm, 20% of wind speed and direction data are missing at three sites (A, B, and C). This missing data can negatively impact the accuracy of power predictions, operational planning, efficiency, safety, and overall system reliability. An LSTM model has been developed to forecast missing wind speed and direction data to address the issue, as mentioned in Section 3 and compared with non-linear autoregressive (NAR) [53], and autoregressive integrated moving average (ARIMA) [54] models. The LSTM model has one input, 200 hidden units, and one output. The model uses learning and dropout rates of 0.005 and 0.2, respectively, for 1000 iterations. Table 2 displays RMSE values for various models used in forecasting missing wind speed and direction data at three sites. The table compares the performance of three different forecasting models: LSTM, NAR, and ARIMA. Based on Table 2, the performance analysis of various forecasting models is as follows:

- The LSTM model exhibits the lowest RMSE values compared to the NAR and ARIMA models for forecasting missing wind speed data across all sites.
- At Site A, the LSTM model achieved the lowest RMSE value of 0.16, followed by NAR with an RMSE of 0.353 and ARIMA with an RMSE of 0.583.
- Similarly, the LSTM model at Site B outperformed the other models with an RMSE of 0.185, while the NAR and ARIMA models showed higher RMSE values of 0.297 and 0.458, respectively.
- Finally, at Site C, the LSTM model exhibited the lowest RMSE of 0.112, followed by ARIMA with an RMSE of 0.387 and NAR with the highest RMSE of 0.457.
- The following analysis is related to missing wind direction data forecasting, where the performance of the models varies across different sites.

- At Site A, the LSTM model had the lowest RMSE of 0.18, followed by ARIMA with an RMSE of 0.386 and NAR with the highest RMSE of 0.442.
- Similarly, at Site B, the LSTM model performed best with an RMSE of 0.425, followed by NAR with an RMSE of 0.185, and ARIMA with the highest RMSE of 0.572.
- Finally, at Site C, the NAR model had the lowest RMSE of 0.395, followed by LSTM with an RMSE of 0.126, and ARIMA with the highest RMSE of 0.454.

Table 2. Performance comparison of various forecasting models for missing data of wind speed and direction at different sites.

Model	Site	Wind Speed (m/s)	Wind Direction (deg)
		RMSE	RMSE
LSTM	Site A	0.18	0.16
	Site B	0.425	0.185
	Site C	0.112	0.126
NAR	Site A	0.353	0.442
	Site B	0.297	0.185
	Site C	0.457	0.395
ARIMA	Site A	0.583	0.386
	Site B	0.458	0.572
	Site C	0.387	0.454

The results show that the LSTM model performs better than the NAR and ARIMA models in forecasting wind speed and wind direction missing data across different sites. However, the performance may vary depending on the specific site and the nature of the wind data. Figures 11 and 12 depict the actual wind speed and direction and those forecast by the LSTM model. Further, the numerical values in Table 2 indicate the model's best performance, with RMSEs of around 0.11 and 0.12, respectively.

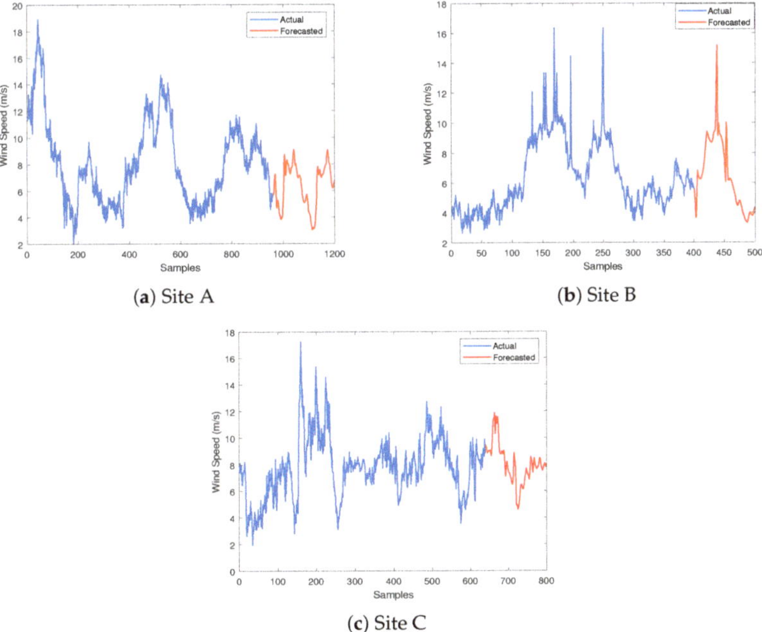

Figure 11. Actual and forecast wind speed data at different sites.

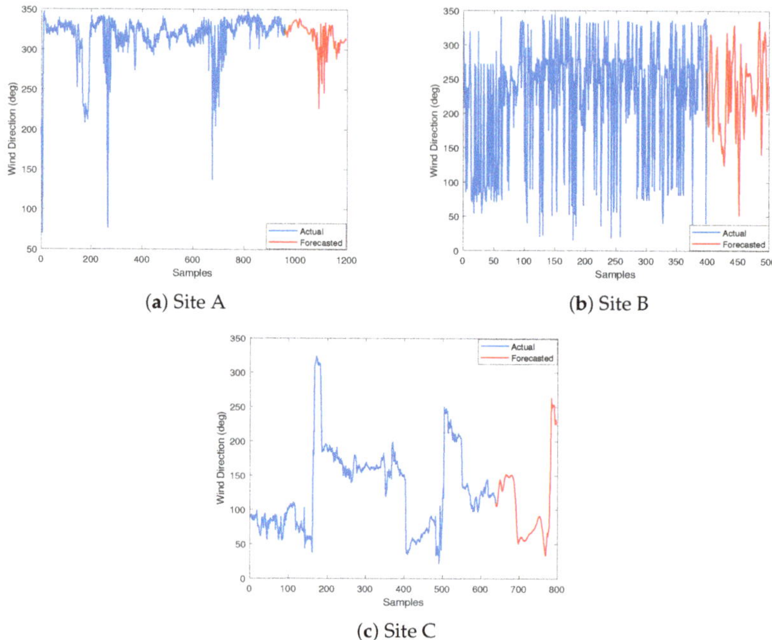

(**a**) Site A (**b**) Site B

(**c**) Site C

Figure 12. Actual and forecast wind direction data at different sites.

4.2. Performance of FONN Model

This section presents the FONN model's performance in predicting wind power using the forecast missing wind speed and direction data with the LSTM model at different sites. As per the previous section, the LSTM model showed the best performance among the forecasting models. There are two case studies considered for predicting wind power. The first case study involves predicting generated wind power using wind direction data and forecast missing wind speed data. The second case study presents generated wind power using wind speed data and forecast missing wind direction data.

4.2.1. Case Study 1

As mentioned, a FONN model with a single hidden layer was used in the first case study, as depicted in Figure 8. This neural network predicts wind power using wind direction data and forecasts missing wind speed data. The architecture comprises 2 nodes in the input layer, 30 in the hidden layer, and 1 in the output layer. The activation function "Purelin" was chosen for the network's output layer. In contrast, the hidden layer's activation function "F" varied between conventional and developed tangential functions to evaluate the performance of the FONN model in terms of R^2 and MSE. The analysis of the performance of the activation function is given in Table 3, based on the results obtained, as follows.

The following analysis was conducted to determine the accuracy of different activation functions on Site A during the training and testing phases. The results indicate that the fractional arctan function had the highest accuracy, with R^2 values of 0.9749 and 0.9831 during the training and testing phases, respectively, with MSE values of 0.0205 and 0.0142. Similarly, the fractional hard tansig function performed the best, with R^2 values of 0.9263 and 0.9369 and MSE values of 0.0424 and 0.0397 during the training and testing phases, respectively. The hard tansig function also performed well, with R^2 values of 0.8954 and 0.9075 in the training and testing phases, respectively, and an MSE value of 0.0521 in both phases. On the other hand, the conventional tansig function had the lowest accuracy, with

R^2 values of 0.8578 and 0.8642 during the training and testing phases, respectively, and MSE values of 0.0753 and 0.0764, respectively.

At Site B, the fractional tansig function performed the best, with R^2 values of 0.9428 and 0.9497 during training and testing phases, respectively, and MSE values of 0.0534 and 0.0529. The hard tansig function also performed well, with R^2 values of 0.9489 and 0.9517 during the training and testing phases, respectively, and MSE values of 0.0583 and 0.0578, respectively. On the other hand, the worst-performing function for Site B was the conventional tansig function, with R^2 values of 0.9328 and 0.9436 in the training and testing phases, respectively, and MSE values of 0.0662 and 0.0652, respectively. Moreover, the fractional hard tansig function performed better during the training and testing phases, with R^2 values of 0.9543 and 0.9609 and MSE values of 0.0464 and 0.0432, respectively. The conventional LiSHT function had R^2 values of 0.9532 and 0.9584 and MSE values of 0.0428 and 0.0414 during the training and testing phases, respectively. The fractional LiSHT function performed better, with R^2 values of 0.9572 and 0.9621 and MSE values of 0.0399 and 0.0386 during the training and testing phases, respectively. For the highest accuracy, the fractional arctan function proved to be the best option, with R^2 values of 0.9929 and 0.9952 in the training and testing phases, respectively, and MSE values of 0.0046 and 0.0032, respectively. Similarly, the conventional arctan function also performed well, with R^2 values of 0.9901 and 0.9948 during the training and testing phases, respectively, and MSE values of 0.0063 and 0.0035, respectively.

For Site C, the fractional tansig function performed better than the conventional tansig function, with R^2 values of 0.8931 and 0.9026 and MSE values of 0.0742 and 0.0629 during the training and testing phases, respectively. Similarly, the fractional hard tansig function performed better than the conventional hard tansig function, with R^2 values of 0.9035 and 0.9163 and MSE values of 0.0598 and 0.0586 during the training and testing phases, respectively. For the LiSHT function, the fractional LiSHT function performed slightly better than the conventional LiSHT function, with R^2 values of 0.8864 and 0.8973, and MSE values of 0.0752 and 0.0745 during the training and testing phases, respectively. The highest accuracy was achieved using the fractional arctan function, with R^2 values of 0.9573 and 0.9635 and MSE values of 0.0123 and 0.0115 during the training and testing phases, respectively. The conventional arctan function also performed well, with R^2 values of 0.9469 and 0.9529 and MSE values of 0.0158 and 0.0134 during the training and testing phases, respectively.

Therefore, from the results across all the sites shown in Table 3, the FONN model's best performance with the conventional arctan function is depicted in Figure 13 and the fractional arctan function is shown in Figure 14 at the hidden layer. The neural network's performance varied across sites and activation functions. The developed arctan arctan function provided improved results compared to other functions, reflected in higher R^2 values and lower MSE values for training and testing across different sites.

4.2.2. Case Study 2

The following analysis presents the second case study, which uses the same network as the previous case, with an identical node count and activation functions. However, this network uses wind speed data and forecast wind direction data as inputs to predict the generated wind power. The obtained results are shown in Table 4, and the analysis of the performance of the activation functions at the three sites is as follows.

Table 4 shows that conventional and fractional functions performed well across all the sites during the training and testing phases. Among the conventional functions, hard tansig and arctan outperformed the others in terms of R^2 and MSE values during both training and testing phases. On the other hand, the fractional function performed better overall, with higher R^2 values and lower MSE values than the corresponding conventional functions. For instance, for Site A, arctan had the highest R^2 value of 0.9898 and the lowest MSE value of 0.0081 for training, while for testing, it had the highest R^2 value of 0.9931 and the lowest MSE value of 0.0059. Similarly, the fractional arctan function had the highest R^2 value of 0.9899 and the lowest MSE value of 0.0081 for training, while for testing, it had the highest R^2 value of 0.9946 and the lowest MSE value of 0.0048. These values indicate that arctan and its corresponding fractional function were the best-performing functions for Site A. The second-best performing function for Site A was hard tansig and its corresponding fractional function. For training, hard tansig had the highest R^2 value of 0.9264 and the lowest MSE value of 0.0372, while for testing, it had the highest R^2 value of 0.9378 and the lowest MSE value of 0.0346. Similarly, the fractional hard tansig function had the highest R^2 value of 0.9726 and the lowest MSE value of 0.0218 for training, while for testing, it had the highest R^2 value of 0.9832 and the lowest MSE value of 0.0169. On the other hand, the tansig function had the lowest R^2 value of 0.8973 and the highest MSE value of 0.0621 during training. Similarly, during testing, it had the lowest R^2 value of 0.9043 and the highest MSE value of 0.0594. Similarly, the fractional tansig function had the lowest R^2 value of 0.9264 and the highest MSE value of 0.0519 during training, while it had the lowest R^2 value of 0.9329 and the highest MSE value of 0.0497 when tested. These values indicate that tansig and its corresponding fractional function were the worst-performing functions at Site A.

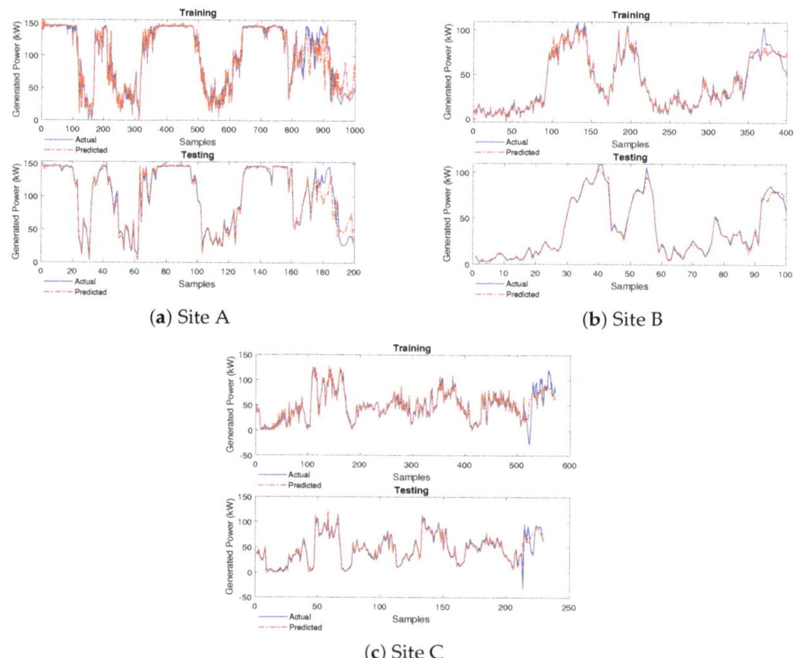

(a) Site A (b) Site B

(c) Site C

Figure 13. Performance of conventional neural network model during training and testing with forecast missing wind speed.

At Site B, the activation function arctan performed the best in both training and testing phases, with R^2 values of 0.9826 and 0.9875, respectively, and MSE values of 0.0129 and 0.0094, respectively. The arctan arctan function also performed well, with R^2 values of 0.9835 and 0.9867 and MSE values of 0.0124 and 0.0094, respectively. The best-performing activation function was arctan hard tansig, with R^2 values of 0.8864 and 0.8949 in the training and testing phases, respectively, and MSE values of 0.0682 and 0.0617, respectively. The tansig and LiSHT activation functions performed well but not as well as the arctan and arctan hard tansig functions. In conclusion, the arctan and arctan hard tansig activation functions performed the best for Site B, with arctan being slightly better regarding the R^2 value. Similarly, for Site C, the activation function arctan performed the best in both training and testing phases, with R^2 values of 0.9793 and 0.9866, respectively, and MSE values of 0.0085 and 0.0054, respectively. The arctan arctan function also performed well, with R² values of 0.9816 and 0.9865 and MSE values of 0.0076 and 0.0052, respectively. The second-best-performing activation function was arctan hard tansig, with R² values of 0.9273 and 0.9526 in the training and testing phases, respectively, and MSE values of 0.0341 and 0.0252, respectively. The tansig and LiSHT activation functions also performed well but not as well as the arctan and arctan hard tansig functions. Thus, the arctan and arctan hard tansig activation functions performed the best for Site C, with arctan being slightly better regarding the R² value. The worst-performing activation function was tansig for both conventional and fractional functions.

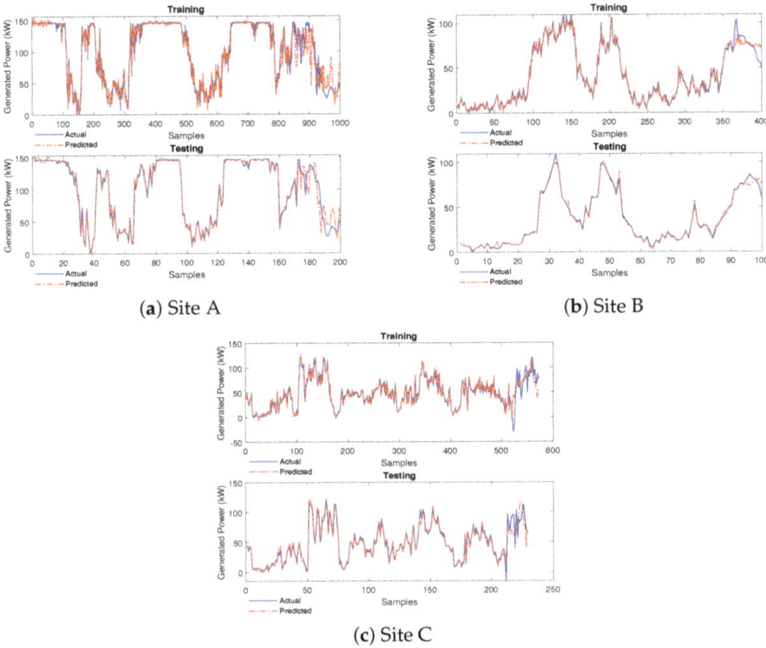

(a) Site A (b) Site B

(c) Site C

Figure 14. Performance of FONN model with forecast missing wind speed.

Table 3. Performance comparison of different functions in training and testing phases for various sites under case study 1.

Site	Conventional Function	Training		Testing		Fractional Function	Training		Testing	
		R^2	MSE	R^2	MSE		R^2	MSE	R^2	MSE
Site A	Tansig	0.8578	0.0753	0.8642	0.0764	Tansig	0.8739	0.0628	0.8864	0.0612
	Hard tansig	0.8954	0.0521	0.9075	0.0516	Hard tansig	0.9263	0.0424	0.9369	0.0397
	LiSHT	0.8749	0.0683	0.8873	0.0621	LiSHT	0.9025	0.0612	0.9173	0.0598
	Arctan	0.9727	0.0227	0.9733	0.0207	Arctan	0.9749	0.0205	0.9831	0.0142
Site B	Tansig	0.9328	0.0662	0.9436	0.0652	Tansig	0.9428	0.0534	0.9497	0.0529
	Hard tansig	0.9489	0.0583	0.9517	0.0578	Hard tansig	0.9543	0.0464	0.9609	0.0432
	LiSHT	0.9532	0.0428	0.9584	0.0414	LiSHT	0.9572	0.0399	0.9621	0.0386
	Arctan	0.9901	0.0063	0.9948	0.0035	Arctan	0.9929	0.0046	0.9952	0.0032
Site C	Tansig	0.8216	0.0853	0.8362	0.0817	Tansig	0.8931	0.0742	0.9026	0.0629
	Hard tansig	0.8453	0.0732	0.8564	0.0695	Hard tansig	0.9035	0.0598	0.9163	0.0586
	LiSHT	0.8762	0.0789	0.8758	0.0778	LiSHT	0.8864	0.0752	0.8973	0.0745
	Arctan	0.9469	0.0158	0.9529	0.0134	Arctan	0.9573	0.0123	0.9635	0.0115

Table 4. Performance comparison of different functions in training and testing phases for various sites under case study 2.

Site	Conventional Function	Training		Testing		Fractional Function	Training		Testing	
		R^2	MSE	R^2	MSE		R^2	MSE	R^2	MSE
Site A	Tansig	0.8973	0.0621	0.9043	0.0594	Tansig	0.9264	0.0519	0.9329	0.0497
	Hard tansig	0.9264	0.0372	0.9378	0.0346	Hard tansig	0.9726	0.0218	0.9832	0.0169
	LiSHT	0.9163	0.0583	0.9289	0.0542	LiSHT	0.9517	0.0487	0.9619	0.0453
	Arctan	0.9898	0.0081	0.9931	0.0059	Arctan	0.9899	0.0081	0.9946	0.0048
Site B	Tansig	0.8245	0.0982	0.8463	0.0968	Tansig	0.8562	0.0841	0.8678	0.0832
	Hard tansig	0.8674	0.0721	0.8689	0.0708	Hard tansig	0.8864	0.0682	0.8949	0.0617
	LiSHT	0.8462	0.0819	0.8573	0.0798	LiSHT	0.8693	0.0739	0.8715	0.0716
	Arctan	0.9826	0.0129	0.9875	0.0094	Arctan	0.9835	0.0124	0.9867	0.0094
Site C	Tansig	0.9041	0.0528	0.9146	0.0512	Tansig	0.9317	0.0425	0.9462	0.0419
	Hard tansig	0.9089	0.0481	0.9163	0.0479	Hard tansig	0.9273	0.0341	0.9526	0.0252
	LiSHT	0.8932	0.0514	0.9023	0.0506	LiSHT	0.9172	0.0459	0.9251	0.0445
	Arctan	0.9793	0.0085	0.9866	0.0054	Arctan	0.9816	0.0076	0.9865	0.0052

Therefore, the results presented in Table 4, the best performance of the FONN model with the conventional arctan function, is shown in Figure 15, and the developed arctan arctan activation function is depicted in Figure 16 at the hidden layer during training and testing at all three sites. The conventional arctan and the developed arctan arctan functions exhibit strong predictive abilities compared to other functions at all three sites. Minor variations in the R^2 and MSE values demonstrate the consistent and dependable performance of both functions for predicting generated wind power using the provided input data.

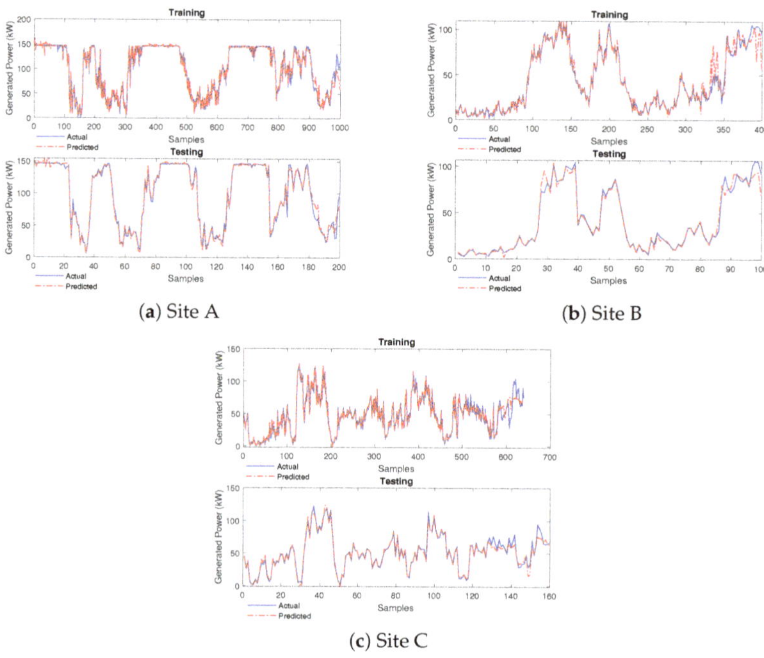

(a) Site A

(b) Site B

(c) Site C

Figure 15. Performance of conventional neural network model during training and testing with forecast missing wind direction.

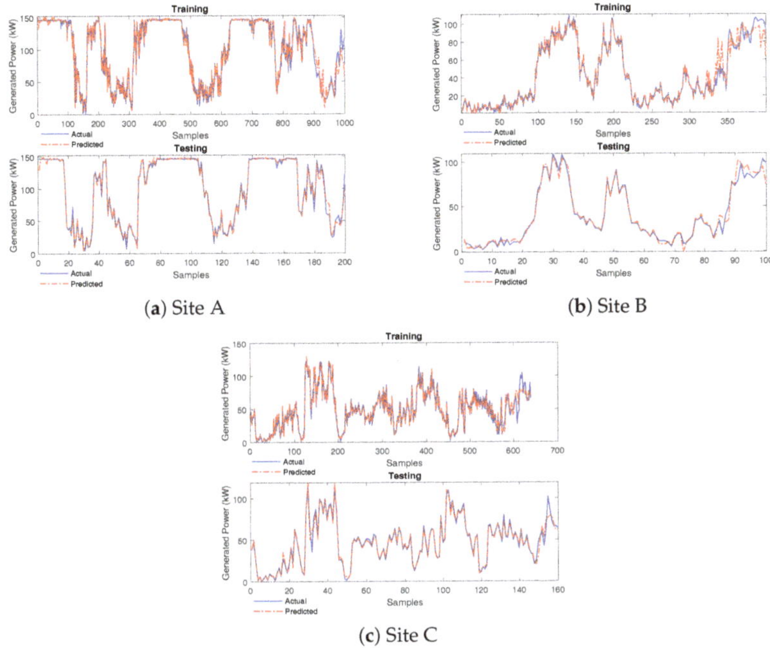

(**a**) Site A (**b**) Site B

(**c**) Site C

Figure 16. Performance of FONN model with forecast missing wind direction.

The results in both case studies compared the performance of the arctan tangential functions and the conventional tangential functions in predicting wind power across various sites. The arctan arctan function consistently achieved higher R^2 values and lower MSE values than the other functions, indicating better predictive capabilities of the FONN model. These findings have important implications for fields that rely on predictive modelling, such as finance, economics, and engineering.

5. Conclusions

A hybrid approach combining LSTMs and FONNs has been presented in this paper to forecast data missing from wind parameters and predict generated wind power across all the sites in the Jeju Island wind farm. An LSTM model was employed to forecast missing wind speed and direction data, obtaining RMSE values of approximately 0.11 and 0.12, respectively. In addition, the FONN model was used to predict wind power with forecast missing wind parameters data through two case studies. In the first case, using wind direction and forecast wind speed data, the developed arctan arctan activation function outperformed the conventional arctan function in the neural network, with high R^2 and low MSE values, around 0.97 and 0.003, respectively, during training and testing. Similarly, both activation functions exhibited strong predictive capabilities in predicting wind power using wind speed. During training and testing, the forecast wind direction in the second case achieved high R^2 and low MSE values, around 0.98 and 0.004, respectively. The results highlight the potential of the developed arctan arctan function, which consistently proved its effectiveness in enhancing predictive capabilities compared to the conventional arctan function and among all the tangential functions in both case studies. The study provides valuable insights into predicting generated wind power and fills gaps in missing data, demonstrating the potential of advanced neural networks in renewable energy applications. The developed arctan tangential activation functions have improved predictive capabilities compared to the conventional tangential functions, but their increased complexity may limit their practical implementation. In future work, there is a possibility of expanding the analysis carried out on fractional activation functions at $\alpha = 0.1$ to determine the optimal α

value. This extension of α could potentially increase the predictive accuracy of power in wind farms.

Author Contributions: Conceptualisation, B.R. and K.B.; methodology, B.R. and V.R.K.; software, B.R.; validation, B.R., K.B., and V.R.K.; formal analysis, B.R.; investigation, B.R.; resources, V.R.K.; data curation, B.R.; writing—original draft preparation, B.R.; writing—review and editing, K.B. and V.R.K.; visualisation, B.R.; supervision, K.B.; project administration, K.B.; funding acquisition, K.B. All authors have read and agreed to the published version of the manuscript.

Funding: This research was funded by Short-Term Internal Research Funding (STIRF) with grant number 015LA0-048.

Data Availability Statement: Datasets related to this article can be found at https://figshare.com/articles/dataset/J_2014_01_C_txt/8330285 (accessed on 14 November 2023).

Acknowledgments: The authors further thank the Vellore Institute of Technology in Vellore, India, for their assistance.

Conflicts of Interest: The authors declare no conflicts of interest.

References

1. Kariniotakis, G.; Stavrakakis, G.; Nogaret, E. Wind power forecasting using advanced neural networks models. *IEEE Trans. Energy Convers.* **1996**, *11*, 762–767. [CrossRef]
2. Han, Z.; Jing, Q.; Zhang, Y.; Bai, R.; Guo, K.; Zhang, Y. Review of wind power forecasting methods and new trends. *Power Syst. Prot. Control* **2019**, *47*, 178–187.
3. Cui, Y.; Li, L.; Chen, D. Ultra-Short-Term Wind Power Load Forecast Based on Least Squares SVM. *Electr. Autom. Pap.* **2014**, *5*, 35–37.
4. Zhang, Y.; Wang, P.; Ni, T.; Cheng, P.; Lei, S. Wind power prediction based on LS-SVM model with error correction. *Adv. Electr. Comput. Eng.* **2017**, *17*, 3–8. [CrossRef]
5. Pinson, P.; Kariniotakis, G. Wind power forecasting using fuzzy neural networks enhanced with on-line prediction risk assessment. In Proceedings of the 2003 IEEE Bologna Power Tech Conference Proceedings, Bologna, Italy, 23–26 June 2003; Volume 2, p. 8.
6. Shi, J.; Lee, W.J.; Liu, Y.; Yang, Y.; Wang, P. Short term wind power forecasting using Hilbert-Huang Transform and artificial neural network. In Proceedings of the 2011 4th International Conference on Electric Utility Deregulation and Restructuring and Power Technologies (DRPT), Weihai, China, 6–9 July 2011; pp. 162–167.
7. Sideratos, G.; Hatziargyriou, N.D. Probabilistic wind power forecasting using radial basis function neural networks. *IEEE Trans. Power Syst.* **2012**, *27*, 1788–1796. [CrossRef]
8. Hong, Y.Y.; Yu, T.H.; Liu, C.Y. Hour-ahead wind speed and power forecasting using empirical mode decomposition. *Energies* **2013**, *6*, 6137–6152. [CrossRef]
9. Lotfi, E.; Khosravi, A.; Akbarzadeh-T, M.; Nahavandi, S. Wind power forecasting using emotional neural networks. In Proceedings of the 2014 IEEE International Conference on Systems, Man, and Cybernetics (SMC), San Diego, CA, USA, 5–8 October 2014; pp. 311–316.
10. Chen, N.; Qian, Z.; Nabney, I.T.; Meng, X. Wind power forecasts using Gaussian processes and numerical weather prediction. *IEEE Trans. Power Syst.* **2013**, *29*, 656–665. [CrossRef]
11. Çevik, H.H.; Acar, Y.E.; Çunkaş, M. Day ahead wind power forecasting using complex valued neural network. In Proceedings of the 2018 International Conference on Smart Energy Systems and Technologies (SEST), Seville, Spain, 10–12 September 2018; pp. 1–6.
12. Naik, J.; Dash, S.; Dash, P.K.; Bisoi, R. Short term wind power forecasting using hybrid variational mode decomposition and multi-kernel regularized pseudo inverse neural network. *Renew. Energy* **2018**, *118*, 180–212. [CrossRef]
13. Higashiyama, K.; Fujimoto, Y.; Hayashi, Y. Feature extraction of NWP data for wind power forecasting using 3D-convolutional neural networks. *Energy Procedia* **2018**, *155*, 350–358. [CrossRef]
14. Abesamis, K.; Ang, P.; Bisquera, F.I.; Catabay, G.; Tindogan, P.; Ostia, C.; Pacis, M. Short-Term Wind Power Forecasting Using Structured Neural Network. In Proceedings of the 2019 IEEE 11th International Conference on Humanoid, Nanotechnology, Information Technology, Communication and Control, Environment, and Management (HNICEM), Laoag, Philippines, 29 November–1 December 2019; pp. 1–4.
15. Ansari, S.; Sampath Vinayak Kumar, T.G.; Dhillon, J. Wind Power Forecasting using Artificial Neural Network. In Proceedings of the 2021 4th International Conference on Recent Developments in Control, Automation & Power Engineering (RDCAPE), Noida, India, 7–8 October 2021; pp. 35–37. [CrossRef]
16. Khelil, K.; Berrezzek, F.; Bouadjila, T. DWT-based Wind Speed Forecasting Using Artificial Neural Networks in the region of Annaba. In Proceedings of the 2020 1st International Conference on Communications, Control Systems and Signal Processing (CCSSP), El Oued, Algeria, 16–17 May 2020; pp. 508–512.

17. Jørgensen, K.L.; Shaker, H.R. Wind power forecasting using machine learning: State of the art, trends and challenges. In Proceedings of the 2020 IEEE 8th International Conference on Smart Energy Grid Engineering (SEGE), Oshawa, ON, Canada, 12–14 August 2020; pp. 44–50.
18. Lipu, M.H.; Miah, M.S.; Hannan, M.; Hussain, A.; Sarker, M.R.; Ayob, A.; Saad, M.H.M.; Mahmud, M.S. Artificial intelligence based hybrid forecasting approaches for wind power generation: Progress, challenges and prospects. *IEEE Access* **2021**, *9*, 102460–102489. [CrossRef]
19. Peiris, A.T.; Jayasinghe, J.; Rathnayake, U. Forecasting wind power generation using artificial neural network:"Pawan Danawi"—A case study from Sri Lanka. *J. Electr. Comput. Eng.* **2021**, *2021*, 5577547. [CrossRef]
20. He, Y.; Li, H. Probability density forecasting of wind power using quantile regression neural network and kernel density estimation. *Energy Convers. Manag.* **2018**, *164*, 374–384. [CrossRef]
21. Wu, Y.X.; Wu, Q.B.; Zhu, J.Q. Data-driven wind speed forecasting using deep feature extraction and LSTM. *IET Renew. Power Gener.* **2019**, *13*, 2062–2069. [CrossRef]
22. Ramadevi, B.; Bingi, K. Chaotic time series forecasting approaches using machine learning techniques: A review. *Symmetry* **2022**, *14*, 955. [CrossRef]
23. Vaswani, A.; Shazeer, N.; Parmar, N.; Uszkoreit, J.; Jones, L.; Gomez, A.N.; Kaiser, Ł.; Polosukhin, I. Attention is all you need. *arXiv* **2017**, arXiv:1706.03762. [CrossRef]
24. Zhu, Q.; Li, H.; Wang, Z.; Chen, J.; Wang, B. Ultra-short-term prediction of wind farm power generation based on long-and short-term memory networks. *Power Grid Technol.* **2017**, *41*, 3797–3802.
25. Wang, S.; Li, B.; Li, G.; Yao, B.; Wu, J. Short-term wind power prediction based on multidimensional data cleaning and feature reconfiguration. *Appl. Energy* **2021**, *292*, 116851. [CrossRef]
26. Khochare, J.; Rathod, J.; Joshi, C.; Laveti, R.N. A short-term wind forecasting framework using ensemble learning for indian weather stations. In Proceedings of the 2020 IEEE International Conference for Innovation in Technology (INOCON), Bangluru, India, 6–8 November 2020; pp. 1–7.
27. Kumar, D.; Abhinav, R.; Pindoriya, N. An ensemble model for short-term wind power forecasting using deep learning and gradient boosting algorithms. In Proceedings of the 2020 21st National Power Systems Conference (NPSC), Gandhinagar, India, 17–19 December 2020; pp. 1–6.
28. Zhou, M.; Wang, B.; Guo, S.; Watada, J. Multi-objective prediction intervals for wind power forecast based on deep neural networks. *Inf. Sci.* **2021**, *550*, 207–220. [CrossRef]
29. Zhang, Y.; Li, Y.; Zhang, G. Short-term wind power forecasting approach based on Seq2Seq model using NWP data. *Energy* **2020**, *213*, 118371. [CrossRef]
30. Kisvari, A.; Lin, Z.; Liu, X. Wind power forecasting—A data-driven method along with gated recurrent neural network. *Renew. Energy* **2021**, *163*, 1895–1909. [CrossRef]
31. Lin, W.H.; Wang, P.; Chao, K.M.; Lin, H.C.; Yang, Z.Y.; Lai, Y.H. Wind power forecasting with deep learning networks: Time-series forecasting. *Appl. Sci.* **2021**, *11*, 10335. [CrossRef]
32. Cali, U.; Sharma, V. Short-term wind power forecasting using long-short term memory based recurrent neural network model and variable selection. *Int. J. Smart Grid Clean Energy* **2019**, *8*, 103–110. [CrossRef]
33. Zhang, K.; Jin, H.; Jin, H.; Wang, B.; Yu, W. Gated Recurrent Unit Neural Networks for Wind Power Forecasting based on Surrogate-Assisted Evolutionary Neural Architecture Search. In Proceedings of the 2023 IEEE 12th Data Driven Control and Learning Systems Conference (DDCLS), Xiangtan, China, 12–14 May 2023; pp. 1774–1779.
34. Arora, P.; Jalali, S.M.J.; Ahmadian, S.; Panigrahi, B.; Suganthan, P.; Khosravi, A. Probabilistic Wind Power Forecasting Using Optimized Deep Auto-Regressive Recurrent Neural Networks. *IEEE Trans. Ind. Inform.* **2022**, *19*, 2814–2825. [CrossRef]
35. Miele, E.S.; Ludwig, N.; Corsini, A. Multi-Horizon Wind Power Forecasting Using Multi-Modal Spatio-Temporal Neural Networks. *Energies* **2023**, *16*, 3522. [CrossRef]
36. Wu, N.; Green, B.; Ben, X.; O'Banion, S. Deep transformer models for time series forecasting: The influenza prevalence case. *arXiv* **2020**, arXiv:2001.08317.
37. Han, K.; Xiao, A.; Wu, E.; Guo, J.; Xu, C.; Wang, Y. Transformer in transformer. *Adv. Neural Inf. Process. Syst.* **2021**, *34*, 15908–15919.
38. Ren, J.; Yu, Z.; Gao, G.; Yu, G.; Yu, J. A CNN-LSTM-LightGBM based short-term wind power prediction method based on attention mechanism. *Energy Rep.* **2022**, *8*, 437–443. [CrossRef]
39. Zhou, X.; Liu, C.; Luo, Y.; Wu, B.; Dong, N.; Xiao, T.; Zhu, H. Wind power forecast based on variational mode decomposition and long short term memory attention network. *Energy Rep.* **2022**, *8*, 922–931. [CrossRef]
40. Wang, L.; He, Y.; Li, L.; Liu, X.; Zhao, Y. A novel approach to ultra-short-term multi-step wind power predictions based on encoder–decoder architecture in natural language processing. *J. Clean. Prod.* **2022**, *354*, 131723. [CrossRef]
41. Wei, H.; Wang, W.s.; Kao, X.x. A novel approach to ultra-short-term wind power prediction based on feature engineering and informer. *Energy Rep.* **2023**, *9*, 1236–1250. [CrossRef]
42. Ramadevi, B.; Kasi, V.R.; Bingi, K. Fractional ordering of activation functions for neural networks: A case study on Texas wind turbine. *Eng. Appl. Artif. Intell.* **2024**, *127*, 107308. [CrossRef]
43. Esquivel, J.Z.; Vargas, J.A.C.; Lopez-Meyer, P. Fractional adaptation of activation functions in neural networks. In Proceedings of the 2020 25th International Conference on Pattern Recognition (ICPR), Milan, Italy, 10–15 January 2021; pp. 7544–7550.

44. Son, N.; Yang, S.; Na, J. Hybrid forecasting model for short-term wind power prediction using modified long short-term memory. *Energies* **2019**, *12*, 3901. [CrossRef]
45. Dubey, S.R.; Singh, S.K.; Chaudhuri, B.B. A comprehensive survey and performance analysis of activation functions in deep learning. *arXiv* **2021**, arXiv:2109.14545.
46. Ding, B.; Qian, H.; Zhou, J. Activation functions and their characteristics in deep neural networks. In Proceedings of the 2018 Chinese Control and Decision Conference (CCDC), Shenyang, China, 9–11 June 2018; pp. 1836–1841.
47. Nwankpa, C.; Ijomah, W.; Gachagan, A.; Marshall, S. Activation functions: Comparison of trends in practice and research for deep learning. *arXiv* **2018**, arXiv:1811.03378.
48. Lederer, J. Activation functions in artificial neural networks: A systematic overview. *arXiv* **2021**, arXiv:2101.09957.
49. Job, M.S.; Bhateja, P.H.; Gupta, M.; Bingi, K.; Prusty, B.R. Fractional Rectified Linear Unit Activation Function and Its Variants. *Math. Probl. Eng.* **2022**, *2022*, 1860779. [CrossRef]
50. Sharma, S.; Sharma, S.; Athaiya, A. Activation functions in neural networks. *Towards Data Sci.* **2017**, *6*, 310–316. [CrossRef]
51. Adhikari, R.; Agrawal, R.K. An introductory study on time series modeling and forecasting. *arXiv* **2013**, arXiv:1302.6613.
52. Bingi, K.; Prusty, B.R.; Kumra, A.; Chawla, A. Torque and temperature prediction for permanent magnet synchronous motor using neural networks. In Proceedings of the 2020 3rd International Conference on Energy, Power and Environment: Towards Clean Energy Technologies, Shillong, India, 5–7 March 2021; pp. 1–6.
53. Ramadevi, B.; Bingi, K. Time Series Forecasting Model for Sunspot Number. In Proceedings of the 2022 International Conference on Intelligent Controller and Computing for Smart Power (ICICCSP), Hyderabad, India, 21–23 July 2022; pp. 1–6. [CrossRef]
54. Siami-Namini, S.; Tavakoli, N.; Siami Namin, A. A Comparison of ARIMA and LSTM in Forecasting Time Series. In Proceedings of the 2018 17th IEEE International Conference on Machine Learning and Applications (ICMLA), Orlando, FL, USA, 17–20 December 2018; pp. 1394–1401. [CrossRef]

 *fractal and fractional*

Article

Smooth and Efficient Path Planning for Car-like Mobile Robot Using Improved Ant Colony Optimization in Narrow and Large-Size Scenes

Likun Li [1], Liyu Jiang [3], Wenzhang Tu [4], Liquan Jiang [2,5,*] and Ruhan He [6]

1   School of Management, Huazhong University of Science and Technology, 1037 Luoyu Road,
    Wuhan 430074, China; lilikun0725@gmail.com
2   The State Key Laboratory of New Textile Materials and Advanced Processing Technologies, Wuhan Textile
    University, 1 Yangguang Avenue, Wuhan 430200, China
3   Hubei Institute of Measurement and Testing Technology, 2 Maodianshan Middle Road,
    Wuhan 430223, China
4   Hubei Province Fibre Inspection Bureau, 6 Gongping Road, Wuhan 430064, China; tuwenzhang@me.com
5   Hubei Key Laboratory of Digital Textile Equipment, Wuhan Textile University, 1 Yangguang Avenue,
    Wuhan 430200, China
6   School of Computer Science and Artificial Intelligence, Wuhan Textile University, 1 Yangguang Avenue,
    Wuhan 430200, China; heruhan@wtu.edu.cn
*   Correspondence: lqjiang@wtu.edu.cn

Citation: Li, L.; Jiang, L.; Tu, W.; Jiang, L.; He, R. Smooth and Efficient Path Planning for Car-like Mobile Robot Using Improved Ant Colony Optimization in Narrow and Large-Size Scenes. *Fractal Fract.* **2024**, *8*, 157. https://doi.org/10.3390/ fractalfract8030157

Academic Editors: Norbert Herencsar, Kishore Bingi and Abhaya Pal Singh

Received: 31 January 2024
Revised: 4 March 2024
Accepted: 7 March 2024
Published: 10 March 2024

Abstract: Car-like mobile robots (CLMRs) are extensively utilized in various intricate scenarios owing to their exceptional maneuverability, stability, and adaptability, in which path planning is an important technical basis for their autonomous navigation. However, path planning methods are prone to inefficiently generate unsmooth paths in narrow and large-size scenes, especially considering the chassis model complexity of CLMRs with suspension. To this end, instead of traditional path planning based on an integer order model, this paper proposes fractional-order enhanced path planning using an improved Ant Colony Optimization (ACO) for CLMRs with suspension, which can obtain smooth and efficient paths in narrow and large-size scenes. On one hand, to improve the accuracy of the kinematic model construction of CLMRs with suspension, an accurate fractional-order-based kinematic modelling method is proposed, which considers the dynamic adjustment of the angle constraints. On the other hand, an improved ACO-based path planning method using fractional-order models is introduced by adopting a global multifactorial heuristic function with dynamic angle constraints, adaptive pheromone adjustment, and fractional-order state-transfer models, which avoids easily falling into a local optimum and unsmooth problem in a narrow space while increasing the search speed and success rate in large-scale scenes. Finally, the proposed method's effectiveness is validated in both large-scale and narrow scenes, confirming its capability to handle various challenging scenarios.

Keywords: car-like mobile robot; path planning; ant colony optimization; fractional-order; narrow and large-size scene

1. Introduction

Path planning is a critical technology in the field of mobile robotics, enabling a mobile robot to efficiently navigate from its starting point to a designated target while circumventing obstacles within a given environment. It serves as a fundamental component of autonomous navigation and intelligent decision-making in mobile robot systems [1–3]. Meanwhile, Car-Like Mobile Robots (CLMRs) play an important role in the fields of warehousing and logistics, inspection, and distribution, etc. [4–6], and their chassis is equipped with a steering mechanism and suspension system [7], which makes CLMRs have good load capacity, passability, and flexibility. CLMRs are commonly utilized in environments characterized by a combination of large-scale areas and narrow spaces. Examples include neighborhoods with narrow alleys, dynamic manufacturing plants, or wild landscapes with dense vegetation [8,9]. These types of scenes impose greater demands on the path planning capabilities of CLMRs, requiring them to efficiently and accurately plan smooth paths.

Recently, path planning methods have emerged as a highly prominent area of research, captivating the attention of scholars worldwide, and researchers have extensively explored and developed innovative techniques across a diverse range of scenarios and for various types of robots [10–12]. In general, research on path planning usually focuses on two major aspects: the construction of the robot chassis model and the optimization of path planning methods. Fortunately, the fractional-order method is commonly used for modelling and optimization, which is a mathematical tool dealing with non-integer order calculus [13–16]. It extends the traditional integer order calculus by allowing derivatives or integrals of non-integer orders to exist in the model. Fractional-order methods have gained significant attention in capturing the behavior of complex nonlinear systems. These methods offer a more accurate representation of system dynamics by incorporating fractional-order differential equations, which enable the modelling of properties like nonlocal dependence and nonsmooth behavior, allowing fractional-order models to better fit the behavior of real systems and provide more accurate predictions and analyses [17–20]. Therefore, this paper aims to utilize the fractional-order approach to extend the conventional path planning method based on an integer-order model, to devise a path planning scheme that is not only smoother but also more efficient.

In terms of robot chassis model construction, it can usually be categorized into kinetic model and kinematic model construction [21]. In path planning, kinematic model construction is widely used in mobile robot path planning because it is efficient and practical [22], unless robots involving special loads or structures need to consider kinetic models [23]. Traditional kinematic model construction assumes that the steering and drive mechanisms of the vehicle are rigid bodies and uses an integer order approach for model construction [24]. This approach simplifies the modelling and computational process but also poses the problem that once the structure and parameters of the robot chassis have been determined, the angle constraints are fixed. However, for the kinematic modelling of a CLMR, the traditional approach is not applicable because CLMRs are usually equipped with shock-absorbing suspensions on the drive and steering mechanisms to enhance passability and stability [7]. This leads to changes in the chassis structure when steering or crossing obstacles, which makes the angle constraints in the kinematic model time-varying. Only by more accurately describing the time-varying angle constraints can the CLMR's ability to move in a narrow space be improved. The accurate description of time-varying systems using fractional-order methods offers a valuable opportunity to enhance the CLMR's maneuverability in narrow spaces. In this regard, this paper aims to make a significant contribution by incorporating dynamic factors, such as chassis suspension, into the precise construction of the fractional-order kinematic model.

There are many different path planning methods available [25], mainly including graph-search-based methods (e.g., A* algorithm [26]), stochastic path planning methods (e.g., Rapidly-Exploring Random Tree, RRT [27]), and optimization algorithms (e.g., Ant Colony Optimization, ACO [28], and Genetic Algorithm, GA [29]). The above path planning methods are usually used for global planning, but for dynamic obstacles, they are combined with local planning in practical applications, such as dynamic window approaches [30]. Graph-search-based methods utilize a heuristic function to assess the priority of nodes within a graph, enabling the identification of an optimal path by traversing the nodes. This approach is known for its high search efficiency and accuracy, making it particularly suitable for small-scale path planning problems. Besides, stochastic path planning methods employ random sampling and tree expansion techniques to swiftly explore feasible paths and gradually approach the desired goal position. These methods excel in high-dimensional environments and complex terrains but are susceptible to planning failures in narrow scenarios. Alternatively, optimization algorithms iteratively search for either the global optimal solution or a near-optimal solution through an optimization process [31]. These algorithms aim to find the most optimized path by iteratively refining the solution. With the advantages of a global search ability, complex scene adaptability and learning ability, ACO has a strong solving ability in path planning problems and is widely used in real-world scenarios [10].

A lot of good research has been done to use ACOs in a better way, and usually, their efficiency and smoothing are the focus [32]. The pheromone concentration settings and heuristic mechanisms of ACO are the classical means around the efficiency improvement aspect. Liu et al. propose an enhanced heuristic mechanism for Ant Colony Optimization (ACO) that incorporates adaptive pheromone concentration settings and a heuristic mechanism with directional judgments, which increases the purposefulness of planned paths and reduces turn times [33]. However, ACO usually realizes real-time planning in a small search space, and its experimental scene is generally less than a 50×50 grid map for algorithm verification [34], which still falls short of the demand for fine path planning in actual large-scale application scenarios. Path smoothing techniques commonly involve incorporating angle or path curvature constraints into the planning method and utilizing spline interpolation to refine the path. For instance, Ali et al. introduce a Markov decision process trajectory evaluation model that considers arc-length parameterization. This model effectively filters and reduces the sharpness of global paths, thereby enhancing path smoothness [35]. Tight constraints on steering angle or path curvature for the sake of smoothing can limit the robot's ability to move, especially in narrow spaces. Feng et al. put forward a path planning algorithm based on immune ACO and B-spline interpolation, which introduces a B-spline curve smoothing strategy based on the optimal solution to make the obtained path shorter and smoother [36]. Nonetheless, in narrow environments, the paths derived using spline interpolation are not necessarily usable, and they may collide with obstacles. In light of large-scale and narrow environments, further investigation of existing ACO algorithms is warranted. To address this, the integration of fractional-order models in path planning holds promise due to their advantages, including flexible and accurate parameter optimization as well as faster convergence. This paper aims to leverage fractional-order models to enhance path planning efficiency and smoothness, which represents a key highlight of the research.

Overall, the path planning performance of CLMRs in large-scale and narrow environments is still limited by inaccurate kinematic models as well as inefficient, insecure, and unsmooth planning methods. To tackle the aforementioned challenges, this paper presents fractional-order enhanced path planning for CLMRs in narrow and large-scale environments, which combines the benefits of fractional-order modelling and optimization techniques to enhance both the kinematic modelling of CLMRs and the ACO algorithm, thereby improving the efficiency of path planning and achieving smoother paths compared to traditional integer-order-based methods. The key contributions of this paper can be summarized as follows:

(1) To enhance the accuracy of kinematic model construction for CLMRs equipped with suspension systems, an innovative fractional-order-based kinematic modelling method is proposed. This method takes into account the dynamic adjustment of angle constraints to address the issue caused by the time-varying position of the steering wheel's virtual center due to suspension changes. By considering these constraints, the proposed method improves the kinematic capabilities of CLMRs, especially in limit steering states, which lays a solid foundation for subsequent efficient and smooth path planning.

(2) To address the issue of unsmooth and inefficient planning paths in narrow and large-scale scenes, an improved Ant Colony Optimization (ACO) based path planning method that incorporates fractional-order models is presented, which overcomes the limitations of traditional approaches by establishing a global multifactorial heuristic function, utilizing dynamic angle constraints in fractional-order-based kinematic modelling, incorporating adaptive pheromone adjustment rules, and adopting fractional-

order descriptive state-transfer models. These enhancements enable the algorithm to quickly acquire smooth paths and mitigate the problem of the algorithm getting trapped in local optima in narrow spaces, ultimately enhancing the searching speed and success rate of the algorithm in large-scale scenes.

(3)   Several experiments are conducted in narrow and large-size sceneries, and the effectiveness of the proposed path planning method is proved by comparison with advanced path planning methods.

The rest of this paper is organized as follows. In Section 2, system modelling and problem formulation are described. Section 3 gives the accurate fractional-order-based kinematic modeling of a CLMR. Then, improved ACO-based path planning using fractional-order models is introduced in Section 4. Experimental results are provided in Section 5, followed by the conclusions and future outlook in Section 6.

2. System Modelling and Problem Formulation

2.1. System Modelling

Constructing accurate kinematic models is essential as a prerequisite for effective path planning. However, to improve the passability of CLMRs, it is insufficient to treat the CLMR as a simple rigid structure. This is because the kinematic constraints imposed on CLMRs during their movement can vary significantly depending on the specific structure of its wheel system. As shown in Figure 1a,b, for CLMRs, limiting the minimum radius of curvature has now become a mainstream method of constructing kinematic constraints, and traditional kinematic models that do not consider suspension can be expressed in the following form:

$$v_x^2 + v_y^2 - \rho_{\max}\omega^2 \geq 0 \tag{1}$$

$$v_x \sin\theta - v_y \cos\theta = 0 \tag{2}$$

$$\frac{1}{\rho_{\max}} = \frac{1}{l}\tan(\varphi_{\max}) \tag{3}$$

where v_x and v_y are the velocity components in the direction of the x- and y-axis in the global coordinate system, respectively, ω is the angle velocity of the steering of the mobile robot, $\rho_{\max}$ refers to the maximum curvature of the running path of the mobile robot, l is the axis distance of the robot, θ denotes the angle of the mobile robot in the global coordinate system, and $\varphi_{\max}$ is the maximum steering angle of the virtual wheel system.

However, the condition for the Equations (1)–(3) to hold is that the center of rotation of the kinematic is on the extension of the rear wheels. As depicted in Figure 1c,d, the four wheels of CLMRs are usually designed in independent suspension mode to ensure the abilities of obstacle crossing and shock absorption. As a result, the center of the circle of the turn is usually not on the extension line of the rear wheels, in which case the maximum steering angle and the maximum curvature are variable quantities, and the constraints of the robot need to be recalculated. Considering the one-to-one mapping relationship between the robot's direction angle and the path taken, the feasible path needs to take into account the robot's kinematic constraints. Therefore, in this paper, we will use the fractional-order technique to construct a more accurate kinematic model for a CLMR with suspension, which will be introduced in detail in Section 3.

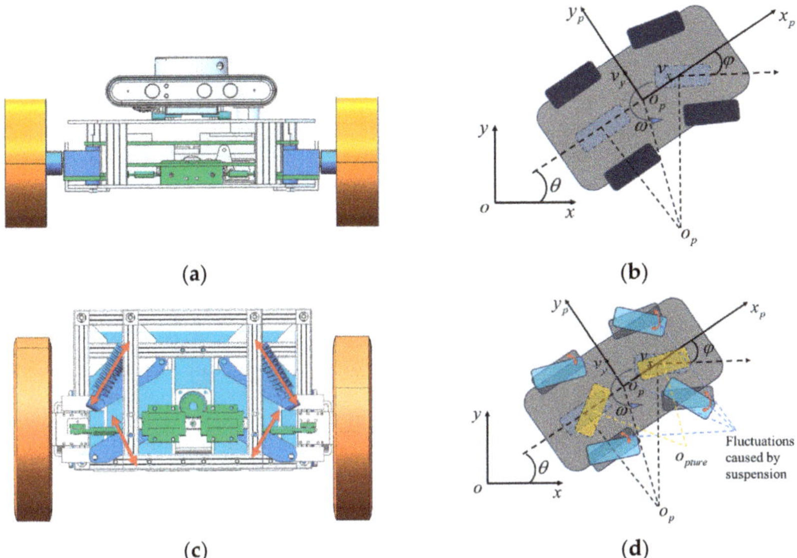

Figure 1. The kinematic model of CLMRs. (**a**) CLMR that does not consider suspension; (**b**) Traditional kinematic model that does not consider suspension; (**c**) CLMR that considers suspension; (**d**) Actual kinematic model that consider suspension.

2.2. Fractional-Order Modelling

Fractional-order calculus, with its rich mathematical properties and characteristics, is an important tool for studying and analyzing complex systems. Considering that fractional order has obvious advantages in processing and modelling real data in nonlinear systems, it can be used in constructing complex kinematic models in path planning, and local characteristic constraints more accurately, and thus close to the real situation. The commonly used fractional-order definitions are the Grunwald–Letnikov definition, Riemann–Liouville definition, and Caputo definition [17–20]. Among them, the Grunwald–Letnikov definition provides an expression for the $\alpha - th$ derivative, which allows for the consideration of the so-called short-memory principle. The Grunwald–Letnikov fractional derivative is based on discrete data points, which transform the continuity of a function into a discrete differential form. Therefore, Grunwald–Letnikov fractional derivatives apply to discrete data. This applies to the description of discrete path points in this article. Specifically, in defining the fractional-order factor $\alpha > 0$ and continuous functions $f(t)$, we have the following:

$$D^{\alpha}[f(t)] = \lim_{h \to 0} \frac{1}{h^{\alpha}} \sum_{n=0}^{\infty} (-1)^n \binom{\alpha}{n} f(t - nh) \tag{4}$$

where

$$\binom{\alpha}{n} = \frac{\Gamma(\alpha + 1)}{\Gamma(n + 1)\Gamma(\alpha - n + 1)} = \frac{\alpha(\alpha - 1)(\alpha - 2)\dots(\alpha - n + 1)}{n!} \tag{5}$$

where $D^{\alpha}(\cdot)$ denotes the GL fractional derivative of order α, $\Gamma(\cdot)$ is the Gamma function, h is the time step, and $(\alpha, n)^{\mathrm{T}}$ represents binomial coefficient.

3. Accurate Fractional-Order-Based Kinematic Modeling of CLMR

As illustrated in Figure 1c,d, the CLMRs can dampen the vibration and improve the ability to cross the ditch by installing the suspension, which also leads to the unpredictability of the steering angle during the cornering process. Based on the parameters of the damping and hydraulic cylinders, the current steering angle constraints of the robot can be obtained, which provides the kinematic constraints for path acquisition. The path acquired in this way can satisfy the obstacle avoidance while improving the tracking accuracy. To simplify the calculation process, the steering wheel can be defined as a freewheel, i.e., the wheel can rotate freely around the axle. In the calculation process, a uniform local coordinate system is defined $x_p o_p y_p$, with the robot center point as the origin of the local coordinate system and the direction perpendicular to the front of the vehicle as the x-axis. Then, the velocity of the virtual wheel in the global coordinate system is defined as:

$$\begin{bmatrix} v_{xp} \\ v_{yp} \end{bmatrix} = \begin{bmatrix} (-x_m \sin\theta - y_m \cos\theta)\omega + v_{oxg} \\ (x_m \cos\theta - y_m \sin\theta)\omega + v_{oyg} \end{bmatrix} \tag{6}$$

where, v_{xp} and v_{yp} are the velocities of the virtual wheel in the global coordinate system in the x and y directions, respectively, x_m and y_m denote the coordinates of the virtual wheel in the local coordinate system in the x and y directions, respectively, and v_{oxg} and v_{oyg} are the velocity of the origin of the local coordinate system in the global coordinate system. Further, the acceleration expression can be obtained as:

$$\begin{bmatrix} \dot{v}_{xp} \\ \dot{v}_{yp} \end{bmatrix} = \begin{bmatrix} (-x_m \cos\theta + y_m \sin\theta)\omega^2 + \\ (-x_m \sin\theta - y_m \cos\theta)\dot{\omega} + \dot{v}_{oxg} \\ (-x_m \sin\theta - y_m \cos\theta)\omega^2 + \\ (x_m \cos\theta - y_m \sin\theta)\dot{\omega} + \dot{v}_{oyg} \end{bmatrix} \tag{7}$$

According to [37], it can be known that changes in the steering angle of the wheel system can cause dynamic torque distribution. In non-rigid suspension structures, torque fluctuation can cause wheel system displacement. From the torque distribution law, the deformation of the suspension near the inner side of the arc is greater than that on the outer side of the arc, resulting in a change of angle constraint. Fortunately, onboard sensors can accurately capture the current state information during the CLMR's movement, allowing real-time constraint information to be calculated. Therefore, the variation of the virtual wheel direction angle ϕ_c for the CLMR's movement is calculated as:

$$\begin{aligned} \phi_c &= \phi_m - \theta \\ \phi_m &= \arctan\left(\frac{x_{cm}}{y_{cm}}\right) \end{aligned} \tag{8}$$

Assuming that the posterior axis is fixed and parallel to the y-axis of the defined local coordinate system, there is no change in the point of the posterior axis. Consider that the velocity relation can be represented as:

$$v^2 = v_{xp}^2 + v_{yp}^2 \tag{9}$$

therefore, it can be concluded that:

$$x_v \omega = v_{xp} \sin\theta - v_{yp} \cos\theta \tag{10}$$

This leads to a general equation for the relationship between the CLMR's attitude angle, velocity, and position, and a general constraint equation for the first-order derivatives. The relationship between the effects of velocity, attitude, and steering angle on path planning should be further clarified considering that the robot moves along a curve at different velocities. The running path (the planned path is obtained in the following section) is defined as:

$$y = f(x) \tag{11}$$

Next, the slopes at the virtual wheels are calculated and the offset of the wheel system is taken into account. Conventional equations of kinematics do not correctly express the correctness of the system's kinematic process, and fractional-order models offer the possibility of accurate system modelling. Therefore, we derive the trajectory equations and bring them into the above equation to obtain the following:

$$D^{\rho}\theta = \frac{1}{x_v}(\sin\theta - f'(x))D^{\alpha}(x) \tag{12}$$

where α and ρ are predefined fractional-order operators.

The steering angle of the CLMR imposes a constraint on the maximum curvature of the planned path, considering the dynamic characteristics of the CLMR's kinematics. However, directly calculating the curvature constraints proves challenging. From Equation (8), ϕ_m can be obtained from onboard sensors. Therefore, calculating the real-time rate of θ becomes the key to the solution. By combining Equations (11) and (12) we have the following:

$$\frac{D^{\rho}\theta}{D^{\alpha}(x)} = \frac{(\sin\theta - D^{1-\alpha}(x)f'(x)\cos\theta)}{x} \tag{13}$$

and bringing Equation (7) into Equation (8), the corner constraint can be obtained as:

$$\varphi_{max} = \arctan\left(\frac{(x_m\cos\theta - y_m\sin\theta)\frac{(\sin\theta - D^{1-\alpha}(x)f'(x)\cos\theta)}{x} + f'(x)}{(-x_m\sin\theta - y_m\cos\theta)\frac{(\sin\theta - D^{1-\alpha}(x)f'(x)\cos\theta)}{x} + 1}\right) - \theta \tag{14}$$

Considering the fluctuation of suspension in different environments, fractional-order-based kinematic modelling provides precise and dynamic angle constraints. This improves the success rate of path planning for a CLMR in narrow and difficult-to-pass scenarios.

From Figure 1, the adjustment of the angular constraints mainly relies on the suspension adjustment of the wheel system in two degrees of freedom. However, through the change of the wheel system structure, the maximum constraint angle is also changed. At this point, the circular extension of the steering is not on the rear wheel system, which is of greater relevance to the planning of the path considering the mapping of the direction angle to the path. From Equation (8), it can be seen that the wheel system angle constraint varies with the change of the wheel system angle of rotation and the initial calibration position. For computational convenience, this paper focuses on the summation constraints of the virtual wheel system to improve computational and planning efficiency. With the calculation of the maximum constraint angle and the acquisition of the current steering angle from the sensing module, we can calculate the change in the maximum constraint angle.

4. Improved ACO Based Path Planning Using Fractional-Order Model

The traditional ACO usually uses the path length as the heuristic function term when solving the path planning problem; however, the environment faced during robot operation is more complex. Path planning, as a key module of mobile robot operation, plays a vital role in the safety and smoothness of robot operation. The pseudocode of the proposed ACO method is shown in Algorithm 1. In the algorithm, lines 1 to 3 are the initialization phase of the algorithm, which completes the initialization of the weight factors and pheromones. Lines 4 to 20 are the iterative part of the algorithm. Specifically, line 6 gives the initial position of the ant colony. Lines 8 to 16 are the ant colony search under the current iteration cycle, and the next moment position of the ant is obtained by transferring the probability model, recording the status of the ant colony, and determining the relationship with the target point. After the completion of the current iteration loop, the pheromone values $\tau_{ij}(t+1)$ and path optimums L_k for the scenario are updated. In Line 21, the optimal values for each loop are compared and the optimal path L_n is selected.

Algorithm 1 The pseudocode of the improved ACO.

1	/*Initialization*/
2	Initialize the parameters, including λ^1 λ^2 φ^1 L_{tr} ε η β_1 β_2 β_3
3	Calculate initialize pheromone matrix $\tau_{ij}(0)$
4	/*main Loop*/
5	**While** iteration number n does not arrive at the target $N_{\max}$ **do**:
6	Place all ants at the start point;
7	/*inner loop*/
8	**For** k = 1 to K **do**
9	Calculate the $p_{ij}^k(t)$ using Formula (25) and confirm the next node
10	**If** Ant k reach the target point do
11	Goto step 15
12	**Else**
13	Goto step 9
14	**End if**
15	Select the optimal ant path for this round according to Equation (15)
16	**End for**
17	Update the $\tau_{ij}(t+1)$ by Formulas (23)–(25)
18	$n = n + 1$, k = 0
19	Select the optimal path L_n
20	**End while**
21	Return final optimal path L_k

To obtain a safe and feasible path, the safety, smoothness, and path distance of the path need to be considered comprehensively, so the improved multi-factor heuristic function is as follows:

$$J = \varphi^1\left\{\lambda^1\omega + \lambda^2(D + \frac{1}{d})\right\} \tag{15}$$

where J is the path planning heuristic function, φ^1 refers to the heuristic function that ensures the safe operation of the CLMR, λ^1 and λ^2 denote the weighting factors, respectively, $\omega_{ij}(k)$ refers to the curvature smoothing factor, $D(k)$ is the modified path heuristic function, and $d(k)$ implies the standard path heuristic function, which is required by the planning method to obtain the minimum value of the cost function.

4.1. Factorization of the Cost Function with Fractional-Order Model

4.1.1. Safety Functions with Local Region Preprocessing

The operational safety of the mobile robot is the first factor to be considered for path planning. As shown in Figure 2, considering the existence of tracking errors, the planning module needs to leave enough redundant space. To facilitate the process, a common approach is to uniformly inflate the static map with the CLMR's radius; however, in large-scale or highly dynamic scenarios, the optimal or relatively optimal paths are difficult to obtain and the length of the planned paths increases dramatically. Treating robots as a fixed matrix reduces the passability of a CLMR and leads to lower search efficiency. For this reason, this paper proposes a safety factor function based on ACO storage information, defined as follows:

$$\varphi_{ij}^1(k) = \begin{cases} 1, L_{tr} \times S(i,j) \cap imdilate(M_{A \times B}(i,j), L_{tr} \times S(i,j)) \\ \cap L_{pix} \times mod(Dir(i,j), 2) == 0 \neq Inf; \\ Inf, Others \end{cases} \tag{16}$$

where

$$S(i,j) = R_{C \times D} \times f(\theta) \tag{17}$$

where L_{tr} is the safety threshold constant, $S(i,j)$ denotes the intermediate function, $imdilate(\cdot)$ refers to the map expansion function, $M_{A \times B}(i,j)$ implies the map information stored by the ACO in the map information, $A \times B$ is expressed as the information dimension matrix, L_{pix}

denotes the length of the map pixel point, $mod(\cdot)$ refers to the residual function, $Dir(i,j)$ is the searching direction raster labelling, $R_{C \times D}$ stands for the CLMR's matrix under the CLMR's coordinate system, $C \times D$ denotes the robot's matrix dimensions, $f(\theta)$ is the coordinate system transfer matrix, and θ refers to the robot direction angle in global coordinates.

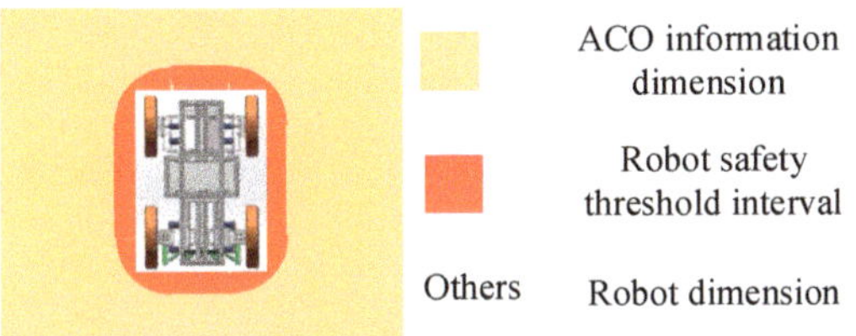

Figure 2. Planning Safety Thresholds of the CLMR.

4.1.2. Smoothing Function Based on Dynamic Angle Constraints

The ACO iterates towards the final heuristic function during the planning process, while the smoothness and feasibility of the paths are not given much attention. However, the angle constraints of the robot impose new requirements on the planning of paths, while excessive corners reduce the feasibility of paths. Improving the smoothness of the path and eliminating excessive corners will help reduce the travelling time and improve the smoothness of the path. To address these issues, considering the dynamic characteristics of the dynamic angle constraints in fractional-order-based kinematic modelling, a dynamic smoothing factor is introduced to reduce the integrated angle probability and improve the comprehensive performance of the algorithm, and the corner smoothing function is:

$$\omega_{ij}(k) = \begin{cases} \varepsilon G(i,j) \frac{\varphi_{ij}(N_{\max}-N_k)}{\varphi_c N_k} & \varphi_{ij} \leq \varphi_c \\ Inf & \varphi_{ij} > \varphi_c \end{cases} \tag{18}$$

where φ_c denotes the computed wheel system corner constraint, φ_{ij} is the planning corner at point i to point j, ε denotes the path angle adjustment factor, $G(i,j)$ represents the robot straight travelling function, $N_{\max}$ stands for the maximum number of iterations, and N_k refers to the current number of iterations. Further, as shown in Figure 3, the robot straight line function is expressed as:

$$G(i,j) = \begin{cases} \frac{\varphi_{m-1}(l_{m-2}+l_{m-1})+\varphi_m(l_{m-1}+l_m)+\varphi_{m+1}(l_m+l_{m+1})}{2(l_{m-2}+l_{m-1}+l_m+l_{m+1})} & m \geq 4 \\ 1 & m < 4 \end{cases} \tag{19}$$

where l_{m-2}, l_{m-1}, l_m and l_{m+1} are the four consecutive trajectories planned by the colony at the current point, and $\varphi_{m-1}\varphi_m$ and φ_{m+1} represent the three consecutive corners consisting of these four trajectories.

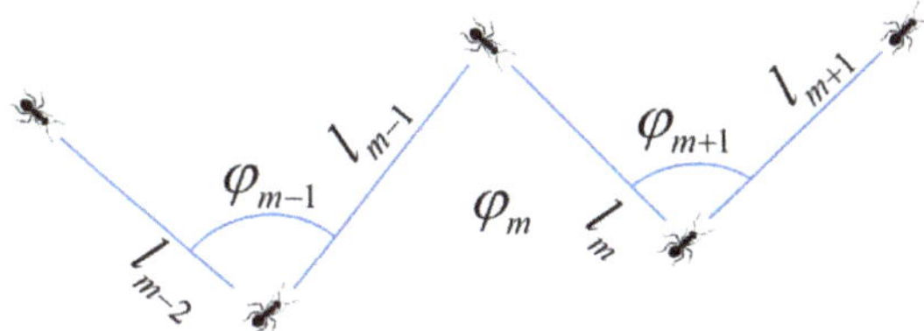

Figure 3. Planning Straight-line Constraints of the CLMR.

The feasibility of the path is improved by the smoothing function with angle constraints. In the function, the smoothing factor of the angle is added to ensure the smoothness of the planned path, which is more favorable to the operation of the CLMR.

4.1.3. Path Functions by Adding Adjusting Factor

In actual operation, the length of the path is still an important factor to be considered, which is closely related to the CLMR's work efficiency and energy utilization. For the traditional ACO, at the beginning of the iteration, the very small distance difference easily causes search confusion. At the late stage of convergence, there is a certain probability of falling into a local minimum. For this reason, we need to amplify the very small factor of fluctuation in the early stage in the distance factor to accelerate the convergence speed. In the later stages of iteration, we need to reduce the influence brought by the path, so that the path obtained is comprehensively optimal. The modified path factor function is:

$$D_{ij}(k) = \left\{ \frac{\eta\left(L(max(P_{ij}, P_{Goal})) - d(P_{ij}, P_{Goal})\right)}{L_{pix} + L(max(P_{ij}, P_{Goal})) - L(min(P_{ij}, P_{Goal}))} \frac{(N_{max} - N_k)}{N_k} \right. \tag{20}$$

where η is the path coefficient, $L(max(P_{ij}, P_{Goal}))$ denotes the longest path from the current point P_{ij} to the target point P_{Goal} planned by the ACO, $L(min(P_{ij}, P_{Goal}))$ denotes the shortest path from the current point P_{ij} to the target point P_{Goal}, and $d(P_{ij}, P_{Goal})$ refers to the Euclidean distance from the current point P_{ij} to the target point P_{Goal}.

4.2. Adaptive Pheromone Update Rules

Traditional ACO algorithms are usually set to a constant C in the initial stage, which leads to a blind search mainly relying on the heuristic function at the initial stage, and it is very easily falls into a local minimum in large scene maps. To solve this problem and improve the search efficiency, the initial pheromone is redistributed in the initial stage of the map with the help of the convergence method of the initial A* algorithm to speed up the subsequent path replanning in large scenes. The initial pheromone is recorded as:

$$\tau_{ij}(0) = \left\{ \begin{array}{l} nc, j \in l_p \\ c, j \in otherwise \end{array} \right. \tag{21}$$

In the actual operation of the CLMR, path planning is influenced by multiple factors, and the goal is to find an optimal path that considers all of these conditions collectively. Currently, efforts are focused on improving the amount of pheromone changes at different points along the path. The specific follow-up rules are as follows:

$$\tau_{ij}(t + 1) = (1 - \zeta(t))\tau_{ij}(t) + \triangle\tau_{ij}(t) \tag{22}$$

$$\triangle\tau_{ij}(t) = \left\{ \begin{array}{l} \frac{\kappa_1 Ph}{W} + \frac{\kappa_2 Ph}{L_A}, (i, j) \in allowed \\ 0, others \end{array} \right. \tag{23}$$

where $\zeta(t)$ is the dynamic volatilization factor of pheromone, $\tau_{ij}(t+1)$ denotes the pheromone matrix at the current moment, Ph represents the pheromone concentration, κ_1 and κ_2 refer to the conditioning factors, W implies the mean squared deviation value of the walking angle, and L_A denotes the cumulative path length from the starting point to the target point.

The iterative values of $\triangle\tau_{ij}(t)$ are also dynamically adjusted through the changes of angle W and distance values L_A. By setting the magnitude of the values of weight coefficients κ_1 and κ_2, the acquisition of effective paths that are more compatible with the scene is facilitated.

The dynamic pheromone volatilization factor is designed as:

$$\zeta(t) = \left\{ \begin{array}{l} a^{\frac{N_{max}}{N_k}} \zeta(t - 1), t \neq 0, (0 < a < 1) \\ \zeta_{init}, t \neq 0 \end{array} \right. \tag{24}$$

where a and ζ_{init} are self-defined constants, and $\zeta(t)$ can be adjusted adaptively with the search.

From Equation (24), it can be seen that the pheromone volatilization function $\zeta(t)$ reaches the maximum value in the pre-search period of the ACO which is in the period of the fastest change of the pheromone volatilization value. This increases the uncertainty factor in the early stage of the algorithm during the optimization process, which is more conducive to obtaining the globally optimal feasible solution. As the number of iterations increases, the pheromone volatility function $\zeta(t)$ tends to stabilize, and the local search process is more frequent, which helps to improve the quality of the path. Therefore, the iterative process accomplishes the adaptive regulation of pheromone concentration, which facilitates the realization of rapid path planning and optimization.

4.3. Fractional-Order Transfer Probability Rules

To obtain the feasible path faster and ensure the quality of the path, this paper improves the transfer probability of the algorithm. It makes the target probability increase the angle factor and distance factor. It is expected to obtain the shortest path under the premise of ensuring a smooth path. The improved state transfer probability is:

$$p_{ij}^k(t) = \begin{cases} \dfrac{[\tau_{ij}(t)]^{\beta_1}[D^{\alpha_1}R_{ij}(t)]^{\beta_2}[D^{\alpha_2}Q_{ij}(t)]^{\beta_3}}{\sum\limits_{s \in allowed}[\tau_{is}(t)]^{\beta_1}[D^{\alpha_1}R_{is}(t)]^{\beta_2}[D^{\alpha_2}Q_{is}(t)]^{\beta_3}}, \\ 0 \end{cases} \tag{25}$$

where β_1, β_2, and β_3 denote the heuristic term factor, respectively, and $R_{ij}(t)$ and $Q_{ij}(t)$ are defined as follows:

$$D^{\alpha_1}R_{ij}(t) = D^{\alpha_1}\omega_{ij}(t) \tag{26}$$

$$D^{\alpha_2}Q_{ij}(t) = D^{\alpha_1}(D_{ij}(t) + d_{ij}(t)) \tag{27}$$

The fractional reciprocal of the angle factor and the distance factor is calculated to improve the sensitivity of the transition probability to its change, to ensure the timeliness of the path change and to improve the passability of the path.

A fractional-order state-transfer model can more accurately adjust the exploration probability of ant colonies in unexplored areas, which is beneficial for ant colony algorithms to jump out of the current local optimal solution and search for the global optimal solution with a greater probability, thereby improving the success rate of the search in large-scale scenarios. At the same time, due to the high dependence of the pheromone concentration on the optimal path, the modification of the transfer model increases the search breadth and the search probability of the optimal path, avoiding the acquisition of the optimal path, accelerating the search process around the optimal path, and thus improving the convergence speed.

5. Experimental Validations

5.1. Experimental Implementation

The narrow and large-size experimental scenes and self-developed CLMRs are shown in Figure 4. The CLMR consists of an industrial computer (Intel(R) Core (TM) i7-6500U CPU @2.50 GHz, 8 GB of RAM, 64-bit operating system), LiDARs, motor encoders, and some related sensors, such as an ultrasonic transducer and IMU. More specifically, with an impressive range of 150 m and a scanning rate of 10 Hz, the Velodyne VLP-16 LiDAR provides the CLMR with a broad field of view, which guarantees that the CLMR has enough field of view to ensure safety and real-time mapping and path planning. As illustrated in Figure 4a, to improve the stability and passability of the CLMRs in complex environments, the drive and steering mechanisms are fitted with suspensions. As we can see from Figure 4b–d, the entire neighborhood is quite expansive, covering an area of sixteen thousand square meters. However, the alleyways within the neighborhood are remarkably narrow. These tight spaces are often filled with temporarily parked cars and bustling pedestrians, which severely limits the available space for CLMRs to navigate through. In particular, shown in Figure 4b, to obtain more detailed and practical paths, we chose a grid map with a resolution of 0.2 m, making the map size 800×500, which is a great challenge for the path planning algorithm.

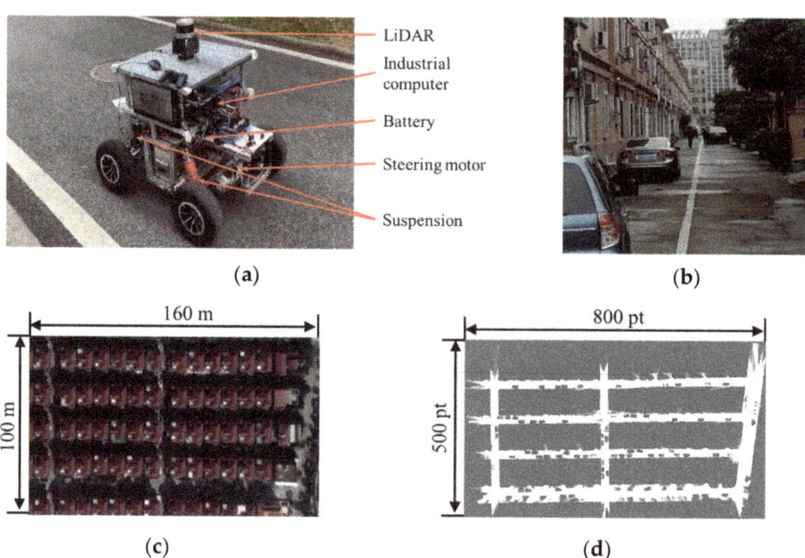

Figure 4. Experimental scene and platform. (**a**) Self-developed CLMR; (**b**) Real-world scene; (**c**) Google map of the narrow and large-size scene; (**d**) Grid map of the narrow and large-size scene.

5.2. Experimental Results and Discussions

In the experiment, it is necessary to obtain real-time vehicle positioning data and wheel steering angle data. Real-time recording and storage of experimental data was performed on the PC. The planning, calculation, allocation, and execution process are as follows: The current data are processed by the data processing unit and returned to MATLAB. Then, MATLAB is used to complete the calculation of the planning algorithm. Finally, the path instructions generated by the planning are sent to the control unit, completing the current path planning process. The initial state of the considered robot is the same, all parameters are optimally adjusted, and experiments are conducted under the same operating conditions. In the process of parameter tuning, parameter stabilization is achieved through the use of a nature-inspired optimization algorithm called Artificial Bee Colony [38], which reduces the sensitivity of the parameters to the environment and also ensures the fairness of the algorithm comparison process. The values of the parameters obtained are as follows: $\lambda^1 = 0.6$, $\lambda^2 = 0.2$, $\varepsilon = 0.1$, $N_{\max} = 50$, $\eta = 1.1$, $\kappa_1 = 0.8$, $\kappa_2 = 0.2$, and $a = 0.4$. The superiority of the proposed fractional-order ACO (FACO) is verified by comparing it with the traditional A*, improved A* (IA) [26], ACO [28], improved ACO combined with path fitting (ACOF) [39], Genetic Algorithm (GA) [29], and the GA method combined with A* (AGA). We used different algorithms to run each of them ten times in large-scale narrow scenes. To better validate the effectiveness and advantages of the proposed method, we drew on the comparative methods in literature [40] and selected common path lengths, times, and success rates of trajectory planning (including planning failures and collisions with obstacles) to accurately describe the process of path planning. The comparison results are shown in Table 1, and the experimental results of the planned path are shown in Figures 5–11.

Table 1. Performance comparison of different path planning methods.

Methods	Path Length (m)	Times (s)	Success Rate (%)
A*	91.60	857.26	100
IA	93.57	591.73	100
GA	101.24	100.61	100
AGA	93.36	123.67	100
ACO	89.20	350.66	100
ACOF	94.97	153.71	0
FACO	92.98	63.74	100

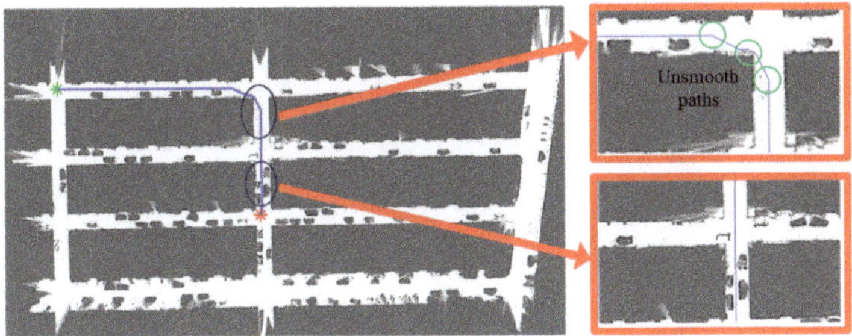

Figure 5. Path planning result of A*.

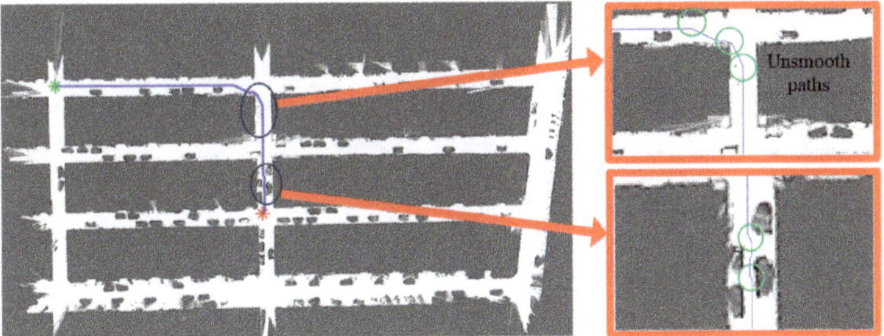

Figure 6. Path planning result of IA.

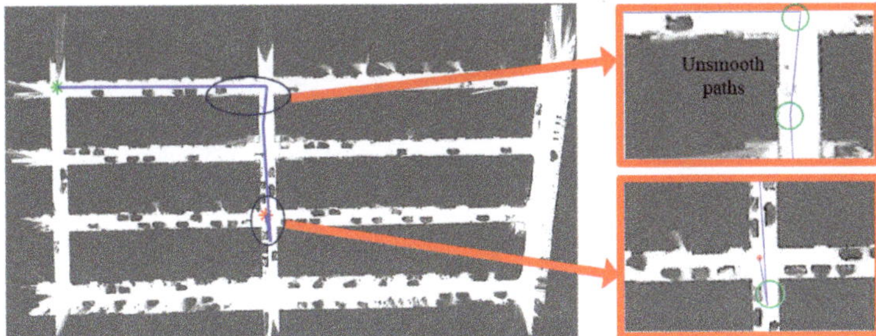

Figure 7. Path planning result of GA.

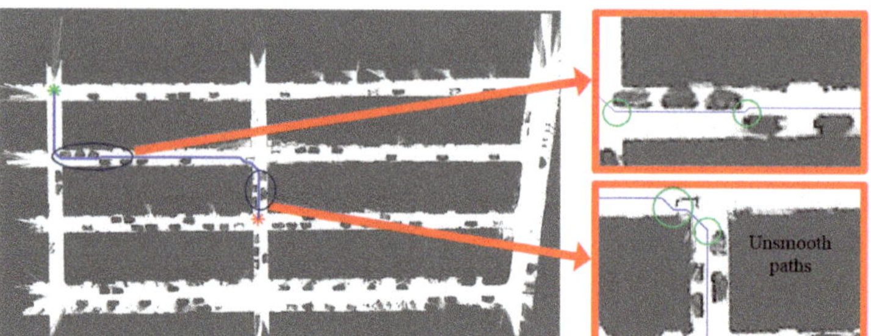

Figure 8. Path planning result of AGA.

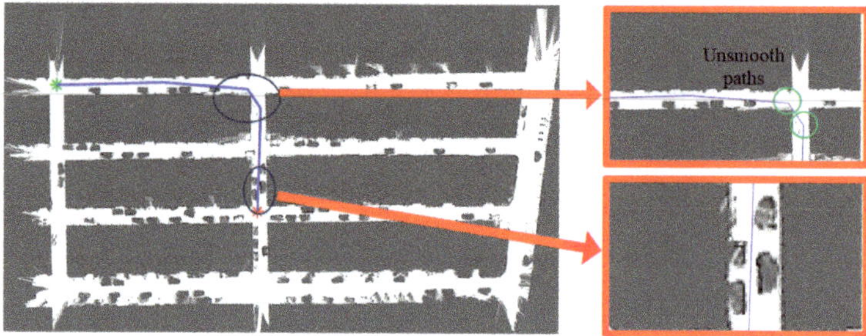

Figure 9. Path planning result of ACO.

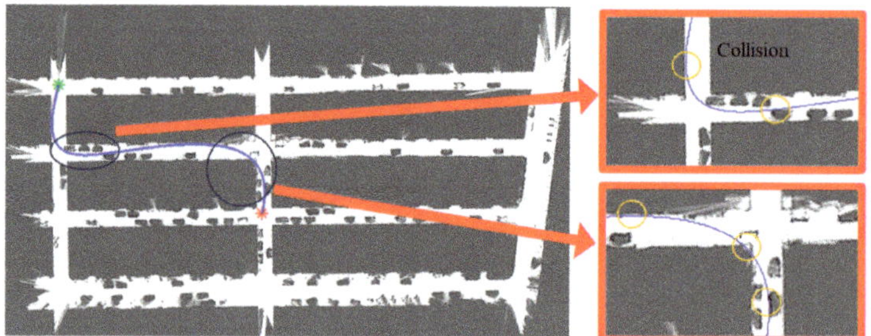

Figure 10. Path planning result of ACOF.

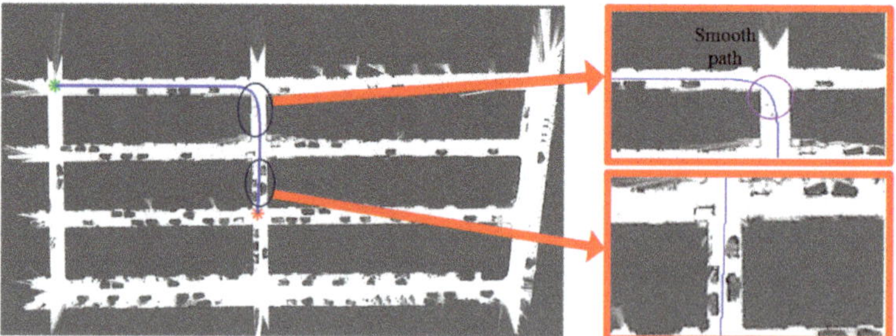

Figure 11. Path planning result of FACO.

From Table 1, it can be seen that the A* algorithm takes the longest time, which is because the A* algorithm completes the traversal of the surrounding space before finding the target point, and due to the high-precision attribute of the map, more and more computational resources are consumed and the computational speed decreases dramatically in the planning process. In the improved A* method, the search direction is guided to improve the planning efficiency, but to ensure the success rate of path planning, the efficiency improvement for a large-space search is not obvious. GA has excellent performance in the field of optimization, and in the planning process GA achieves random sampling in the space by repeated cross-compilation, which greatly reduces the planning time, and by combining it with the A* algorithm, there is an increase in the planning time, but there is an increase in the stability of the path. The ACO-based planning algorithm demonstrates reduced dependency on parameters in path planning. However, employing pure ACO preprocessing alone increases the time required. By combining the ACO algorithm with A*, the overall time decreases further. In the proposed method, the additional search burden caused by the large space is mitigated through local space weighting in the ACO search. This reduction in computational burden leads to improved planning efficiency. Experimental results show a significant enhancement in efficiency with the proposed method, achieving improvements of 92.56%, 57.84%, and 81.82% compared to traditional A*, GA, and ACO methods, respectively. These improvements have great significance for real-world scenarios.

In terms of path length, the A* method shows a greater advantage due to the objective of the method to obtain shorter paths, and with the increase in constraints, the paths of the improved methods based on A*, GA, and ACO all increase to varying degrees. Further analysis shows that GA shows significant non-randomness of paths due to the random sampling in space and the path length appears to increase to a greater extent, whereas the ACO algorithm shows better stability of paths due to having an advanced spatial search and shorter path lengths. The proposed method needs to meet the actual operational requirements and the dynamic constraints make the planned paths increase, but the length of the planned paths decreases by 0.63%, 0.40%, and 2.10% compared to the improved methods of A*, GA, and ACO, respectively. When comparing the various path planning, all of them show better results in terms of path length. However, the proposed method is more advantageous under the constraints.

In Table 2, the smoothness indicator is given, which reflects the proportion of different steering angles between path points in path planning. It has been proven that a smoother trajectory is achieved when there is a lower proportion of large steering angles. However, optimization-based planning algorithms, such as GA, AGA, ACO, and ACOF, often prioritize a single performance improvement, resulting in a path consisting of a relatively simple finite set of points. For example, the GA algorithm represents the path using only five coordinate points, and the proportion of large steering angles reaches 66.67%. Consequently, these optimization algorithms introduce numerous angle mutation points, making it challenging to ensure a smooth path. Moreover, traditional path planning methods like A* and IA struggle to incorporate dynamic angle constraints at corners, leading to scattered abrupt changes in the path. Although smoothing the obtained path can prevent the occurrence of mutation points, it may also result in high spatial requirements after the smoothing process. To address these challenges, we propose a novel path planning algorithm that fully leverages the constraint characteristics of angles. This algorithm performs real-time smoothing during the planning phase, avoiding abrupt changes in the path's angle, the proportion of small steering angles exceeds 99%, which means that the proposed method is essentially free of steering mutations. As a result, it offers significant benefits for planning in narrow spaces while still considering angle constraints.

Table 2. The proportion of different steering angles between path points.

Methods	0–5°	5–10°	>10°
A*	98.39%	1.15%	0.46%
IA	96.77%	2.30%	0.93%
GA	0	33.33%	66.67%
AGA	0	0	100%
ACO	33.33%	0	66.67%
ACOF	100%	0	0
FACO	99.4%	0.6%	0

According to the results of path planning from Figures 5–11, it is evident that both the A* algorithm and its improved version can produce relatively smooth paths. However, it is noticeable that the paths tend to excessively prioritize shorter routes at corner nodes and articulation points. This pursuit of shorter paths increases operational risks and reduces feasibility. In the GA method, random sampling points introduce a certain possibility of encountering large corners, resulting in longer paths, and placing higher demands on the CLMR's maneuverability. To address this issue, the combination of A* and GA methods improves the path's smoothness and provides guidance. Nonetheless, the presence of more corners and narrow areas in the path poses greater challenges for the robot's capabilities and model. In the planning of ACO, it follows the pursuit of the shortest path when the path shows better smoothness, but the passage is not considered in the method. Given this, after combining with A*, the smoothing of the path is proposed, and here the B-spline is used for processing because the smoothing of the path is usually accompanied by a change in the path points, and it can be seen from the figure that, due to the space being relatively small, the path status quo is changed, and a larger set of points appeared to be in contact with the obstacles, which poses a greater threat to the operational safety, making it difficult for the traditional path smoothing to be sufficient. In the proposed method, a transfer model of fractional-order is established by building a fractional-order model with constraints in the method, which improves the path smoothing, avoids the path changes brought about by the additional path smoothing, and improves the path safety, while the obtained smoothed paths ensures the path feasibility. At the same time, we can see that the paths of ACOF are corrected to increase smoothness, but in a narrow space, and this late correction makes it easy for the planned paths to run into obstacles, resulting in a drastic decrease in their success rate. The proposed method, on the other hand, maintains smoothness and at the same time has a high success rate, which proves the excellent performance of the planning method based on the fractional-order model.

Figure 12 provides the mean and current lowest values of the objective function for various optimization algorithms: GA—both Mean GA (MGA) and Current Lowest GA (CLGA), AGA—both Mean AGA (MAGA) and Current Lowest AGA (CLAGA), ACO—both Mean ACO (MACO) and Current Lowest ACO (CLACO), ACOF—both Mean ACOF (MACOF) and Current Lowest ACOF (CLACOF), and FACO—both Mean FACO (MFACO) and Current Lowest FACO (CLFACO). GA is characterized by fast convergence in the initial stage; however, with the depth of the iteration, the GA is more likely to fall into premature maturity. In contrast to the ACO calculation, the method is more dependent on the initial value, and the convergence is slower in the early stage; however, with the help of the method's adaptation to nonlinear and complex problems it has an advantage in dealing with the local optimal solution. To improve the search efficiency, the traversal based on the A* method is used in this paper: AGA, ACOF, and FACO. This shows great advantages in the fast convergence of the mean value and the optimization of the minimum value. In the iterative search for the minimum value, AGA achieves faster convergence, while ACOF and FACO can still jump out of the current local optimality conditions and further search for the global optimal solution. Therefore, the proposed method has a great advantage in realizing the optimal value of the cost function in this paper.

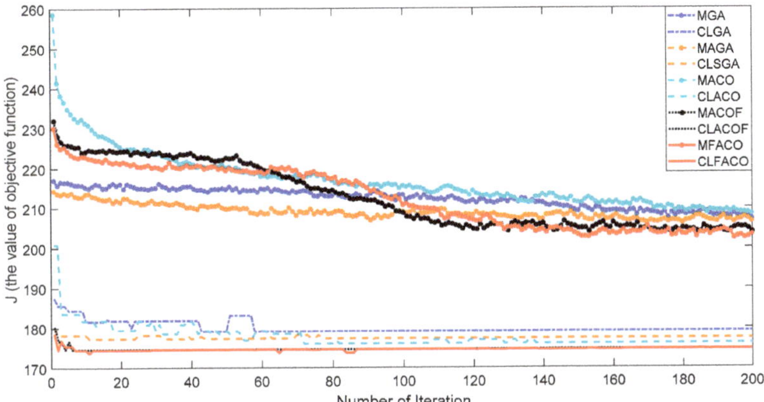

Figure 12. Comparison of the objective function value.

6. Conclusions and Outlook

This paper presents a novel ACO approach for the path planning of CLMRs with suspension in narrow and large-scale environments, which combines fractional-order enhanced path planning with an improved ACO algorithm to achieve smooth and efficient paths. To improve the accuracy of the kinematic model construction for CLMRs with suspension, a precise fractional-order-based kinematic modeling method is introduced. This method takes into account the dynamic adjustment of angle constraints, resulting in a more accurate representation of the CLMR's motion. Furthermore, the path planning algorithm is further enhanced by incorporating fractional-order transfer-probability modelling into the ACO framework. This extension effectively addresses the challenges associated with local optima and lack of smoothness in narrow spaces. Additionally, it improves the search speed in large-scale scenes, ensuring more efficient and optimized path planning.

It is worth noting that the proposed method adopts a fixed fractional order although it is more accurate and flexible than the integer order, but the exact value of the fractional order is still a challenge. In future work, we will introduce sensor observation, further implement variable fractional order to improve the accuracy of model construction and explore the potential application of fractional-order models in other path planning methods. In addition, we will subsequently refine the path planning methods and further investigate local planning algorithms based on the existing global path planning to cope with highly dynamic scenarios.

Author Contributions: Conceptualization, L.L. and L.J. (Liquan Jiang); methodology, L.L. and L.J. (Liquan Jiang); software, L.J. (Liyu Jiang) and L.J. (Liquan Jiang); validation, L.J. (Liquan Jiang); formal analysis, W.T.; investigation, L.L. and W.T.; resources, L.J. (Liquan Jiang)); data curation, W.T.; writing—original draft preparation, L.L., R.H. and L.J. (Liquan Jiang); writing—review and editing, L.J. (Liquan Jiang); visualization, L.J. (Liyu Jiang) and R.H.; supervision, L.J. (Liquan Jiang)); project administration, L.J. (Liquan Jiang); funding acquisition, L.J. (Liquan Jiang). All authors have read and agreed to the published version of the manuscript.

Funding: This research was funded in part by the Hubei Provincial Engineering Research Center for Intelligent Textile and Fashion (Wuhan Textile University) under Grant no. 2023HBITF03, in part by State Key Laboratory of New Textile Materials and Advanced Processing Technologies no. FZ20230023, in part by Hubei Key Laboratory of Digital Textile Equipment no. DTL2023023 and in part by Fund of National Engineering Research Center for Water Transport Safety under Grant no. A202303.

Data Availability Statement: The data presented in this study may be available on request from the corresponding author.

Conflicts of Interest: The authors declare no conflicts of interest.

References

1. Liu, L.; Wang, X.; Yang, X.; Liu, H.; Li, J.; Wang, P. Path planning techniques for mobile robots: Review and prospect. *Expert Syst. Appl.* **2023**, *227*, 120254. [CrossRef]
2. Meng, J.; Wang, S.; Xie, Y.; Li, G.; Zhang, X.; Jiang, L.; Liu, C. A safe and efficient LIDAR-based navigation system for 4WS4WD mobile manipulators in manufacturing plants. *Meas. Sci. Technol.* **2021**, *32*, 045203. [CrossRef]
3. Wu, L.; Huang, X.; Cui, J.; Liu, C.; Xiao, W. Modified adaptive ant colony optimization algorithm and its application for solving path planning of mobile robot. *Expert Syst. Appl.* **2023**, *215*, 119410. [CrossRef]
4. Li, G.; Liu, C.; Wu, L.; Xiao, W. A mixing algorithm of ACO and ABC for solving path planning of mobile robot. *Appl. Soft. Comput.* **2023**, *148*, 110868. [CrossRef]
5. Ben-Asher, J.Z.; Rimon, E.D. Time optimal trajectories for a car-like mobile robot. *IEEE Trans. Robot.* **2021**, *38*, 421–432. [CrossRef]
6. Shui, Y.; Zhao, T.; Dian, S.; Hu, Y.; Guo, R.; Li, S. Data-driven generalized predictive control for car-like mobile robots using interval type-2 T-S fuzzy neural network. *Asian J. Control* **2022**, *24*, 1391–1405. [CrossRef]
7. Theunissen, J.; Tota, A.; Gruber, P.; Dhaens, M.; Sorniotti, A. Preview-based techniques for vehicle suspension control: A state-of-the-art review. *Annu. Rev. Control* **2021**, *51*, 206–235. [CrossRef]
8. Meng, J.; Wang, S.; Jiang, L.; Hu, Z.; Xie, Y. Accurate and Efficient Self-localization of AGV Relying on Trusted Area Information in Dynamic Industrial Scene. *IEEE Trans. Veh. Technol.* **2023**, *72*, 7148–7159. [CrossRef]
9. Qie, T.; Wang, W.; Yang, C.; Li, Y.; Liu, W.; Xiang, C. A path planning algorithm for autonomous flying vehicles in cross-country environments with a novel TF-RRT* method. *Green Energy Intell. Trans.* **2022**, *1*, 100026. [CrossRef]
10. Miao, C.; Chen, G.; Yan, C.; Wu, Y. Path planning optimization of indoor mobile robot based on adaptive ant colony algorithm. *Comput. Ind. Eng.* **2021**, *156*, 107230. [CrossRef]
11. Mac, T.T.; Copot, C.; Tran, D.T.; De Keyser, R. Heuristic approaches in robot path planning: A survey. *Robot. Auton. Syst.* **2016**, *86*, 13–28. [CrossRef]
12. Song, B.; Wang, Z.; Zou, L. An improved PSO algorithm for smooth path planning of mobile robots using continuous high-degree Bezier curve. *Appl. Soft. Comput.* **2021**, *100*, 106960. [CrossRef]
13. Sang, H.; You, Y.; Sun, X.; Zhou, Y.; Liu, F. The hybrid path planning algorithm based on improved A* and artificial potential field for unmanned surface vehicle formations. *Ocean Eng.* **2021**, *223*, 108709. [CrossRef]
14. Nassef, A.M.; Abdelkareem, M.A.; Maghrabie, H.M.; Baroutaji, A. Metaheuristic-Based Algorithms for Optimizing Fractional-Order Controllers—A Recent, Systematic, and Comprehensive Review. *Fractal Fract.* **2023**, *7*, 553. [CrossRef]
15. Jiang, L.; Wang, S.; Xie, Y.; Xie, S.; Zheng, S.; Meng, J.; Ding, H. Decoupled fractional supertwisting stabilization of interconnected mobile robot under harsh terrain conditions. *IEEE Trans. Ind. Electron.* **2021**, *69*, 8178–8189. [CrossRef]
16. Jiang, L.; Wang, S.; Xie, Y.; Xie, S.Q.; Zheng, S.; Meng, J. Fractional robust finite time control of four-wheel-steering mobile robots subject to serious time-varying perturbations. *Mech. Mach. Theory* **2022**, *169*, 104634. [CrossRef]
17. Xie, Y.; Zhang, X.; Meng, W.; Zheng, S.; Jiang, L.; Meng, J.; Wang, S. Coupled fractional-order sliding mode control and obstacle avoidance of a four-wheeled steerable mobile robot. *ISA Trans.* **2021**, *108*, 282–294. [CrossRef] [PubMed]
18. Patnaik, S.; Hollkamp, J.P.; Semperlotti, F. Applications of variable-order fractional operators: A review. *Proc. R. Soc. London Ser. A-Math. Phys. Eng. Sci.* **2020**, *476*, 20190498. [CrossRef]
19. Zhao, C.; Dai, L.; Huang, Y. Fractional-order Sequential Minimal Optimization Classification Method. *Fractal Fract.* **2023**, *7*, 637. [CrossRef]
20. Muresan, C.I.; Birs, I.; Ionescu, C.; Dulf, E.H.; De Keyser, R. A review of recent developments in autotuning methods for fractional-order controllers. *Fractal Fract.* **2022**, *6*, 37. [CrossRef]
21. Tzafestas, S.G. Mobile robot control and navigation: A global overview. *J. Intell. Robot. Syst.* **2018**, *91*, 35–58. [CrossRef]
22. Jiang, H.; Xu, G.; Zeng, W.; Gao, F. Design and kinematic modeling of a passively-actively transformable mobile robot. *Mech. Mach. Theory* **2019**, *142*, 103591. [CrossRef]
23. Mohammadpour, M.; Kelouwani, S.; Gaudreau, M.A.; Zeghmi, L.; Amamou, A.; Bahmanabadi, H.; Graba, M. Energy-efficient motion planning of an autonomous forklift using deep neural networks and kinetic model. *Expert Syst. Appl.* **2024**, *237*, 121623. [CrossRef]
24. Song, B.; Wang, Z.; Zou, L.; Xu, L.; Alsaadi, F.E. A new approach to smooth global path planning of mobile robots with kinematic constraints. *Int. J. Mach. Learn. Cybern.* **2019**, *10*, 107–119. [CrossRef]
25. Rubio, F.; Valero, F.; Llopis-Albert, C. A review of mobile robots: Concepts, methods, theoretical framework, and applications. *Int. J. Adv. Robot. Syst.* **2019**, *16*, 1729881419839596. [CrossRef]
26. Niu, C.; Li, A.; Huang, X.; Li, W.; Xu, C. Research on Global Dynamic Path Planning Method Based on Improved A* Algorithm. *Math. Probl. Eng.* **2021**, *2021*, 1–13. [CrossRef]
27. Wang, J.; Chi, W.; Li, C.; Wang, C.; Meng, M.Q.H. Neural RRT*: Learning-based optimal path planning. *IEEE Trans. Autom. Sci. Eng.* **2020**, *17*, 1748–1758. [CrossRef]
28. Lyridis, D.V. An improved ant colony optimization algorithm for unmanned surface vehicle local path planning with multi-modality constraints. *Ocean Eng.* **2021**, *241*, 109890. [CrossRef]
29. Nazarahari, M.; Khanmirza, E.; Doostie, S. Multi-objective multi-robot path planning in continuous environment using an enhanced genetic algorithm. *Expert Syst. Appl.* **2019**, *115*, 106–120. [CrossRef]

30. Li, Y.; Zhao, J.; Chen, Z.; Xiong, G.; Liu, S. A robot path planning method based on improved genetic algorithm and improved dynamic window approach. *Sustainability* **2023**, *15*, 4656. [CrossRef]
31. Aggarwal, S.; Kumar, N. Path planning techniques for unmanned aerial vehicles: A review, solutions, and challenges. *Comput. Commun.* **2020**, *149*, 270–299. [CrossRef]
32. Khedr, A.M.; Al Aghbari, Z.; Khalifa, B.E. Fuzzy-based multi-layered clustering and ACO-based multiple mobile sinks path planning for optimal coverage in WSNs. *IEEE Sens. J.* **2022**, *22*, 7277–7287. [CrossRef]
33. Liu, C.; Wu, L.; Xiao, W.; Li, G.; Xu, D.; Guo, J.; Li, W. An improved heuristic mechanism ant colony optimization algorithm for solving path planning. *Knowl. Based Syst.* **2023**, *271*, 110540. [CrossRef]
34. Ajeil, F.H.; Ibraheem, I.K.; Azar, A.T.; Humaidi, A.J. Grid-based mobile robot path planning using aging-based ant colony optimization algorithm in static and dynamic environments. *Sensors* **2020**, *20*, 1880. [CrossRef] [PubMed]
35. Ali, H.; Gong, D.; Wang, M.; Dai, X. Path planning of mobile robot with improved ant colony algorithm and MDP to produce smooth trajectory in grid-based environment. *Front. Neurorobotics* **2020**, *14*, 44. [CrossRef] [PubMed]
36. Feng, K.; He, X.; Wang, M.; Chu, X.; Wang, D.; Yue, D. Path Optimization of Agricultural Robot Based on Immune Ant Colony: B-Spline Interpolation Algorithm. *Math. Probl. Eng.* **2022**, *2022*, 2585910. [CrossRef]
37. Zhang, X.; Xie, Y.; Jiang, L.; Li, G.; Meng, J.; Huang, Y. Fault-tolerant dynamic control of a four-wheel redundantly-actuated mobile robot. *IEEE Access* **2019**, *7*, 157909–157921. [CrossRef]
38. Kiran, M.S.; Hakli, H.; Gunduz, M.; Uguz, H. Artificial bee colony algorithm with variable search strategy for continuous optimization. *Inf. Sci.* **2015**, *300*, 140–157. [CrossRef]
39. Abdulakareem, M.; Raheem, F.A. Development of path planning algorithm using probabilistic roadmap based on ant colony optimization. *Engin. Techn. J.* **2020**, *38*, 343–351. [CrossRef]
40. Gao, W.; Tang, Q.; Ye, B.; Yang, Y.; Yao, J. An enhanced heuristic ant colony optimization for mobile robot path planning. *Soft Comput.* **2020**, *24*, 6139–6150. [CrossRef]

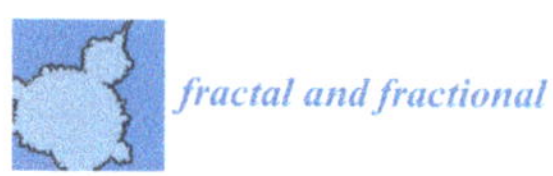

fractal and fractional

MDPI

Article

Fractional Active Disturbance Rejection Positioning and Docking Control of Remotely Operated Vehicles: Analysis and Experimental Validation

Weidong Liu, Liwei Guo, Le Li *, Jingming Xu and Guanghao Yang

School of Marine Science and Technology, Northwestern Polytechnical University, Xi'an 710072, China; liuwd@nwpu.edu.cn (W.L.); 2019100504@mail.nwpu.edu.cn (L.G.); 2018260584@mail.nwpu.edu.cn (J.X.); 2022260761@mail.nwpu.edu.cn (G.Y.)
* Correspondence: leli@nwpu.edu.cn

Abstract: In this paper, a fractional active disturbance rejection control (FADRC) scheme is proposed for remotely operated vehicles (ROVs) to enhance high-precision positioning and docking control in the presence of ocean current disturbances and model uncertainties. The scheme comprises a double closed-loop fractional-order $PI^\lambda D^\mu$ controller (DFOPID) and a model-assisted finite-time sliding-mode extended state observer (MFSESO). Among them, DFOPID effectively compensates for non-matching disturbances, while its fractional-order term enhances the dynamic performance and steady-state accuracy of the system. MFSESO contributes to enhancing the estimation accuracy through the integration of sliding-mode technology and model information, ensuring the finite-time convergence of observation errors. Numerical simulations and pool experiments have shown that the proposed control scheme can effectively resist disturbances and successfully complete high-precision tasks in the absence of an accurate model. This underscores the independence of this control scheme on accurate model data of an operational ROV. Meanwhile, it also has the advantages of a simple structure and easy parameter tuning. The FADRC scheme presented in this paper holds practical significance and can serve as a valuable reference for applications involving ROVs.

Keywords: fractional active disturbance rejection control (FADRC); double closed-loop fractional-order $PI^\lambda D^\mu$ controller (DFOPID); model-assisted finite-time sliding-mode extended state observer (MFSESO); remotely operated vehicle (ROV); remotely operated vehicle (ROV)

Citation: Liu, W.; Guo, L.; Li, L.; Xu, J.; Yang, G. Fractional Active Disturbance Rejection Positioning and Docking Control of Remotely Operated Vehicles: Analysis and Experimental Validation. *Fractal Fract.* **2024**, *8*, 354. https://doi.org/10.3390/fractalfract8060354

Academic Editors: Kishore Bingi and Abhaya Pal Singh

Received: 4 May 2024
Revised: 9 June 2024
Accepted: 10 June 2024
Published: 14 June 2024

1. Introduction

ROVs play a significant role in the realm of underwater robotics due to their cost-effectiveness, safety features, and robust operational capacities. These devices have found extensive application in diverse fields, including marine environmental surveillance, seabed topography assessments, underwater search-and-rescue operations, marine resource collection, etc. [1]. The above tasks require ROVs to have excellent control performance. This enables them to carry out high-precision docking operations with underwater devices to facilitate ROV submarine operations, underwater recovery, underwater device data backhaul, fault inspection, and power supply replacement, among other functions. However, challenges arise in achieving high-precision control of ROVs due to difficulties in accurately acquiring the ROV model and external disturbances like ocean currents in the marine environment.

In recent years, many researchers have dedicated themselves to studying the positioning and docking control of underwater vehicles. Hiroshi proposed the linear parameter-varying model predictive control (MPC) method for the docking operation, and simulation results show that this control method can effectively handle the influence of various ocean current disturbances [2]. Ohrem designed a nonlinear robust adaptive backstepping controller to ensure the dynamic positioning of ROVs in environments with model uncertainty and unknown disturbances. Extensive field trials in aquaculture applications have been successfully conducted using this controller [3]. Xie proposed a 3D mobile docking control method based on backstepping sliding mode control (SMC), which efficiently completed the underactuated autonomous underwater vehicle (AUV) mobile docking task in the presence of unknown ocean current disturbances [4]. Wu proposed a hybrid proportional integral derivative (PID) controller for a work-class ROV to achieve high-performance maneuvering [5]. Song developed an improved model-based PI robust controller using a nominal model for the precise positioning control of a hexagonal multi-vector propulsion ROV with communication time-delay constraints [6]. Li utilized a linear ADRC scheme that combines a reduced-order extended state observer and approximate time-optimal control; simulation results confirmed its effective control performance [7]. Zhang developed a model-free docking controller using deep reinforcement learning to complete three-dimensional docking tasks under disturbances [8]. Wang proposed a two-step adaptive control method to solve the planar-type docking problem, seamlessly combining horizontal dynamic positioning and visual servo docking [9].

From the above discussion, it can be seen that the control methods for the positioning and docking of underwater vehicles can be roughly divided into model-based control paradigms, such as SMC, MPC, and backstepping, and data-based control paradigms, such as PID. Model-based control paradigms are mathematically rigorous and demonstrate excellent theoretical control performance, but their application is limited. The key point is that the mathematical model of the system object may not be entirely accurate in most scenarios. Operational ROVs are often subject to unknown disturbances from umbilical cables and ocean currents. Additionally, depending on the operation's content, ROVs often require the replacement of manipulators and other work tools, making it more difficult to obtain an accurate dynamic model. A data-based control paradigm has a simple structure, allows easy parameter tuning, and is economical and practical. Employing data-based control paradigms is still the most widely used strategy in control. Traditional PID control is based on feedback error correction, which inherently exhibits a hysteresis effect and a limited anti-disturbance capability [10]. This characteristic makes it less ideal for control scenarios demanding high accuracy. Therefore, the proposal of a control paradigm that can combine the advantages of both is urgently needed.

As a new nonlinear robust control technique, ADRC can unify the above two control paradigms by incorporating the nominal model of the system in the observer design. Nonlinear ADRC mainly consists of a tracking differentiator, an extended state observer, and a nonlinear-state error feedback control law. The core idea is to consider the nominal model or integral series type of the system as the standard type. Simultaneously, the components of the system dynamics that differ from the standard type, such as system uncertainty and external disturbances, are considered as total disturbances. An observer is used to estimate the total disturbance in real time and eliminate it. Finally, the error is eradicated through the application of a nonlinear-state error feedback control law [11]. This technique exhibits good robustness, gives a fast response, has a strong anti-disturbance ability, and does not rely on the accurate mathematical model of the controlled object. It can be used when the model is completely unknown or when some information about the model is known. In light of the limitations of nonlinear ADRC due to its complex structure, numerous control parameters, and challenging tuning process, the linear ADRC method simplifies the structure by converting all controllers and extended state observers into a linear form. This approach allows for individual adjustments to be made to the controller bandwidth and observer bandwidth, thereby enhancing the effectiveness of ADRC in engineering applications [12]. Therefore, ADRC has attracted the attention of many researchers in the field of underwater vehicle motion control. Liu introduced the ADRC technique to achieve depth control for autonomous underwater vehicle (AUV). He utilized an improved speed saturation tracking differentiator to enhance the controller's adaptability to control instructions [13]. Wang utilized ADRC-based dynamic controllers in AUV formations to ensure that followers and leaders consistently maintained the desired distance [14]. Gao proposed an ADRC method based on dynamic inversion to achieve motion control of underwater vehicle-manipulator systems (UVMSs) [15]. Zhou designed a robust dynamic heading-tracking control method based on an improved ADRC method and an enhanced anti-convolution compensator. Zhou's study confirmed that the proposed control method can achieve high accuracy in heading tracking [16]. Li utilized ADRC technology to develop a tandem-level ADRC controller for a water–air multi-rotor vehicle. Additionally, Li introduced the particle swarm optimization (PSO) algorithm to efficiently adjust the controller parameters, ensuring that the controller meets the performance criteria in challenging underwater environments [17]. Liu proposed a depth-tracking method for underactuated AUVs, using an ADRC framework to compensate for the complex unknown pitch dynamics by approximating them into an integral series; the effectiveness and strong disturbance rejection capabilities of the proposed method were verified with field comparison experiments [18]. Nevertheless, many of the research studies referenced here fail to consider the effects of non-matching disturbances, while the conventional ADRC's PID controller encounters challenges in achieving precise control performance at a high level.

Fractional calculus is an extension of traditional calculus that describes the fractal dimension of a space. Podlubny first applied the concept of fractional order to controller design and proposed the fractional-order $PI^{\lambda}D^{\mu}$ controller [19]. Compared with the integer PID controller, a fractional-order $PI^{\lambda}D^{\mu}$ controller has two additional adjustable parameters, namely integral order λ and differential order μ, which can obtain more flexible amplitude–phase characteristics, so as to achieve high-precision and fast-response control performance. At the same time, the fractional-order $PI^{\lambda}D^{\mu}$ controller exhibits greater adaptability to parameter changes in the controlled object of the system. When the parameter of the controlled object changes within a certain allowable range, the system characteristics remain basically unchanged, indicating that the fractional-order $PI^{\lambda}D^{\mu}$ controller exhibits strong robustness. Fractional-order $PI^{\lambda}D^{\mu}$ controllers have been extensively researched in the field of underwater vehicle control. For AUV heading control, Liu designed a robust fractional-order $PI^{\lambda}D^{\mu}$ controller that effectively resists parametric uncertainty and demonstrates good robustness and dynamic performance [20]. Zhu proposed a fractional-order control method based on fuzzy logic and achieved good dynamic and steady-state characteristics through an AUV depth control simulation [21]. Li proposed an adaptive fractional-order non-singular terminal sliding-mode trajectory-tracking controller for an underwater robot, which can achieve fast switching gain, avoid over-tuning, and effectively improve the accuracy and robustness [22]. Zhang proposed a nonlinear fractional-order PD^{μ} controller based on saturation, which exhibits good dynamic performance and robustness. Additionally, it offers the advantages of a simple structure and easy implementation [23]. Cui designed a single-input fractional fuzzy logic controller for an unmanned underwater vehicle (UUV) motion control system. Simulation results demonstrate that Cui's control algorithm exhibits good stability and transient performance [24]. Liu proposed a fractional-order PI^{λ} controller for UUVs that guarantees both frequency-domain and time-domain behavior, offering greater flexibility in enhancing the system robustness and transient performance [25]. Hansan designed an adaptive neural network with a nonlinear fractional-order $PI^{\lambda}D^{\mu}$ controller for the path-tracking problem of underwater vehicles [26]. ROV positioning and docking operations have high requirements for the robustness and dynamic performance of the control system. The application of fractional-order $PI^{\lambda}D^{\mu}$ control in this task has not been reported. Simultaneously, the faster dynamic response of fractional-order $PI^{\lambda}D^{\mu}$ can more effectively reduce the observation error of the observer in ADRC. Therefore, FADRC can combine the advantages of ADRC and fractional-order $PI^{\lambda}D^{\mu}$ control, resulting in superior control performance. At the same time, it also ensures the simple structure of the controller and is easy to implement in practice.

Aiming to address the challenges of high-precision positioning and docking control of ROVs under ocean current disturbances and model uncertainties, this paper proposes an FADRC scheme. The proposed scheme consists of a double closed-loop fractional-order

$PI^\lambda D^\mu$ controller and a model-assisted finite-time sliding-mode extended state observer. Regarding the existing research on ADRC for ROVs, compared to previous studies focusing on improving the observer and the tracking differentiator algorithms themselves, this paper innovatively introduces a fractional-order $PI^\lambda D^\mu$ controller and adds the observation of the velocity term in the kinematic channel, enhancing the robustness of high-precision ROV operations and effectively reducing the impact of matching and non-matching disturbances. Its main contributions are as follows:

1. In order to better compensate for the non-matching disturbance caused by ocean currents on the kinematics of ROVs and to generate a smooth and ideal transition process, a double closed-loop control structure composed of a position control loop and a velocity control loop is adopted. In order to effectively enhance the robustness and dynamic performance of high-precision positioning and docking control of ROVs, a fractional-order $PI^\lambda D^\mu$ controller is introduced in the velocity control loop. Its integral and differential orders can be arbitrarily selected, providing more flexibility than an integer-order controller. At the same time, the fractional-order controller exhibits strong robustness to changes in the parameters of the controlled object;

2. The ROV nominal model is integrated into the extended state observer, and a model-assisted finite-time sliding-mode extended state observer is designed to eliminate the dependence on the accurate model. A Lyapunov function is formulated to demonstrate the finite-time convergence of the observation error. The introduction of this nominal model can effectively reduce the gain of the observer and improve the estimation accuracy. The sliding-mode technology can enhance the robustness of the observer, accelerate error convergence [27], and further improve the performance of the ROV positioning and docking control;

3. Numerical simulations and pool experiments are conducted on the ROV to perform positioning and docking tasks in the presence of ocean current disturbances and model uncertainties. Compared to the currently most widely used PID and ADRC method, the control scheme proposed in this paper has advantages in high-precision operations.

The remainder of this paper is arranged as follows: Section 2 introduces the kinematic and dynamic models of an operational ROV in the presence of ocean currents and describes the control objectives. Section 3 introduces the FADRC scheme, discusses the double closed-loop fractional-order $PI^\lambda D^\mu$ controller, and elaborates on the model-assisted finite-time sliding-mode extended state observer. Section 4 elaborates on numerical simulations and pool experiments, which verify the advantages of the proposed scheme. The conclusions are provided in Section 5.

2. ROV Modeling and Problem Formulation

This section provides a detailed analysis of the kinematics and dynamics of an operational ROV in an ocean current environment. It also outlines the control objectives for positioning and docking.

2.1. ROV Kinematics

As shown in Figure 1, the inertial coordinate system {I} and the body coordinate system {B} are established to describe the ROV's spatial motion. Among them, $\eta = [x, y, z, \phi, \theta, \psi]^T \in \mathbb{R}^6$ represents the position and direction angle of the ROV in the inertial coordinate system {I} and the body coordinate system {B}, and $v = [u, v, w, p, q, r]^T \in \mathbb{R}^6$ represents the linear velocity and angular velocity of the ROV under the body coordinate system {B}, while $\tau = [X, Y, Z, K, M, N]^T \in \mathbb{R}^6$ represents the external force and moment acting on the ROV under the body coordinate system {B}. The kinematic model of the ROV considering the current field is

$$\dot{\eta} = J(\eta)v_r + v_f \tag{1}$$

In Equation (1), $J(\eta) \in \mathbb{R}^{6\times6}$ represents the velocity transformation matrix of the ROV between the inertial coordinate system and the body coordinate system, and $v_r \in \mathbb{R}^6$ represents the velocity vector of the ROV relative to the ocean current under the body coordinate system {B}, while $v_f \in \mathbb{R}^6$ represents the velocity vector of the ocean current under the inertial coordinate system {I}. The relationship between v_r, v_f, and v is described as follows:

$$\begin{cases} \dot{\eta} = J(\eta)v_r + v_f \\ v = v_r + v_c \\ v_f = J(\eta)v_c \end{cases} \tag{2}$$

$v_c \in \mathbb{R}^6$ represents the velocity vector of the ocean current in the body coordinate system {B}.

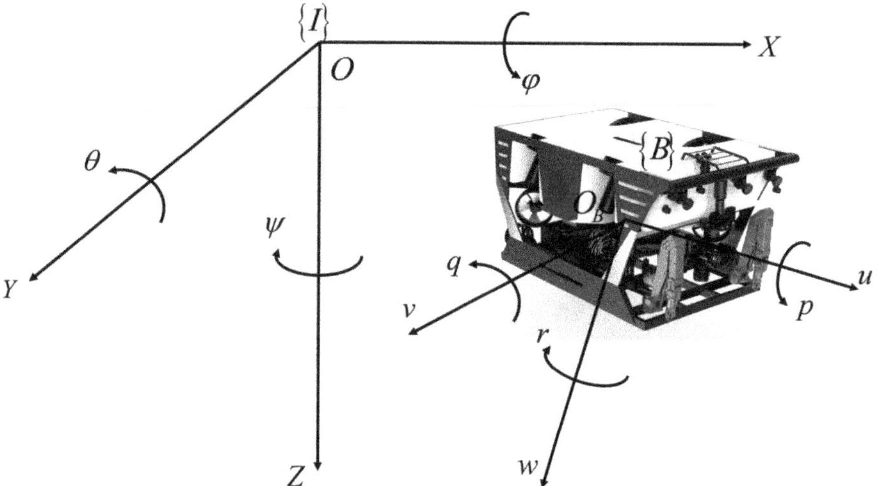

Figure 1. The inertial coordinate system {I} and the body coordinate system {B}.

ROVs can obtain v_r using their own inertial guidance equipment, but it is usually difficult to obtain v_f. When an ROV performs high-precision tasks, v_f needs to be estimated and compensated. Since v_f and $\dot{v}_f$ satisfy the law of conservation of fluid energy [28], the following reasonable assumption can be made:

Assumption 1. *The ocean current velocity v_f and its derivative $\dot{v}_f$ satisfy the bounded condition $\|v_f\|_2 \leq k_f$, $\|\dot{v}_f\|_2 \leq k_{df}$, where $\|\cdot\|_2$ represents the Euclidean norm. k_f and k_{df} are definite constants.*

2.2. ROV Dynamics

The dynamic model of the ROV is depicted in Equation (3):

$$M\dot{v}_r + C(v_r)v_r + D(v_r)v_r + g(\eta) = \tau_T + \tau_D \tag{3}$$

$M \in \mathbb{R}^{6\times6}$, $C \in \mathbb{R}^{6\times6}$, $D \in \mathbb{R}^{6\times6}$, $g \in \mathbb{R}^6$, $\tau_T \in \mathbb{R}^6$, and $\tau_D \in \mathbb{R}^6$ represent the inertial matrix; the Coriolis and centripetal force matrix; the damping matrix; the restoring force matrix; the control force and moment vector; and the lumped disturbance vector under the nominal model, respectively. Among them, $\tau_D = \tau_F + \tau_E$. $\tau_F \in \mathbb{R}^6$ represents the disturbance vector caused by ocean currents. Since it is difficult to obtain an accurate model of the ROV, $\tau_E \in \mathbb{R}^6$ represents the system model uncertainty caused by umbilical cables, etc., which is the error between the accurate model and the nominal model. τ_F is given by calculating Equation (4):

$$\tau_F = -M_{RB}\dot{v}_c - C_{RB}(v_r + v_c)v_c - C_{RB}(v_c)v_r \tag{4}$$

where $M_{RB} \in R^{6\times6}$ represents the rigid-body inertia matrix and $C_{RB} \in \mathbb{R}^{6\times6}$ represents the rigid-body Coriolis force and centripetal force matrix.

Assumption 2. *The lumped disturbance τ_D and its derivative $\dot{\tau}_D$ satisfy the bounded condition $\|\tau_D\|_2 \leq k_D$, $\|\dot{\tau}_D\|_2 \leq k_{dD}$, where k_D and k_{dD} are definite constants.*

2.3. Distribution of Thrust Forces

The operational ROV studied in this paper is equipped with four horizontal thrusters and four vertical thrusters. The distribution relationship between the control force and moment τ_T and the thrust of each thruster is

$$\tau_T = BU, \tag{5}$$

where $U \in \mathbb{R}^8$ represents the thrust vector generated by the thruster and $B \in \mathbb{R}^{6\times8}$ represents the ROV thrust distribution matrix. The thrusters can provide a thrust range of ± 4000 N.

2.4. Control Objectives

The control objectives of this paper are to design a high-precision motion control scheme for an operational ROV that is affected by ocean current disturbance and cannot obtain an accurate model. The aim is to enable the ROV to achieve precise positioning and docking with the underwater tool platform. Due to the structural requirements of the underwater tool platform, the positioning error of the ROV should be less than 0.05 m, and the error of each attitude angle should be less than $1°$.

3. Model-Assisted Finite-Time Sliding-Mode Extended State Observer

This section provides a detailed description and proof of the model-assisted finite-time sliding-mode extended state observer. First, the operational ROV nominal model is integrated into the extended state observer. The known model information can reduce the computational burden of the extended state observer and improve the estimation accuracy of disturbance and uncertainty. At the same time, sliding-mode technology is introduced to further enhance the robustness and convergence velocity of the observer.

3.1. Design of MFSESO

From the above discussion on modeling the kinematics and dynamics of the operational ROV, it can be seen that the ocean current velocity v_f and the lumped disturbance τ_D are unknown disturbance quantities that need to be estimated. Therefore, these two variables are set as the extended state variables of the system, and the ROV extended state equation is established as follows:

$$\begin{cases} \dot{x}_1 = x_2 + f_1(x_1) + hu(t) \\ \dot{x}_2 = \dot{x}_2 \\ y = x_1 \end{cases} \tag{6}$$

In Equation (6), $x = \left[x_1^T, x_2^T\right]^T \in \mathbb{R}^{24}$. $x_1 = \left[\eta^T, v_r^T\right]^T \in \mathbb{R}^{12}$ represents the state vector of the system, which can be measured with sensors; $x_2 = \left[v_f^T, \tau_{MD}^T\right]^T \in \mathbb{R}^{12}$ represents the extended state vector of the system, which needs to be estimated using the observer. The lump-like disturbance vector $\tau_{MD} = M^{-1}\tau_D \in \mathbb{R}^6$. $f_1(x_1) = \left[J(\eta)v_r, -M^{-1}f(v_r, \eta)\right]^T \in \mathbb{R}^{12}$ is the known function vector of the system, where $f(v_r, \eta) = C(v_r)v_r + D(v_r)v_r +$

$g(\eta) \in \mathbb{R}^6$. $h = \left[0_{6\times6}, M^{-1}\right]^T \in \mathbb{R}^{12\times6}$ is the known function matrix of the system. $u(t) = \tau_T \in \mathbb{R}^6$ indicates the system control input and $y \in \mathbb{R}^{12}$ indicates the system output. According to the actual operational conditions of the ROV, it can be assumed that each element in the state vectors x_1 and x_2 of the system is bounded.

In order to estimate the ocean current velocity v_f and lumped disturbance τ_D, the equation for the model-assisted finite-time sliding-mode extended state observer is as follows:

$$\begin{cases} \dot{\hat{x}}_1 = \hat{x}_2 + f_1(x_1) + hu(t) + \omega l_1(x_1 - \hat{x}_1) + \kappa_1|x_1 - \hat{x}_1|^{\frac{p}{q}}\mathrm{sgn}(x_1 - \hat{x}_1) \\ \dot{\hat{x}}_2 = \omega^2 l_2(x_1 - \hat{x}_1) + \omega\kappa_2|x_1 - \hat{x}_1|^{\frac{p}{q}}\mathrm{sgn}(x_1 - \hat{x}_1) \end{cases} \tag{7}$$

In Equation (7), $\hat{x} = \left[\hat{x}_1^T, \hat{x}_2^T\right]^T \in \mathbb{R}^{24}$ represents the state variable of MFSESO, which estimates the state variable x, where $\hat{x}_1 = \left[\hat{\eta}^T, \hat{v}_r^T\right]^T \in \mathbb{R}^{12}$, $\hat{x}_2 = \left[\hat{v}_f^T, \hat{\tau}_{MD}^T\right]^T \in \mathbb{R}^{12}$, and $\hat{\tau}_D = M\hat{\tau}_{MD} \in \mathbb{R}^6$. ω is the observer scale parameter; l_1 and l_2 are the observer gain parameters; and κ_1 and κ_2 are the observer sliding-mode gain parameters, both of which are definite positive real numbers. sgn represents the sign function, and p and q are the observer quasi-sliding-mode parameters, both of which are positive odd numbers, and $p < q$. In the aforementioned parameters, the correlation between $\hat{x}_1$ and $\hat{x}_2$ can be modified through the adjustment of ω. By selecting appropriate values for l_1 and l_2, the poles of the MFSESO characteristic equation can be determined, thereby affecting the convergence performance of the observer. κ_1 and κ_2 enhance the robustness of the observer. At the same time, $|x_1 - \hat{x}_1|^{\frac{p}{q}}\mathrm{sgn}(x_1 - \hat{x}_1)$ replaces the traditional sign function, effectively reducing the chattering phenomenon of sliding mode control.

3.2. Convergence Analysis of MFSESO

Theorem 1. *For the ROV extended state equation established above in Equation (6), the observation error of MFSESO designed in Equation (7) can converge to zero in finite time.*

The observation error state equation can be obtained by differentiating the system's extended state equation in Equation (6) and the MFSESO equation in Equation (7), as shown in Equation (8):

$$\begin{bmatrix} \dot{e}_1 \\ \dot{e}_2 \end{bmatrix} = \begin{bmatrix} -\omega l_1 I_{12\times12} & \omega I_{12\times12} \\ -\omega l_2 I_{12\times12} & 0_{12\times12} \end{bmatrix}\begin{bmatrix} e_1 \\ e_2 \end{bmatrix} + \begin{bmatrix} 0_{12\times1} \\ \dot{x}_2/\omega \end{bmatrix} - \begin{bmatrix} \kappa_1|e_1|^{\frac{p}{q}}\mathrm{sgn}(e_1) \\ \kappa_2|e_1|^{\frac{p}{q}}\mathrm{sgn}(e_1) \end{bmatrix} \tag{8}$$

where $e = \left[e_1^T, e_2^T\right]^T \in \mathbb{R}^{24}$; $e_1 = x_1 - \hat{x}_1$ and $e_2 = \frac{x_2}{\omega} - \frac{\hat{x}_2}{\omega}$; $\overline{A} = \omega\begin{bmatrix} -l_1 I_{12\times12} & I_{12\times12} \\ -l_2 I_{12\times12} & 0_{12\times12} \end{bmatrix} \in \mathbb{R}^{24\times24}$; $\overline{D} = \begin{bmatrix} 0_{12\times1} \\ \dot{x}_2/\omega \end{bmatrix} \in \mathbb{R}^{24}$; and $\overline{B} = \begin{bmatrix} \kappa_1|e_1|^{\frac{p}{q}}\mathrm{sgn}(e_1) \\ \kappa_2|e_1|^{\frac{p}{q}}\mathrm{sgn}(e_1) \end{bmatrix} \in \mathbb{R}^{24}$. When the observer gain parameters l_1 and l_2 satisfy $l_1^2 - 4l_2 > 0$, all eigenvalues of $\overline{A}$ have a negative real part. That is, $\overline{A}$ is the Hurwitz matrix.

Lemma 1 (Lyapunov Matrix Equation). *If the matrix $\overline{A}$ is a Hurwitz matrix, then for any given symmetric positive definite matrix $Q \in \mathbb{R}^{24\times24}$, there exists a symmetric positive definite matrix $P \in \mathbb{R}^{24\times24}$, such that $\overline{A}^T P + P\overline{A} = -Q$ is satisfied.*

To facilitate calculation, $Q = I$ is selected, matrix $\overline{A}$ is brought into Lemma 1, and P satisfying the condition is calculated:

$$P = \begin{bmatrix} P_{11} & P_{12} \\ P_{21} & P_{22} \end{bmatrix} = \begin{bmatrix} \frac{1}{2\omega}\frac{l_2+1}{l_1}I_{12\times12} & -\frac{1}{2\omega}I_{12\times12} \\ -\frac{1}{2\omega}I_{12\times12} & \frac{1}{2\omega}\frac{l_1^2+l_2+1}{l_1 l_2}I_{12\times12} \end{bmatrix} \tag{9}$$

From the above, $\|P\|_F = \frac{\sqrt{3}}{\omega}\left[\left(\frac{l_2+1}{l_1}\right)^2 + \left(\frac{l_1^2+l_2+1}{l_1 l_2}\right)^2 + 2\right]^{\frac{1}{2}}$; $\|\cdot\|_F$ is the Fibonacci norm.

To prove the convergence of the MFSESO system, the Lyapunov function of the system is chosen:

$$V = e^T P e. \tag{10}$$

Taking the derivative of the Lyapunov function V,

$$\dot{V} = e^T P \dot{e} + \dot{e}^T P e. \tag{11}$$

Substituting Equation (8) into Equation (11) yields the following:

$$\begin{aligned}
\dot{V} &= e^T\left(P\overline{A} + \overline{A}^T P\right)e + 2e^T P\overline{D} - 2e^T P\overline{B} \\
&= -\|e\|_2^2 + 2e^T P\overline{D} - 2e^T P\overline{B}
\end{aligned} \tag{12}$$

According to Assumption 1 and Assumption 2, it can be seen that $\left\|\dot{v}_f\right\|_2 \leq k_{df}$ and $\left\|\dot{\tau}_{MD}\right\|_2 \leq \left\|M^{-1}\right\|_F \left\|\dot{\tau}_D\right\|_2 \leq k_{dD}\left\|M^{-1}\right\|_F$. Therefore, when $\left\|\overline{D}\right\|_2 \leq \left\|\frac{k_{df}}{k_{dD}\|M^{-1}\|_F}\right\|_2/\omega = M_D$, we obtain the following:

$$2e^T P\overline{D} \leq 2M_D\|P\|_F\|e\|_2. \tag{13}$$

Meanwhile, since P is a symmetric positive definite matrix, there is an orthogonal matrix $O \in \mathbb{R}^{24\times24}$, so that $O^T P O = \Lambda \in \mathbb{R}^{24\times24}$ is a diagonal matrix; thus, we can obtain Equation (14):

$$\begin{aligned}
2e^T P\overline{B} &= 2e^T O^T \Lambda O\overline{B} \\
&= 2(Oe)^T \Lambda(O\overline{B}) \\
&\geq 2\lambda_{\min}\min\{\kappa_1,\kappa_2\}\left(\sum_{i=1}^{24}\left(e_i|e_i|^{\frac{p}{q}}\mathrm{sgn}(e_i)\right)\right) \\
&\geq 2\lambda_{\min}\min\{\kappa_1,\kappa_2\}\left(\sum_{i=1}^{24}|e_i|^{\frac{q+p}{q}}\right) \\
&= 2\lambda_{\min}\min\{\kappa_1,\kappa_2\}\|e\|_{(q+p)/q}^{(q+p)/q} \\
&\geq 2\alpha\lambda_{\min}\min\{\kappa_1,\kappa_2\}\|e\|_2^{(q+p)/q}
\end{aligned} \tag{14}$$

where $\|\cdot\|_{(q+p)/q}$ is the p-norm with exponent $(q+p)/q$, and $\lambda_{\min} = \frac{l_1^2+(l_2+1)^2 - \sqrt{[l_1^2+(l_2+1)^2][l_1^2+(l_2-1)^2]}}{4\omega l_1 l_2}$ is the minimum eigenvalue of P. Due to the equivalence of vector norms, there exists $\|e\|_{(q+p)/q}^{(q+p)/q} \geq \alpha\|e\|_2^{(q+p)/q}$, $\alpha > 0$. If we let $M_B = 2\alpha\lambda_{\min}\min\{\kappa_1,\kappa_2\}$, then $-2e^T P\overline{B} \leq -M_B\|e\|_2^{(q+p)/q}$. To sum up, Equation (15) can be obtained:

$$\dot{V} \leq -\|e\|_2^2 - M_B\|e\|_2^{(q+p)/q} + 2M_D\|P\|_F\|e\|_2 \tag{15}$$

Lemma 2 ([29]). *Consider the following nonlinear systems:*

$$\dot{x} = f(x), \tag{16}$$

where $f(0) = 0$, $x \in \mathbb{R}^n$, $f : U_0 \to \mathbb{R}^n$ is a continuous function in an open neighborhood U_0 containing the origin. Suppose there is a continuous positive definite function $V(x) : U_0 \to \mathbb{R}^n$, and that there are real numbers $a, b, c > 0$ and $d \in (0.5, 1)$, and an open neighborhood $\hat{U} \subseteq U_0$ containing the origin, such that the following equation holds:

$$\dot{V} \leq -aV^d - bV + cV^{\frac{1}{2}}.$$ (17)

Then, the origin of the system in Equation (16) is in fast finite time, uniformly bounded, stable. This implies that x converges to a stable region $Q = \left\{ x : \chi V^{d-\frac{1}{2}} + \delta V^{\frac{1}{2}} < c \right\}$, $\chi \in (0, a)$, $\delta \in (0, b)$. And the stable time T depends on the initial value $x(0)$, satisfying $T \leq \frac{1}{(b-\delta)(1-d)} \ln\left(1 + \frac{(b-\delta)V_{x(0)}^{1-d}}{a-\chi} \right)$.

The following can be seen from Equation (10):

$$\lambda_{\min}\|e\|_2^2 \leq V \leq \|P\|_F\|e\|_2^2.$$ (18)

According to Equations (15) and (18), we obtain

$$\dot{V} \leq -M_B\|P\|_F^{-\frac{q+p}{2q}} V^{\frac{q+p}{2q}} - \|P\|_F^{-1}V + 2M_D\lambda_{\min}^{-\frac{1}{2}}\|P\|_F V^{\frac{1}{2}},$$
$$= -a_0 V^{d_0} - b_0 V + c_0 V^{\frac{1}{2}}$$ (19)

where $a_0 = M_B\|P\|_F^{-\frac{q+p}{2q}}$, $b_0 = \|P\|_F^{-1}$, $c_0 = 2M_D\lambda_{\min}^{-\frac{1}{2}}\|P\|_F$, and $d_0 = \frac{q+p}{2q}$. According to Lemma (2), it can be seen that the observation error e can converge to the stable region $Q_0 = \left\{ e : \chi_0 V^{d_0-\frac{1}{2}} + \delta_0 V^{\frac{1}{2}} < c_0 \right\}$, $\chi_0 \in (0, a_0)$, $\delta_0 \in (0, b_0)$ in a finite time. And the convergence time satisfies $T \leq \frac{1}{(b_0-\delta_0)(1-d_0)} \ln\left(1 + \frac{(b_0-\delta_0)V_{e(0)}^{1-d_0}}{a_0-\chi_0} \right)$, where the value of e at time $t = 0$ is defined as $e(0)$.

The above proves that the observation error e of MFSESO can converge to the stable region within $t = T$ time, indicating that MFSESO can achieve finite-time estimation of velocity disturbance v_f and lumped disturbance τ_D. This completes the proof of Theorem 1.

4. Double Closed-Loop Fractional-Order $\text{PI}^\lambda\text{D}^\mu$ Controller

It can be seen from the previous discussion that ocean current velocity v_f is mainly reflected in the ROV kinematic model and is not in the same channel as the system control input τ_T, which belongs to non-matching disturbance. The lumped disturbance τ_E acts on the dynamic model and belongs to the matching disturbance. It is difficult to directly offset the influence of non-matching disturbance simply with the input τ_T in the dynamic model. A double closed-loop controller should be designed to compensate for the disturbance of ocean current velocity v_f in the position loop and the disturbance of lumped disturbance τ_E in the velocity loop. A fractional-order $\text{PI}^\lambda\text{D}^\mu$ controller not only preserves the advantages of the simplicity, practicality, and easy tuning of the traditional PID controller, but also effectively enhances the robustness and dynamic capability of a dynamic system. So, the fractional-order $\text{PI}^\lambda\text{D}^\mu$ controller is introduced into the velocity loop control.

Design of DFOPID

The commonly used definitions of fractional calculus are the Riemann–Lioucille definition, the Grunwald Letnikov definition, and the Caputo definition [19]. The Riemann–Lioucille definition and the Grunwald Letnikov definition require the value of the fractional derivative of the signal at the initial moment to be known, while the Caputo definition requires the value of the signal and its integer derivative at the initial moment to be known, which is closer to practical applications. Therefore, the double closed-loop fractional-order $\text{PI}^\lambda\text{D}^\mu$ controller designed in this paper adopts the Caputo definition, as follows:

For $\alpha \in \mathbb{R}^+$, with $m - 1 < \alpha \leq m$ and $m \in \mathbb{Z}^+$, the α-order Caputo fractional derivative of the function $y(t)$ defined on $[t_0, t]$ is

$$_{t_0}D_t^\alpha y(t) = \frac{1}{\Gamma(m-\alpha)} \int_{t_0}^t \frac{y^{(m)}(\tau)}{(t-\tau)^{1+\alpha-m}} d\tau. \tag{20}$$

For $\gamma \in \mathbb{R}^+$, the γ-order Caputo fractional integral of function $y(t)$ defined on $[t_0, t]$ is

$$_{t_0}D_t^{-\gamma} y(t) = \frac{1}{\Gamma(\gamma)} \int_{t_0}^t \frac{y(\tau)}{(t-\tau)^{1-\gamma}} d\tau. \tag{21}$$

Among them, the Gamma function $\Gamma(x)$ is defined as

$$\Gamma(x) = \int_0^\infty t^{x-1} e^{-t} dt. \tag{22}$$

The output expression of the fractional-order $\text{PI}^\lambda \text{D}^\mu$ controller in the time domain is

$$u(t) = k_p e(t) + k_{i0} D^{-\lambda} e(t) + k_{d0} D^\mu e(t), \tag{23}$$

where $e(t) = r(t) - y(t)$ is the system error signal, which serves as the input signal of the fractional-order $\text{PI}^\lambda \text{D}^\mu$ controller; $r(t)$ is the reference input signal of the system; and $y(t)$ is the actual input signal of the system. k_p, k_i, and k_d represent the proportional, integral, and differential gains, respectively. λ and μ represent fractional orders of the integral and differential terms, respectively. The ranges of these values are $0 < \lambda < 2$ and $0 < \mu < 2$, respectively. It can be seen from the above discussion that due to the presence of fractional-order operators, the fractional-order $\text{PI}^\lambda \text{D}^\mu$ controller can adjust the low- and high-frequency characteristics of the closed-loop system more flexibly by modifying the values of λ and μ. Simultaneously, they are less sensitive to the parameter changes in the control system. When the control parameters and disturbance vary within a certain interval, the system performance does not change significantly. The system has stronger robustness. Meanwhile, fractional $\text{PI}^\lambda \text{D}^\mu$ is an extension of integer PID, naturally inheriting the advantages of the simple structure and easy tuning of integer PID.

The position loop controller is primarily responsible for eliminating the non-matching disturbance caused by the ocean current velocity v_f and guiding the operational ROV to achieve a smooth transition process. The position loop controller designed in this paper incorporates a saturated nonlinear link at the input of the PID algorithm to facilitate the ROV in reaching the target position at an optimal velocity:

$$v_d(t) = k_{\eta p} \bar{e}_\eta(t) + k_{\eta i} \int_{t_0}^t \bar{e}_\eta(t) dt + k_{\eta d} \frac{d\bar{e}_\eta(t)}{dt}. \tag{24}$$

In Equation (24), $\bar{e}_\eta(i) = sat(e_\eta(i))$, $e_\eta(i) = \eta_d(i) - \eta(i)$, and $i = 1, \ldots, 6$, where η_d is the desired position, $sat(e_\eta(i)) = \begin{cases} \Delta(i), e_\eta(i) > \Delta(i) \\ e_\eta(i), |e_\eta(i)| \leq \Delta(i) \\ -\Delta(i), e_\eta(i) < -\Delta(i) \end{cases}$, $\Delta \in \mathbb{R}^6$ is the boundary-layer vector, and $k_{\eta j} \in \mathbb{R}^{6\times6}$, $j = p, i, d$ is the control gain diagonal matrix.

Due to the increased demands of ROV dynamics for control robustness and dynamic performance, a fractional-order $\text{PI}^\lambda \text{D}^\mu$ controller is designed to serve as the velocity loop controller:

$$\tau_c(t) = k_{vp} e_v(t) + k_{vi0} D^{-\lambda_v} e_v(t) + k_{vd0} D^{\mu_v} e_v(t). \tag{25}$$

In Equation (25), $e_v = v_d - v_r - \hat{v}_c \in \mathbb{R}^6$, $\hat{v}_c = J^{-1}(\eta)\hat{v}_f$, where v_d is desired velocity, $k_{vj} \in \mathbb{R}^{6\times6}$, $j = p, i, d$ is the control gain diagonal matrix, and $_0 D^{i_v} e_j(t) \in \mathbb{R}^{6\times6}$, $i = -\lambda, \mu$ is the fractional calculus diagonal matrix.

The FADRC scheme developed for the ROV to perform high-precision positioning and docking control tasks is illustrated in Figure 2. The FADRC scheme, outlined with a red chain line in the figure, consists of DFOPID and MFSESO. DFOPID comprises a position loop controller and a velocity loop controller. The position loop controller utilizes the position error η to generate the desired velocity v_d. The velocity error e_v is obtained by subtracting the ROV velocity v_r and the ocean current velocity observation $\hat{v}_c$ from the desired velocity v_d. The velocity loop controller determines the DFOPID controller output τ_C based on the velocity error e_v. MFSESO, based on inputs such as the ROV position η, the ROV velocity v_r, and the FADRC scheme output τ_T, calculates the ocean current velocity observation $\hat{v}_f$ and the lumped disturbance observation $\hat{\tau}_D$. The FADRC scheme output τ_T is derived from the disparity between the DFOPID controller output τ_C and the lumped disturbance observation $\hat{\tau}_D$. In summary, the FADRC scheme incorporates the ocean current velocity observation value $\hat{v}_f$ and the lumped disturbance observation value $\hat{\tau}_D$ estimated by MFSESO into the double closed-loop controller, so that the non-matching disturbance is transformed into matching disturbance, making it easier to mitigate their impact. During the transition process, DFOPID ensures accurate and rapid tracking of the ROV's position and velocity in relation to the target value. This enhances the robustness and dynamic characteristics of the entire system.

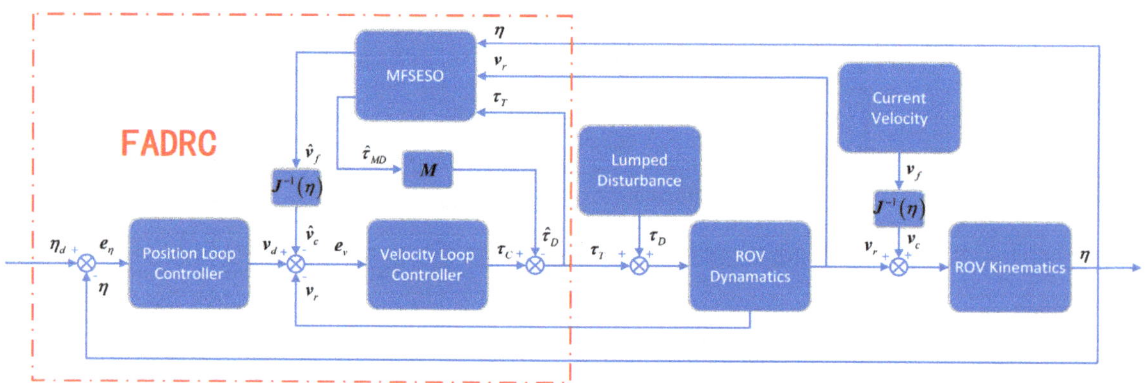

Figure 2. FADRC structural framework. FADRC consists of DFOPID and MFSESO, which is the part outlined with the red chain line. Desired position η_d (defined in line 373); ROV position η (defined in line 195); position error e_η (defined in line 373); velocity error e_v (defined in line 379); desired velocity v_d (defined in line 379); ROV velocity v_r (defined in line 202) under the body coordinate system {B}; ocean current velocity v_c (defined in line 207) and velocity observation $\hat{v}_c$ (defined in line 379) under the body coordinate system {B}; ocean current velocity v_f (defined in line 204) and velocity observation $\hat{v}_f$ (defined in line 271) under the inertial coordinate system {I}; lumped disturbance τ_D (defined in line 220) and lumped disturbance observation $\hat{\tau}_D$ (defined in line 272); lump-like disturbance τ_{MD} (defined in line 262) and lump-like disturbance observation $\hat{\tau}_{MD}$ (defined in line 272); DFOPID controller output τ_C (defined in line 378); FADRC scheme output τ_T (defined in line 222); inertial matrix M (defined in line 220); velocity transformation matrix $J(\eta)$ (defined in line 201).

5. Numerical Simulations and Pool Experiments

In order to verify the effectiveness and advanced nature of the control scheme proposed in this paper, high-precision ROV positioning and docking control experiments were conducted in both simulation and pool environments. Meanwhile, a comparison was carried out using the most widely used traditional method. The high-precision positioning and docking process of the ROV was as follows: In the ocean current environment with a flow velocity of 1 knot, the ROV was guided to the docking position using the visual positioning system and maintained its dynamic position. When the positioning error continued to remain within the required error range for docking, the docking locking mechanism extended downward into the docking hole of the underwater tool platform to complete the docking process.

5.1. ROV Prototype

As shown in Figure 3a, the ROV in this paper has an overall size of 3100 mm × 2000 mm × 1800 mm and a net weight of 4187.5 kg. It is equipped with four horizontal thrusters and four vertical thrusters, and has an omnidirectional driving capability. The bottom protruding part is equipped with two docking rods, which can perform docking operations with docking holes on the underwater tool platform. The parameters of the ROV nominal dynamic model are as follows: the center of gravity coordinate $r_G = [0, 0, 0]^T$, the center of buoyancy coordinate $r_B = [0, 0, -0.493]^T$, the moment of inertia matrix $I = diag(2038, 3587, 3587)$, the additional mass matrix $M_{AM} = -diag(3261.35, 4664.31, 7471.75, 1664, 4118.17, 3708.41)$, the linear damping matrix $D_L = -diag(3610.00, 2462.99, 4566.59, 9810.00, 5220.90, 5841.54)$, the nonlinear damping matrix $D_N = -diag(952|u|, 2442.78|v|, 530.46|w|, 890|p|, 1876|q|, 2085.52|r|)$. As shown in Figure 3b, the underwater tool platform is equipped with various operational tools and a sampling basket necessary for the ROV. Once the ROV dives, it can complete various tasks by changing tools on the tool platform, significantly enhancing the working efficiency. The ROV determines the relative position by identifying the QR code affixed to the tool platform. The tool platform is designed with two central docking holes that work in conjunction with two docking rods to complete the docking operation. According to the design of the docking rods and docking holes, the ROV position error must be less than 0.05 m, and the attitude angle error must be less than $1°$.

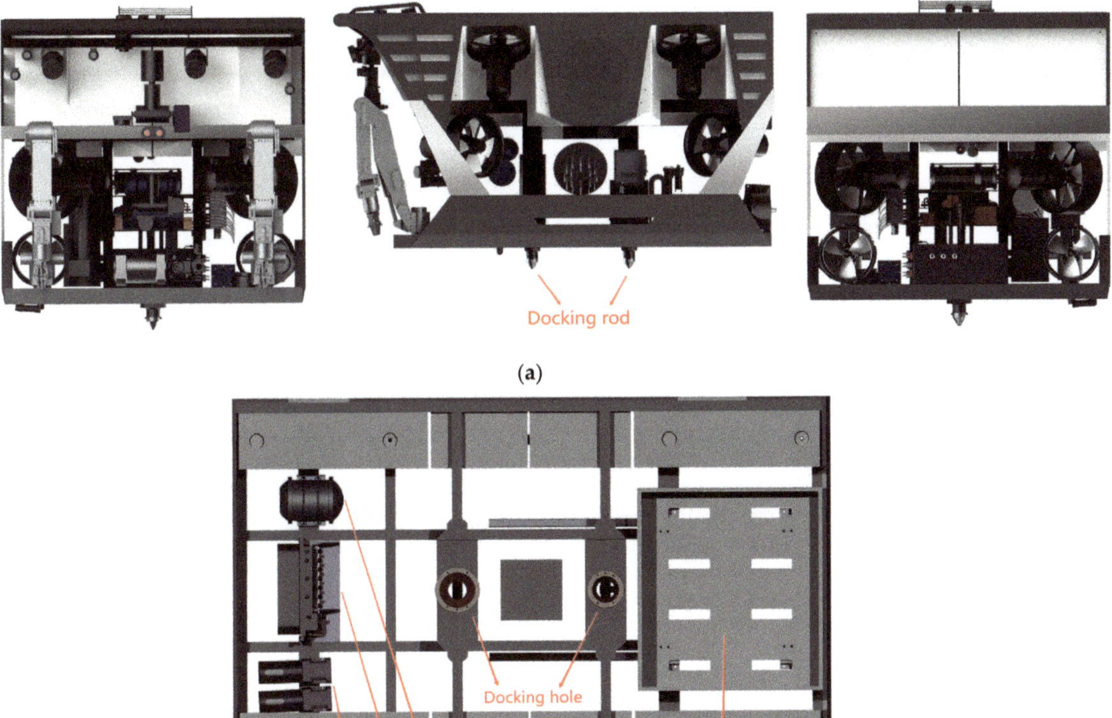

Figure 3. ROV and underwater tool platform structure diagrams. (**a**) ROV structure diagram: front view (left), left view (middle), and rear view (right). (**b**) Underwater tool platform structure diagram.

5.2. Numerical Simulations

In the numerical simulations, the position of the underwater tool platform was set to $\eta_d = [5, 1, 1, 0, 0, 0]^T$. The initial position of the ROV was set to $\eta(0) = [0, 0, 0, 0, 0, 0]^T$ and the initial velocity was set to $v_r(0) = [0, 0, 0, 0, 0, 0]^T$. The ocean current velocity was set to $v_f = [0.3635 \sin(0.05t + \pi/3), 0.3635 \sin(0.05t + \pi/3), 0, 0, 0, 0]^T$, simulating the actual operating conditions of 1 throttling. The disturbance caused by factors other than ocean current such as umbilical cable and the uncertainty of the system model were set to $\tau_E =$

$$\begin{bmatrix} 10\cos(0.05t + \pi/3)\sin(0.05t) \\ 10\cos(0.05t + \pi/4)\cos(0.05t) \\ 10\cos(0.05t + \pi/6)\sin(0.05t + \pi/4) \\ 2\cos(0.05t + \pi/3)\cos(0.05t + \pi/10) \\ 2\sin(0.05t)\sin(0.05t + 2\pi/3) \\ 2\sin(0.05t + 4\pi/3)\cos(0.05t) \end{bmatrix}$$

. The MFSESO observation parameters were set to $\iota_1 = 10$, $\iota_2 = 10$, $\kappa_1 = 2$, and $\kappa_2 = 20$, $\omega = 1$, $p = 3$, $q = 5$. The DFOPID control parameters were set to $k_{\eta p} = diag(1, 1, 1, 1, 1, 1)$, $k_{\eta i} = diag(0, 0, 0, 0, 0, 0)$, $k_{\eta d} = diag(0, 0, 0, 0, 0, 0)$, $\Delta = [0.5, 0.5, 0.5, \pi/180, \pi/180, \pi/180]^T$, $k_{vp} = 100000 diag(1, 1, 1, 10, 10, 30)$, $k_{vi} = diag(0.1, 0.1, 0.1, 0.1, 0.1, 0.1)$, $k_{vd} = diag(10000, 50000, 50000, 0, 0, 0)$, $\lambda_v = [1.1, 1.1, 1.1, 1.1, 1.1, 1.1]^T$ and $\mu_\eta = [0.5, 0.5, 0.5, 0, 0, 0]^T$. An Oustaloup filter [30] was used to implement fractional calculus.

In order to further demonstrate the performance superiority of the FADRC scheme composed of MFSESO and DFOPID, the following simulation scenarios were conducted. The first scenario involved a comparison test between the control scheme based on MFSESO-DFOPID and the control scheme based on MFSESO-DPID. The second test scenario involved comparing the control scheme based on MFSESO-DFOPID with the control scheme based on LESO-DFOPID. The initial conditions of the test remained unchanged.

For simulation scenario 1, the high-precision positioning and docking control simulation results of the control scheme based on MFSESO-DFOPID and the control scheme based on MFSESO-DPID are shown in Figures 4 and 5.

Performance indicators such as the root-mean-square error, adjustment time, and steady-state error are presented in Table 1 to facilitate a more detailed and accurate quantitative comparison between the DFOPID algorithm and the DPID algorithm. Among them, a smaller root-mean-square error indicates that the scheme demonstrates a better control performance, while a shorter adjustment time signifies faster convergence, and the steady-state error reflects the accuracy of the control scheme.

Figure 4 illustrates the position error in the position loop and the velocity error in the velocity loop based on MFSESO-DFOPID and MFSESO-DPID. It can be seen from the figures that the position error and velocity error based on MFSESO-DFOPID show a better dynamic process, with a faster convergence velocity and smaller overshoot. At the same time, when the motion system is stabilized, MFSESO-DFOPID exhibits smaller position steady-state error and velocity steady-state error values. Meanwhile, against ocean current velocity disturbance and lumped disturbance, MFSEPO-DFOPID exhibits smaller error fluctuations and demonstrates greater robustness. Figure 5 shows the thrust curve of the thruster calculated using the schemes mentioned above, which remains generally smooth. This indicates that the control scheme proposed in this paper has good practical application value. The control performance indicators in Table 1 quantitatively support the results presented in Figure 4. MFSESO-DFOPID obviously meets the docking requirements, wherein the ROV position error must be less than 0.05 m and the attitude angle error must be less than $1°$. MFSESO-DFOPID has a significant advantage in most indicators. The above results fully reflect that the DFOPID control algorithm proposed in this paper shows a better control performance than the traditional DPID algorithm. It exhibits a significantly improved dynamic response, reduced steady-state error, and enhanced robustness.

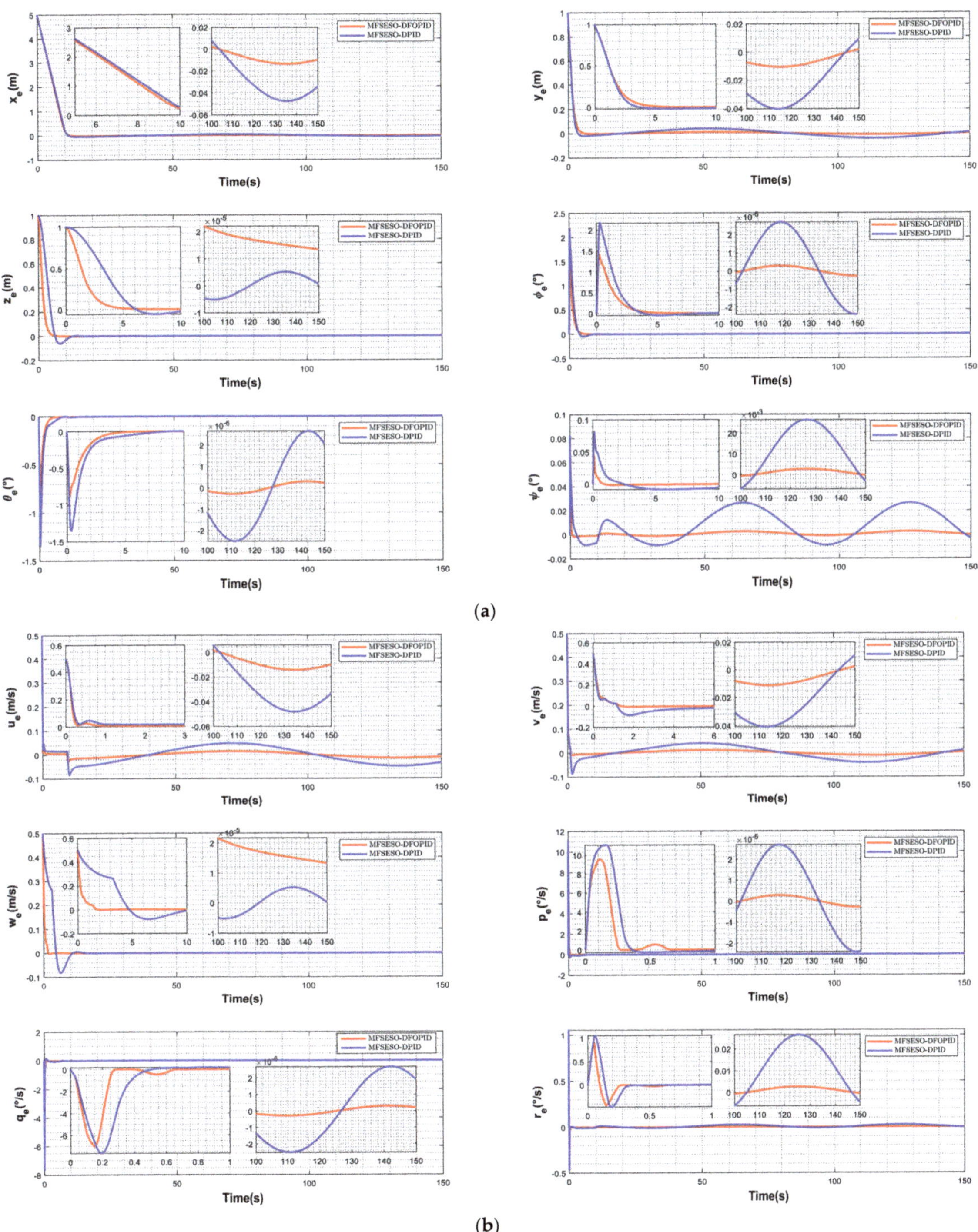

Figure 4. Position error e_η and velocity error e_v under MFSESO-DFOPID (red line) and MFSESO-DPID (blue line). (**a**) Position error e_η (defined in line 373). (**b**) Velocity error e_v (defined in line 379).

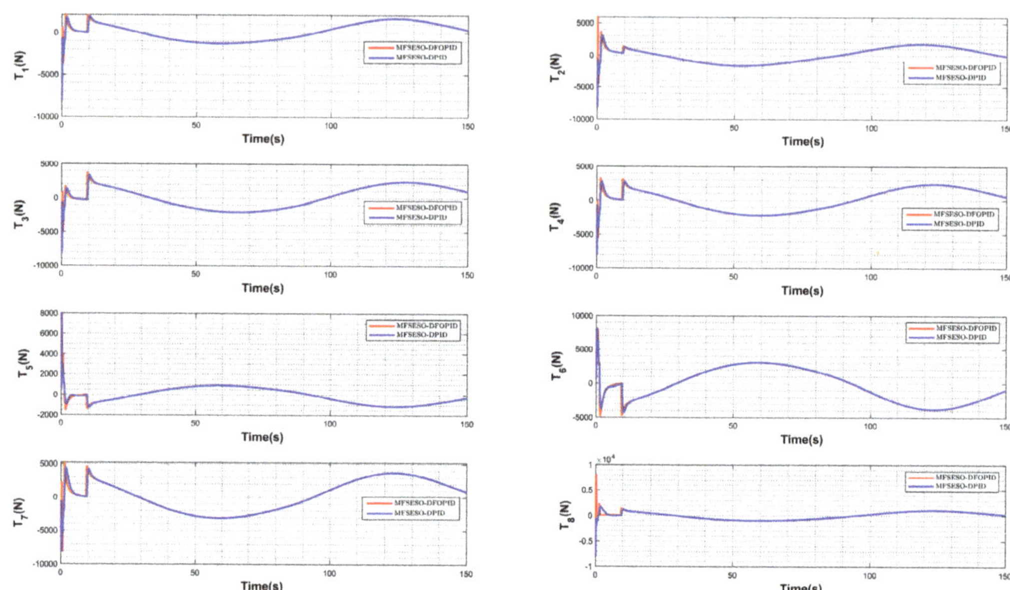

Figure 5. Thruster thrust U (defined in line 236) under MFSESO-DFOPID (red line) and MFSESO-DPID (blue line).

Table 1. Quantitative comparison between MFSESO-DFOPID and MFSESO-DPID control schemes. Better performance indicators are highlighted in red.

Performance Indicator	Control Scheme	x	y	z	ϕ	θ	ψ
Position root-mean-square error (m)	MFSESO-DFOPID	0.7606	0.0762	0.0833	0.0017	0.0010	3.36×10^{-5}
	MFSESO-DPID	0.7716	0.0789	0.1280	0.0028	0.0016	2.57×10^{-4}
Position adjustment time (s)	MFSESO-DFOPID	9.874	3.472	3.628	2.996	2.596	0.162
	MFSESO-DPID	10.030	2.836	5.610	3.118	3.310	0.668
Position steady-state error (m)	MFSESO-DFOPID	-0.0100	0.0019	1.34×10^{-5}	-2.86×10^{-7}	2.18×10^{-7}	-0.0003
	MFSESO-DPID	-0.0347	0.0090	6.78×10^{-7}	-2.41×10^{-6}	2.07×10^{-6}	-0.0032
Velocity root-mean-square error (m/s)	MFSESO-DFOPID	0.0167	0.0167	0.0179	0.0050	0.0033	2.86×10^{-4}
	MFSESO-DPID	0.0371	0.0308	0.0538	0.0067	0.0042	4.78×10^{-4}
Velocity adjustment time (s)	MFSESO-DFOPID	0.224	0.698	1.486	0.250	0.258	0.252
	MFSESO-DPID	12.542	2.916	9.256	0.330	0.422	0.318
Velocity steady-state error (m/s)	MFSESO-DFOPID	-0.0100	0.0022	1.32×10^{-5}	-2.86×10^{-7}	1.98×10^{-7}	-0.0005
	MFSESO-DPID	-0.0335	0.0105	1.66×10^{-7}	-2.39×10^{-6}	1.89×10^{-6}	-0.0045

For simulation scenario 2, the high-precision positioning and docking control simulation results of the control scheme based on MFSESO-DFOPID and the control scheme based on LESO-DFOPID are shown in Figure 6.

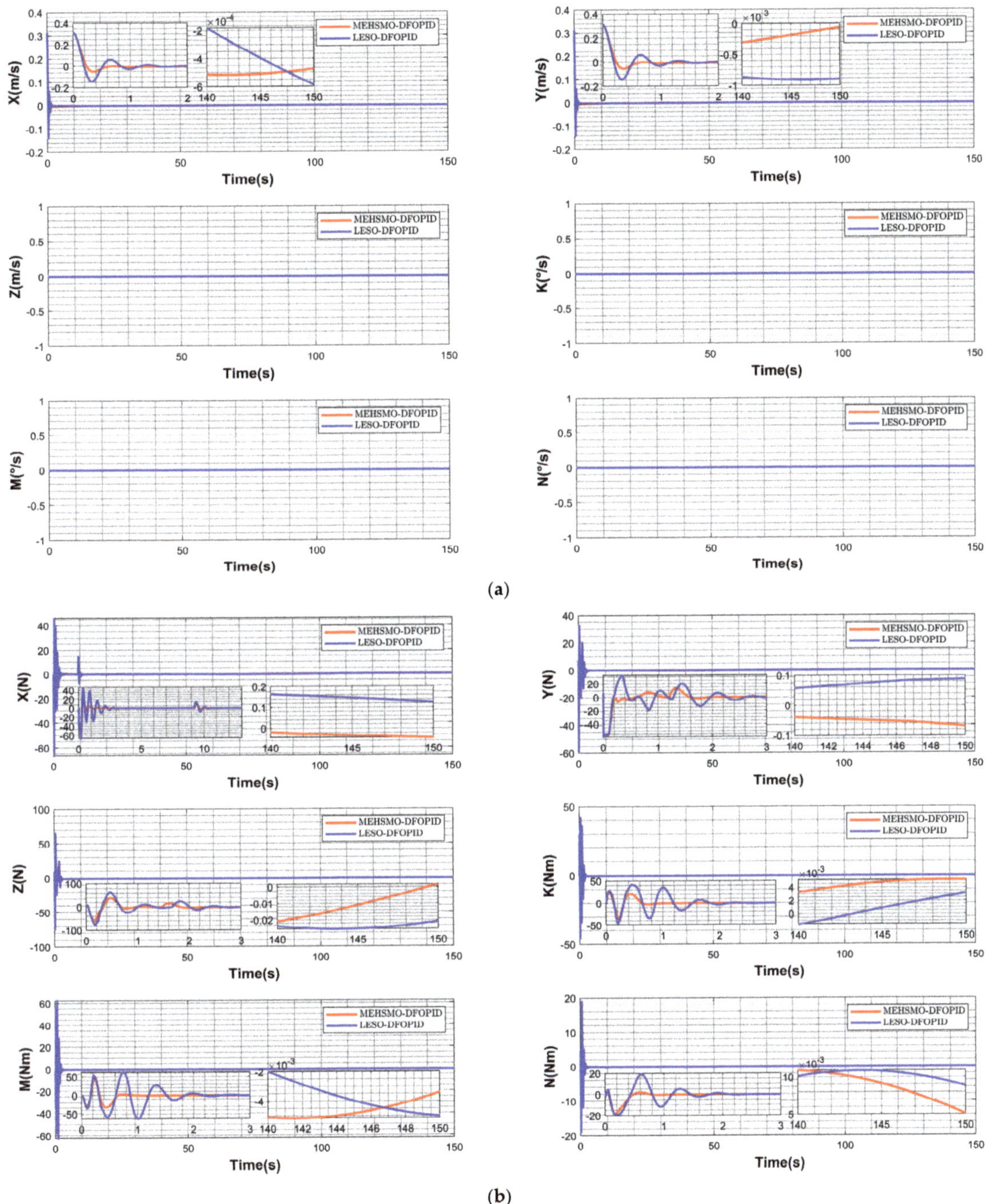

Figure 6. Observation error of ocean current velocity $\hat{v}_f$ and lumped disturbance $\hat{\tau}_D$ under MFSESO-DFOPID (red line) and LESO-DFOPID (blue line). (**a**) Observation error of ocean current velocity $\hat{v}_f$ (defined in line 271). (**b**) Observation error of lumped disturbance $\hat{\tau}_D$ (defined in line 272). Moreover, note that the six degrees of freedom for surge, sway, heave, roll, pitch, and yaw are represented by the letters X, Y, Z, K, M, and N, respectively.

In order to quantitatively compare the observation performance of the MFSESO and LESO algorithms for ocean current disturbance and lumped disturbance, performance indicators such as the root-mean-square error, adjustment time, and steady-state error are introduced in Table 2. The root-mean-square error of observation generally reflects the observation performance, the adjustment time of observation reflects the convergence velocity, and the steady-state error of observation reflects the estimation accuracy.

Table 2. Quantitative comparison between MFSESO-DFOPID and LESO-DFOPID control schemes. Better performance indicators are highlighted in red.

Performance Indicator	Control Scheme	x	y	z	ϕ	θ	ψ
Root-mean-square error of ocean current velocity observation (m/s)	MFSESO-DFOPID	0.0080	0.0080	0	0	0	0
	LESO-DFOPID	0.0091	0.0091	0	0	0	0
Adjustment time of ocean current velocity observation (s)	MFSESO-DFOPID	0.524	0.532	0	0	0	0
	LESO-DFOPID	1.078	1.082	0	0	0	0
Steady-state error of ocean current velocity observation (m/s)	MFSESO-DFOPID	-4.73×10^{-4}	-5.83×10^{-5}	0	0	0	0
	LESO-DFOPID	-5.86×10^{-4}	-8.95×10^{-4}	0	0	0	0
Root-mean-square error of lumped disturbance observation (N)	MFSESO-DFOPID	2.039	1.738	2.193	1.213	1.624	0.597
	LESO-DFOPID	2.953	2.119	3.320	2.517	3.720	1.031
Adjustment time of lumped disturbance observation (s)	MFSESO-DFOPID	10.002	1.888	3.334	0.912	0.942	0.738
	LESO-DFOPID	10.500	3.218	3.604	2.660	2.754	2.036
Steady-state error of lumped disturbance observation (N)	MFSESO-DFOPID	-0.043	0.0022	0.0016	0.0051	0.0034	0.0005
	LESO-DFOPID	0.126	0.0105	-0.0207	0.0031	-0.0050	0.0088

Figure 6a shows the estimation results for the MFSESO and LESO algorithms on ocean current velocity. It is not difficult to observe in the figure that, at X and Y degrees of freedom, MFSESO can approach the ocean current velocity more quickly, while the oscillation amplitude is smaller. With other degrees of freedom, the output of the observer remains 0 because no velocity disturbance is applied. Figure 6b shows the observation results for the MFSESO and LESO algorithms on lumped disturbance. It can be seen from this figure that the convergence process of MFSESO proposed in this paper is faster and smoother, and the steady-state error is smaller. The relevant performance indicators in Table 2 quantitatively describe the two observation algorithms, strongly demonstrating the significant advantages of MFSESO in terms of velocity, stability, and accuracy, and confirming the results in Figure 6. The above statement indicates that MFSESO can yield better observation results compared to the traditional LESO.

5.3. Pool Experiments

To verify the actual operational performance of the FADRC scheme based on MFSESO-DFOPID proposed in this article, a high-precision positioning and docking experiment using an ROV and a tool platform was conducted in a pool. The experimental scenario is depicted in Figure 7, the observation effect of the observer is reflected in Figures 8 and 9, and the position errors of each degree of freedom are illustrated in Figure 10.

As shown in Figure 7, the FADRC scheme based on MFSESO-DFOPID can enable the ROV to quickly and accurately reach the target position and complete the docking operation. Since the ROV is self-stabilized in pitch and roll degrees of freedom through the buoyancy trim, only four degrees of freedom x, y, z, ψ are controlled, where the total distance error $D = \sqrt{x_e{}^2 + y_e{}^2 + z_e{}^2}$. Figures 8 and 9 reflect that the observer in the control scheme effectively compensates for the current velocity disturbance and lumped disturbance at each degree of freedom. As can be seen from Figure 10, when $t > 20.40$ s, the position error $D < 0.05$ m, and when $t > 30.63$ s, the heading angle error $\psi < 1$. This proves that this control scheme can achieve high-precision positioning control and meet the requirements of docking control operations.

Figure 7. High-precision positioning and docking control experimental scenario. From top to bottom, the long image on the left contains frames 1–4, the middle long image includes frames 5–8, and the long image on the right includes frames 9–12. The entire positioning and docking process is clearly shown.

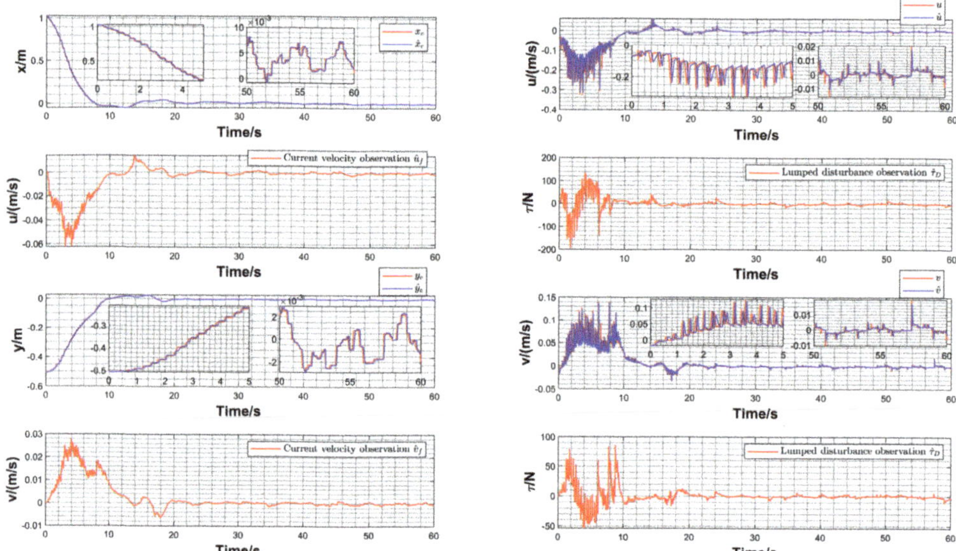

Figure 8. The x and y freedom observation results. Position x_e (red line) and position observation $\hat{x}_e$ (blue line); velocity u (red line) and velocity observation $\hat{u}$ (blue line); current velocity observation $\hat{u}_f$ (red line); lumped disturbance observation $\hat{\tau}_D$ (red line); position y_e (red line) and position observation $\hat{y}_e$ (blue line); velocity v (red line) and velocity observation $\hat{v}$ (blue line); current velocity observation $\hat{v}_f$ (red line); lumped disturbance observation $\hat{\tau}_D$ (red line).

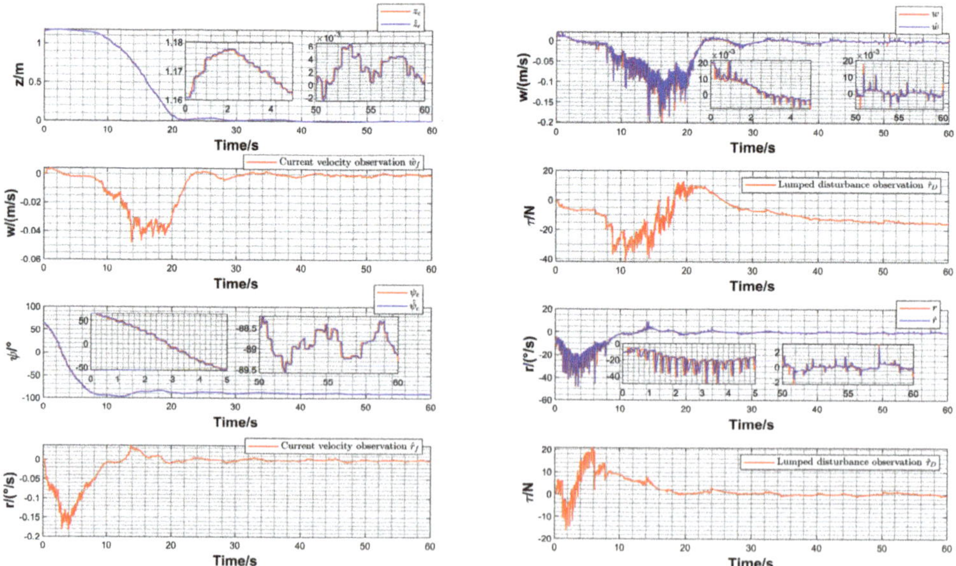

Figure 9. The z and ψ freedom observation results. Position z_e (red line) and position observation $\hat{z}_e$ (blue line); velocity w (red line) and velocity observation $\hat{w}$ (blue line); current velocity observation $\hat{w}_f$ (red line); lumped disturbance observation $\hat{\tau}_D$ (red line); orientation ψ_e (red line) and position observation $\hat{\psi}_e$ (blue line); velocity r (red line) and velocity observation $\hat{r}$ (blue line); current velocity observation $\hat{r}_f$ (red line); lumped disturbance observation $\hat{\tau}_D$ (red line).

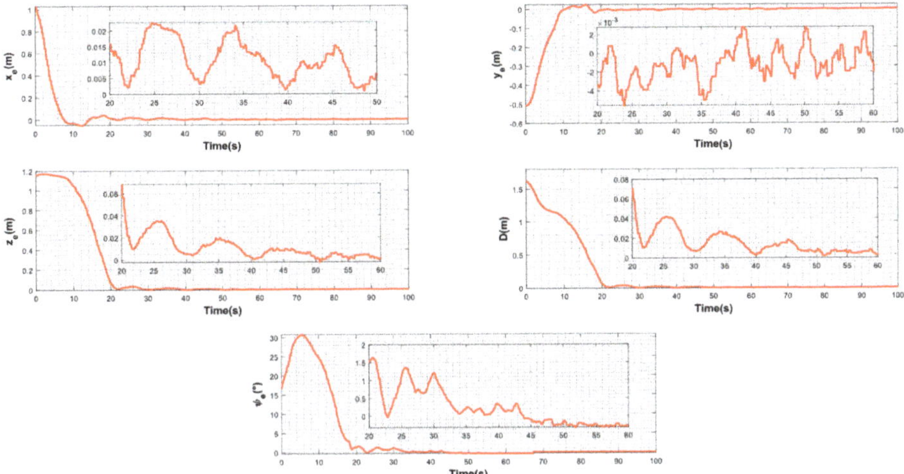

Figure 10. ROV position error. Position x_e (red line); position y_e (red line); position z_e (red line); total distance D (red line); and orientation ψ_e (red line).

6. Conclusions

In this study, an FADRC scheme is introduced, comprising a double closed-loop fractional-order $\text{PI}^\lambda \text{D}^\mu$ controller and a model-assisted finite-time sliding-mode extended state observer. The purpose of this control scheme is to facilitate high-precision positioning and docking tasks for ROVs in the presence of ocean current disturbances and model uncertainties. Specifically, DFOPID effectively addresses non-matched disturbances, with its fractional-order component enhancing the system's dynamic performance and robustness. The MFSESO in this paper further enhances the estimation accuracy by integrating sliding-mode technology and ensuring the finite-time convergence of observation errors. Through numerical simulations and pool experiments, it is demonstrated that the proposed control scheme can effectively mitigate ocean current disturbances and achieve high-precision operations even in the absence of an accurate model. This underscores the scheme's independence from precise model data on the operational ROV, while also highlighting benefits such as its simple structure and easy parameter tuning. Consequently, the FADRC scheme presented in this paper holds significant practical value and can serve as a valuable reference for ROVs engaged in high-precision operations. Future research will focus on exploring adaptive parameter optimization within the control scheme.

Author Contributions: Conceptualization, W.L. and L.G.; methodology, W.L. and L.G.; software, L.G.; validation, W.L. and J.X.; resources, W.L. and L.L.; writing—original draft preparation, L.G. and G.Y.; writing—review and editing, W.L. and L.L.; funding acquisition, W.L. All authors have read and agreed to the published version of the manuscript.

Funding: This research was funded in part by the National Science Foundation of China (project number 61903304), in part by the Fundamental Research Funds for the Central Universities (project number 3102020HHZY030010), and in part by the 111 Project under grant number B18041.

Data Availability Statement: The data presented in this study can be made available by sending a request to 2019100504@mail.nwpu.edu.cn.

Conflicts of Interest: The authors declare no conflicts of interest.

References

1. Makavita, C.D.; Jayasinghe, S.G.; Nguyen, H.D.; Ranmuthugala, D. Experimental Comparison of Two Composite MRAC Methods for UUV Operations with Low Adaptation Gains. *IEEE J. Oceanic. Eng.* **2020**, *45*, 227–246. [CrossRef]
2. Uchihori, H.; Cavanini, L.; Tasaki, M.; Majecki, P.; Yashiro, Y.; Grimble, M.J.; Yamamoto, I.; van der Molen, G.M.; Morinaga, A.; Eguchi, K. Linear Parameter-Varying Model Predictive Control of AUV for Docking Scenarios. *Appl. Sci.* **2021**, *11*, 4368. [CrossRef]
3. Ohrem, S.J.; Amundsen, H.B.; Caharija, W.; Holden, C. Robust adaptive backstepping DP control of ROVs. *Control Eng. Pract.* **2022**, *127*, 105282. [CrossRef]
4. Xie, T.; Li, Y.; Jiang, Y.; Pang, S.; Xu, X. Three-dimensional mobile docking control method of an underactuated autonomous underwater vehicle. *Ocean Eng.* **2022**, *265*, 112634. [CrossRef]
5. Wu, N.; Wang, M.; Ge, T.; Wu, C.; Yang, D.; Yang, R. Experiments on high-performance maneuvers control for a work-class 3000-m remote operated vehicle. *Proc. Inst. Mech. Eng. Part I J. Syst. Control Eng.* **2019**, *233*, 558–569. [CrossRef]
6. Song, D.; Li, L.; Wang, C.; Hou, R.; Li, C. A practical robust yaw servo architecture of ROVs by multi-vector propulsion and nonlinear controller. *Trans. Inst. Meas. Control* **2020**, *42*, 2908–2918. [CrossRef]
7. An, L.; Li, Y.; Cao, J.; Jiang, Y.; He, J.; Wu, H. Proximate time optimal for the heading control of underactuated autonomous underwater vehicle with input nonlinearities. *Appl. Ocean Res.* **2020**, *95*, 102002. [CrossRef]
8. Zhang, T.; Miao, X.; Li, Y.; Jia, L.; Wei, Z.; Gong, Q.; Wen, T. AUV 3D docking control using deep reinforcement learning. *Ocean Eng.* **2023**, *283*, 115021. [CrossRef]
9. Wang, T.; Sun, Z.; Ke, Y.; Li, C.; Hu, J. Two-Step Adaptive Control for Planar Type Docking of Autonomous Underwater Vehicle. *Mathematics* **2023**, *11*, 3467. [CrossRef]
10. Li, S.; Zhang, X. The welding tracking technology of an underwater welding robot based on sliding mode active disturbance rejection control. *Assem. Autom.* **2022**, *42*, 891–900. [CrossRef]
11. Han, J. From PID to Active Disturbance Rejection Control. *IEEE Trans. Ind. Electron.* **2009**, *56*, 900–906. [CrossRef]
12. Gao, Z. Active disturbance rejection control: From an enduring idea to an emerging technology. In Proceedings of the 10th International Workshop on Robot Motion and Control, Poznan University of Technology, Poznan, Poland, 1 July 2015; IEEE: Poznan, Poland, 2015; pp. 269–282.
13. Liu, C.; Xiang, X.; Yang, L.; Li, J.; Yang, S. A hierarchical disturbance rejection depth tracking control of underactuated AUV with experimental verification. *Ocean Eng.* **2022**, *264*, 112458. [CrossRef]
14. Wang, C.; Cai, W.; Lu, J.; Ding, X.; Yang, J. Design, Modeling, Control, and Experiments for Multiple AUVs Formation. *IEEE Trans. Autom. Sci. Eng.* **2021**, *19*, 2776–2787. [CrossRef]
15. Gao, J.; Liang, X.; Chen, Y.; Zhang, L.; Jia, S. Hierarchical image-based visual serving of underwater vehicle manipulator systems based on model predictive control and active disturbance rejection control. *Ocean Eng.* **2021**, *229*, 108814. [CrossRef]
16. Zhou, Y.; Sun, X.; Sang, H.; Yu, P. Robust dynamic heading tracking control for wave gliders. *Ocean Eng.* **2022**, *256*, 111510. [CrossRef]
17. Li, Z.; Liang, S.; Guo, M.; Zhang, H.; Wang, H.; Li, Z.; Li, H. ADRC-Based Underwater Navigation Control and Parameter Tuning of an Amphibious Multirotor Vehicle. *IEEE J. Oceanic. Eng.* **2024**, *13*, 4900. [CrossRef]
18. Liu, C.; Xiang, X.; Duan, Y.; Yang, L.; Yang, S. ADRC-SMC-based disturbance rejection depth-tracking control of underactuated AUV. *J. Field Robot.* **2024**, *41*, 1103–1115. [CrossRef]
19. Podlubny, I. Fractional-order systems and PID controllers. *IEEE Trans. Autom. Control* **1999**, *44*, 208–214. [CrossRef]
20. Liu, L.; Zhang, L.; Pan, G.; Zhang, S. Robust yaw control of autonomous underwater vehicle based on fractional-order PID controller. *Ocean Eng.* **2022**, *257*, 111493. [CrossRef]
21. Zhu, B.; Liu, L.; Zhang, L.; Liu, M.; Duanmu, Y.; Chen, Y.; Dang, P.; Li, J. A Variable-Order Fuzzy Logic Controller Design Method for an Unmanned Underwater Vehicle Based on NSGA-II. *Fractal Fract.* **2022**, *6*, 577. [CrossRef]
22. Li, Q.; Yang, G.; Yu, F.; Chen, Y. Adaptive fractional order non-singular terminal sliding mode controller for underwater soft crawling robots with parameter uncertainties and unknown disturbances. *Ocean Eng.* **2023**, *271*, 113728. [CrossRef]
23. Zhang, L.; Liu, L.; Zhang, S.; Cao, S. Saturation Based Nonlinear FOPD Motion Control Algorithm Design for Autonomous Underwater Vehicle. *Appl. Sci.* **2019**, *9*, 4958. [CrossRef]
24. Cui, Z.; Liu, L.; Zhu, B.; Zhang, L.; Yu, Y.; Zhao, Z.; Li, S.; Liu, M. Spiral Dive Control of Underactuated AUV Based on a Single-Input Fractional-Order Fuzzy Logic Controller. *Fractal Fract.* **2022**, *6*, 519. [CrossRef]
25. Liu, L.; Zhang, L.; Zhang, S. Robust PI controller design for AUV motion control withguaranteed frequency and time domain behaviour. *IET Control Theory A* **2021**, *15*, 784–792. [CrossRef]
26. Hasan, M.W.; Abbas, N.H. An adaptive neural network with nonlinear FOPID design of underwater robotic vehicle in the presence of disturbances, uncertainty, and obstacles. *Ocean Eng.* **2023**, *279*, 114451. [CrossRef]
27. Ali, N.; Tawiah, I.; Zhang, W. Finite-time extended state observer based nonsingular fast terminal sliding mode control of autonomous underwater vehicles. *Ocean Eng.* **2020**, *218*, 108179. [CrossRef]
28. Abdurahman, B.; Savvaris, A.; Tsourdos, A. Switching LOS guidance with speed allocation and vertical course control for path-following of unmanned underwater vehicles under ocean current disturbances. *Ocean Eng.* **2019**, *182*, 412–426. [CrossRef]

29. Hu, Q.; Jiang, B. Continuous Finite-Time Attitude Control for Rigid Spacecraft Based on Angular Velocity Observer. *IEEE Trans. Aerosp. Electron. Syst.* **2018**, *54*, 1082–1092. [CrossRef]
30. Oustaloup, A.; Levron, F.; Mathieu, B.; Nanot, F.M. Frequency-band complex noninteger differentiator: Characterization and synthesis. *IEEE Trans. Circuits Syst. I Fundam. Theory Appl.* **2000**, *47*, 25–39. [CrossRef]

MDPI AG
Grosspeteranlage 5
4052 Basel
Switzerland
Tel.: +41 61 683 77 34

Fractal and Fractional Editorial Office
E-mail: fractalfract@mdpi.com
www.mdpi.com/journal/fractalfract